AF863432

Mechanical Sciences

Uday S. Dixit · Santosha Kumar Dwivedy
Editors

Mechanical Sciences

The Way Forward

Springer

Editors
Uday S. Dixit
Department of Mechanical Engineering
Indian Institute of Technology Guwahati
Guwahati, India

Santosha Kumar Dwivedy
Department of Mechanical Engineering
Indian Institute of Technology Guwahati
Guwahati, India

ISBN 978-981-15-5711-8 ISBN 978-981-15-5712-5 (eBook)
https://doi.org/10.1007/978-981-15-5712-5

© Springer Nature Singapore Pte Ltd. 2021
This work is subject to copyright. All rights are reserved by the Publisher, whether the whole or part of the material is concerned, specifically the rights of translation, reprinting, reuse of illustrations, recitation, broadcasting, reproduction on microfilms or in any other physical way, and transmission or information storage and retrieval, electronic adaptation, computer software, or by similar or dissimilar methodology now known or hereafter developed.
The use of general descriptive names, registered names, trademarks, service marks, etc. in this publication does not imply, even in the absence of a specific statement, that such names are exempt from the relevant protective laws and regulations and therefore free for general use.
The publisher, the authors and the editors are safe to assume that the advice and information in this book are believed to be true and accurate at the date of publication. Neither the publisher nor the authors or the editors give a warranty, expressed or implied, with respect to the material contained herein or for any errors or omissions that may have been made. The publisher remains neutral with regard to jurisdictional claims in published maps and institutional affiliations.

This Springer imprint is published by the registered company Springer Nature Singapore Pte Ltd.
The registered company address is: 152 Beach Road, #21-01/04 Gateway East, Singapore 189721, Singapore

A Compendium to Commemorate the Silver Jubilee of the Mechanical Engineering Department Indian Institute of Technology Guwahati

Editorial Advisory Board

Amitabha Ghosh, Formerly Indian Institute of Technology Kanpur
A. K. Mallik, Formerly Indian Institute of Technology Kanpur
Gautam Biswas, Indian Institute of Technology Kanpur
A. K. Dass, Indian Institute of Technology Guwahati
U. K. Saha, Indian Institute of Technology Guwahati
R. Ganesh Narayanan, Indian Institute of Technology Guwahati
M. S. Shunmugam, Indian Institute of Technology Madras
G. K. Ananthasuresh, Indian Institute of Science
P. K. Das, Indian Institute of Technology Khargapur
Sanjay Mittal, Indian Institute of Technology Kanpur
S. S. Pande, Indian Institute of Technology Bombay
Amit Agrawal, Indian Institute of Technology Bombay
P. M. Pandey, Indian Institute of Technology Delhi
J. Paulo Davim, University of Aveiro
Vadim V. Silberschmidt, Loughborough University

About Indian Institute of Technology Guwahati

Indian Institute of Technology (IIT) Guwahati was established in September 1994 as the sixth IIT of the country. The academic programs of IIT Guwahati started in 1995 with three departments—(1) Computer Science and Engineering, (2) Electronics and Communication Engineering, and (3) Mechanical Engineering; Department of Electronics and Communication Engineering was later renamed as Electronics and Electrical Engineering. At present, the Institute has 11 departments and 5 inter-disciplinary academic centers covering engineering, science, and humanities disciplines, offering B. Tech., B.Des., M.A., M.Des., M.Tech., M.Sc., and Ph.D. programs. Within a quarter-century, IIT Guwahati has been able to build up world class infrastructure for carrying out advanced research and has been equipped with state-of-the-art scientific and engineering instruments. Starting with 64 students in 1995, student strength has crossed 6100 (more than 2700 undergraduate students). There are more than 400 faculty members to nurture the students and carry out high-quality research. Indian Institute of Technology Guwahati's campus is on a 285 hectares plot of land on the north bank of the river Brahmaputra, which is around 19 km from the Guwahati Railway Station and 23 km from airport. With the majestic Brahmaputra on one side, and with hills and vast open spaces on others, the campus provides an ideal setting for learning.

Indian
Institute
of
Technology
Guwahati

About Department of Mechanical Engineering, IIT Guwahati

The Department of Mechanical Engineering is one of the largest and oldest departments of Indian Institute of Technology Guwahati. It started in 1995 with an intake of 16 students. Presently, it has about 400 undergraduate and 450 post-graduate students. There are about 45 faculty members in the department. Department offers a B.Tech. program in Mechanical Engineering and an M.Tech. Program in Mechanical Engineering with specialization in the following five streams: (1) Aerodynamics and Propulsion, (2) Manufacturing Science and Engineering, (3) Computational Mechanics, (4) Machine Design, and (5) Fluids and Thermal. It has about 250 Ph.D. students working in the cutting edge areas of mechanical engineering.

Department of Mechanical Engineering has a state-of-the-art workshop and several laboratories. A mechatronics laboratory was well established in 2000 itself. Presently, a number of projects are being executed in the area of mechatronics and robotics. In the manufacturing area, research activities are going on welding, metal forming, and advanced manufacturing such as laser-based manufacturing and additive manufacturing. There is a strong group of Computational Fluid Dynamics (CFD) in the department; experimental activities are also being pursued to validate CFD models. Research is being carried out in microfluidics also. There are some faculty members actively working in the area of energy; they are also associated with the Energy Center of the institute. Other active research areas are finite element method, nonlinear vibration, composites, and material science.

Preface

Mechanical Engineering may be considered as old as the human civilization. Early humans used various types of tools made of stone and wood. Invention of wheel is considered a landmark event in human civilization; it was extensively used in the fourth millennium BC. Although Archimedes (287 BC-212 BC), inventor of several machines, can be considered as the first Mechanical Engineer, mechanical engineering developed as a distinct practice since the advent of James Watt's steam engine in the late eighteenth century. Institution of Mechanical Engineering was formed in the UK in 1847.

Mechanical Engineering is perhaps the most dynamic disciplines of engineering. It collaborates with almost every engineering discipline and has given birth to several other disciplines. Traditional machines used to run with the help of water, steam, and combustion of fossil fuels. Even today these are used as energy sources. However, it is common nowadays to see active amalgamation of electrical and electronics engineering in engines and machines. Various cyber-physical systems are exploiting the age-old knowledge of mechanical engineering along with the latest knowledge developed in the field of computer science and information technology. In the present century, biology may also play a major role in mechanical engineering.

Department of Mechanical Engineering at Indian Institute of Technology (IIT) Guwahati, the sixth IIT founded in 1994, was established in 1995. It caters to the teaching of traditional mechanical engineering as well as futuristic topics. To commemorate, its Silver Jubilee, it was decided to publish a compendium on mechanical engineering for providing an exposure to research directions in this ever-changing discipline. With the major support from Professor Amitabha Ghosh and other Editorial Advisory Board members, we could compile eleven articles written by the leading experts in mechanical engineering. Articles cover important gamut of mechanical engineering—solid mechanics, fluid mechanics, and manufacturing.

The first article by Amitabha Ghosh throws light on the birth of mechanical engineering and speculates its future. The article argues that early steam-powered prime movers developed in seventeenth–eighteenth century triggered the birth of mechanical engineering. In the future, synthetic biology, molecular engineering, intelligent materials, artificial intelligence, etc. are going to play a major role in mechanical engineering. In the second article, G. K. Ananthasuresh exposes the readers to exciting designs in micro-electro-mechanical systems (MEMS). G. Chakraborty and Nikul Jani describe nonlinear dynamics of MEMS. A glimpse of numerical techniques for studying the chatter has been provided by Chigbogu Ozoegwu and Peter Eberhard; stability study of thin-walled milling process is taken as an example.

It is envisaged that space elevators will replace (or supplement) rockets to transport goods in the space. So far a space elevator has not been developed, but in the next two–three decades, it may become a reality. In the fifth chapter, Arun K. Misra and Stephen Cohen discuss the statics and dynamics of space elevators. Gautam Biswas and Kirti Chandra Sahu have presented a brief survey on free surface flows. They highlight that the numerical study of free surface flows will find application in ink-jet printing, combustion, microfluidics, etc. Kiran Raj M. and Suman Chakraborty discuss the hydrodynamics of deformable micro-channels. With the discussion on bio-fluid mechanics, their article also reveals how the proper study of micro-channels can revolutionize the medical field. The discussion on application of microfluidics in the medical field is continued by Vijai Laxmi, Siddhartha Tripathi, and Amit Agrawal in the eighth chapter; several examples of lab-on-chip devices have been provided in this chapter.

Origami is an ancient Japanese art of folding the paper to produce three-dimensional complex shapes. Derosh George and Marc J. Madou discuss how this art can be used to fabricate MEMS. Finally, in the tenth chapter, Anish Roy, Qiang Liu, U. S. Dixit, and Vadim V Silberschmidt provide a glimpse of the power of simulation in manufacturing. With an example of machining simulation, it is demonstrated how the manufacturing, traditionally an artistic activity, can be scientifically studied and controlled. In the last chapter, Jyoti Mazumder takes us to the exciting world of laser-aided manufacturing.

We hope that this book will provide sufficient exposure to the audience and will also motivate researchers in the futuristic areas of mechanical engineering. It can also be useful for technology managers and policy-makers. We welcome the feedback from the readers.

Guwahati, India

Uday S. Dixit
Santosha Kumar Dwivedy

Editorial Acknowledgments

We sincerely thank all the authors for contributing the state-of-the-art articles. The idea for bringing out this compendium was provided by Professor Amitabha Ghosh. We are very grateful to him. We also thank other members of the Editorial Advisory Board, viz., Professors A. K. Mallik, Gautam Biswas, A. K. Dass, U. K. Saha, R. Ganesh Narayanan, M. S. Shunmugam, G. K. Ananthasuresh, P. K. Das, Sanjay Mittal, S. S. Pande, Amit Agrawal, P. M. Pandey, J. Paulo Davim, and Vadim V. Silberschmidt. We are very grateful to Prof. T. G. Sitharam, Director, IIT Guwahati, for encouraging us to celebrate the Silver Jubilee of Mechanical Engineering Department; this book is a part of that celebration. We are also thankful to all faculty members of the department. We thank Mr. Kaustabh Chatterjee, Research Scholar, for assisting the editorial task. Finally, we sincerely acknowledge the dedicated support provided by Ms. Swati Meherishi and other editorial staff members of Springer Nature.

Contents

About the Editors

Prof. Uday S. Dixit is currently a professor of Mechanical Engineering at the Indian Institute of Technology Guwahati. He has more than two decades' experience in carrying out research in the area of metal forming and machining. Apart from FEM, he uses fuzzy set theory and neural networks in his research. Before taking up a research career, he worked as a machine tool designer in HMT Ltd. Pinjore, India. Prof. Dixit obtained his Bachelor's degree in Mechanical Engineering from the Indian Institute of Technology, Roorkee in 1987. He gained his M.Tech. and Ph.D. in Mechanical Engineering from the Indian Institute of Technology, Kanpur in 1993 and 1998, respectively.

Prof. Santosha Kumar Dwivedy is currently the Head of Department of Mechanical Engineering at Indian Institute of Technology Guwahati. He completed his B.E. (Mechanical Engineering) from University College of Engineering Burla (Odisha) and M.E. (Mechanical Engineering with specialization in Machine Design and Analysis) from Regional Engineering College Rourkela (now NIT Rourkela). He completed his Ph.D. (Mechanical Engineering) from IIT Kharagpur in 1999. His research interests are in the field of nonlinear vibration, robotics and biomechanics. He has guided 6 Ph.D. students, 45 M.Tech. students and around 40 B.Tech. projects and currently 12 Ph.D. students are working under his guidance. He has published more than 150 articles in international journals, conferences and book chapters.

Birth of Mechanical Engineering and a Glimpse into the Future Trends

Amitabha Ghosh

1 Introduction

To begin with it is being emphasized that this article is not on the history of Mechanical Engineering. There exist excellent volumes on the subject and any attempt in this direction in an article is futile. Recently, IIT Guwahati has published a wonderful book on History of Mechanical Engineering. The basic focus of this article is to first identify the primary basic branches of physical science, which form the building blocks of what one calls 'Mechanical Engineering'. Next, it will be examined during what period these areas acquired maturity and were applied for effective engineering use. It is well known that although people have used many clever devices to employ mechanical advantage in accomplishing various tasks in the past, there has been no synthesis of the various scientific principles to evolve into an appropriate contrivance that can be considered as a product of Mechanical Engineering. It will be demonstrated that the word 'birth' instead of 'evolution' is more appropriate as this process of synthesizing various basic components of Mechanical Engineering was accomplished in a relatively short period of time in the human history when the first Prime Mover was invented and improved resulting in the First Industrial Revolution. This happened primarily in the seventeenth century.

The world has drastically changed during the past two centuries, and the human civilization has gone through another major watershed event, the Second Industrial Revolution, in the second half of the twentieth century that has resulted in the current knowledge-based civilization. Computer and communication technology helped by advances in material science has played a key role in this revolution.

As the time progresses, the nature of engineering activity is evolving continuously and, as a result, the subject Mechanical Engineering is also evolving like

A. Ghosh (✉)
Indian National Science Academy, New Delhi, India
e-mail: amitabha@iitk.ac.in

Indian Institute of Engineering Science and Technology Shibpur, Howrah, India

© Springer Nature Singapore Pte Ltd. 2021

U. S. Dixit and S. K. Dwivedy (eds.), *Mechanical Sciences*,
https://doi.org/10.1007/978-981-15-5712-5_1

other branches. In the not-too distant future Mechanical Engineering may look very different from the present-day picture. It is not possible to forecast technological inventions and development accurately. Till the 80s, Internet or mobile phones never appeared even in science fictions, but just after a couple of decades these have become inseparable entities in our daily lives! However, some ideas about the future scenario can be formed by studying the present trends of development in various fields. An effort will be made in this direction, and some ideas about the future of Mechanical Engineering will be presented in this article.

2 Engineering and Mechanical Engineering

Before proceeding further, it may be worthwhile to have a clear idea of the term 'Mechanical Engineering' and to form a concrete idea about what is the scope and objective of this branch of engineering activity. The term 'engineering' originates from the Latin word 'ingeniare' implying 'to innovate', 'to contrive', 'to devise', etc. such innovations or creations coming from clever manipulation of basic scientific knowledge. In Latin, 'ingenium' means cleverness and 'engineering' is expected to be clever application of basic scientific knowledge for useful service to the mankind. As an example, one can consider the case of 'stimulated emission of radiation' as discovered by the physicists. In raw form, it has not much application. But through 'engineering' one can use this phenomenon of 'stimulated emission of radiation' to generate an extremely powerful coherent light beam called 'LASER' that can be applied to many practical tasks. In fact it can be easily demonstrated that the primary objective of engineering is to accomplish useful tasks by taking the help of the knowledge created through research in basic science through appropriate clever and innovative manipulations. The term 'science' originates from the Latin word 'scientia' meaning 'knowledge', 'knowing', and 'creation of new knowledge'. As an activity 'engineering' follows 'science'. Quite often the term 'technology' is mixed up with 'engineering' and very frequently many treat these two to be synonymous. However, 'technology' is the activity that starts after 'engineering'. The term 'technology' comes from the Greek word 'technologia' meaning 'systematic use of craft, art, etc'. Thus when a strong coherent LASER beam is generated its use for a variety of tasks like manufacturing, communication, material processing, etc. the activity goes under the term 'technology'.

Defining 'mechanical engineering' is a more difficult task primarily because of the enormous width of spectrum of activities under this branch of engineering. The diversity of the types of tasks accomplished by 'mechanical engineering' makes it difficult to pinpoint the core issue in this branch. This difficulty becomes clear when one tries to find the definition of 'mechanical engineering' in various English dictionaries as demonstrated below:

According to the Webster dictionary 'mechanical engineering' is described as "A branch of engineering concerned primarily with the generation, transmission and utilization of heat and mechanical power and with production of tools, machinery and their products." In Encyclopedia Britannica, 'mechanical engineering' is described as "The branch of engineering concerned with the design, manufacture, installation and operation of engines, machines and manufacturing processes." It further states "Mechanical Engineering involves the application of the principles of dynamics, control, thermodynamics, heat transfer, fluid mechanics, strength of materials, material science, tribology, mathematics and computation. Increasingly knowledge of electronics and, in particular, of microprocessors are required." In Collier's Encyclopedia, the definition of 'mechanical engineering' takes the following form. "Mechanical Engineering is the application of engineering principles by mechanical means. The scope of Mechanical Engineers' work includes the generation of power and its transmission; the design and production of all kinds of goods from the basic machinery for mining, agriculture and other means of exploiting our natural resources through the machines and tools for heavy and light industry to the everyday consumer products; and the movement of people and goods in all types of transportation and material handling equipment, including marine, rail, automotive, aeronautic, conveying, and pumping equipment." It is agreed that these definitions are elaborated and do not identify the basic core issues of 'mechanical engineering' are a little obscure. The simplest and shortest description of 'mechanical engineering' is given in Lexicon Universal E'pedia. It is "Mechanical Engineering is concerned primarily with means of converting energy to useful mechanical forms. The speciality is machine oriented; Mechanical Engineers' creations involve motion, in contrast to other branches of engineering."

Studying the above definitions of 'mechanical engineering', the basic structure of this branch of engineering can be represented by a diagram as shown in Fig. 1. The figure shows the hierarchy of activities in fundamental tree representing 'Mechanical Engineering'. Some examples of the machines and units for accomplishing the tasks are indicated in the second column. The box shows the basic areas necessary for the tasks encountered in mechanical engineering. It is evident that although there were examples of some isolated applications of a few principles in the past, viz., various mechanisms, water wheels, windmills, etc. There was never any contrivance that used the basic principles of mechanical engineering. The search of a prime mover that led to the invention of the engines by Thomas Newcomen was improved upon by James Watt in the eighteenth century; the mankind first saw the application of most of the basic aspects of mechanical engineering in the developed prime mover. The following brief discussion on the evolution of the early prime movers illustrates the point.

3 Acute Wood Famine and the Search for a Prime Mover

When one examines the old pictures depicting the landscapes of sixteenth-century Europe, a very strange contrast with the present-day scenario can be noticed. Now the landscapes of Europe shows wonderful green cover and the forests over the empty lands, whereas in most old drawings the landscapes are barren and forests are

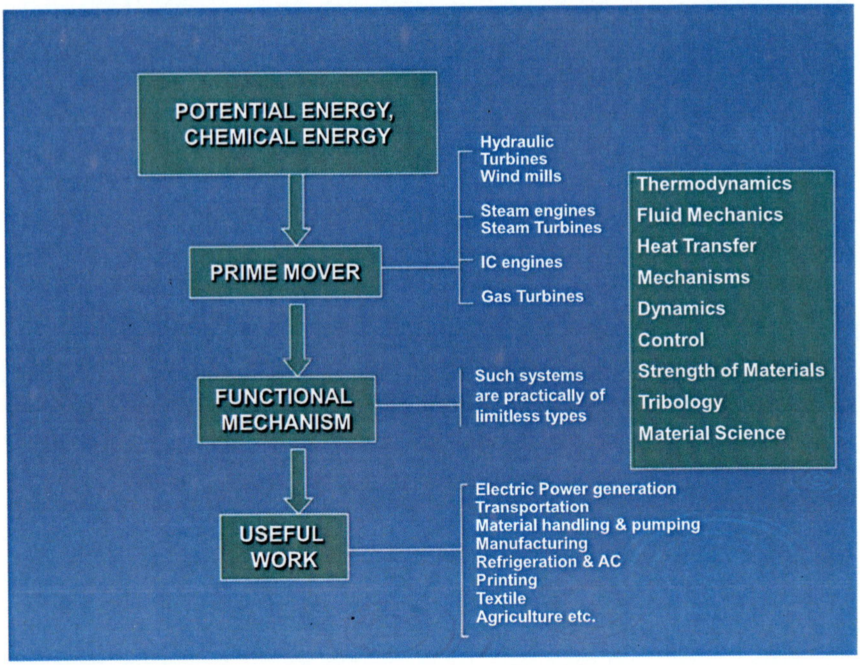

Fig. 1 Basic structure of mechanical engineering activities

conspicuously absent. Figure 2 shows this typical feature in old drawings of Europe. The forest covers are missing as shown. This happened due to the famous 'wood famine'.

In sixteenth- and seventeenth-century Europe, this famine resulted because of excessive use of wood in all spheres of activities. Figure 3 shows the use of wood in those centuries. Thus people started searching for coal as the fuel for heating and other purposes. Soon the coal near the surface got exhausted, forcing people to go deeper that resulted in the need of water pumping from deep mines. Besides, pumping water to higher levels and then getting various services from that raised

Fig. 2 Typical landscape of sixteenth- and seventeenth-century Europe

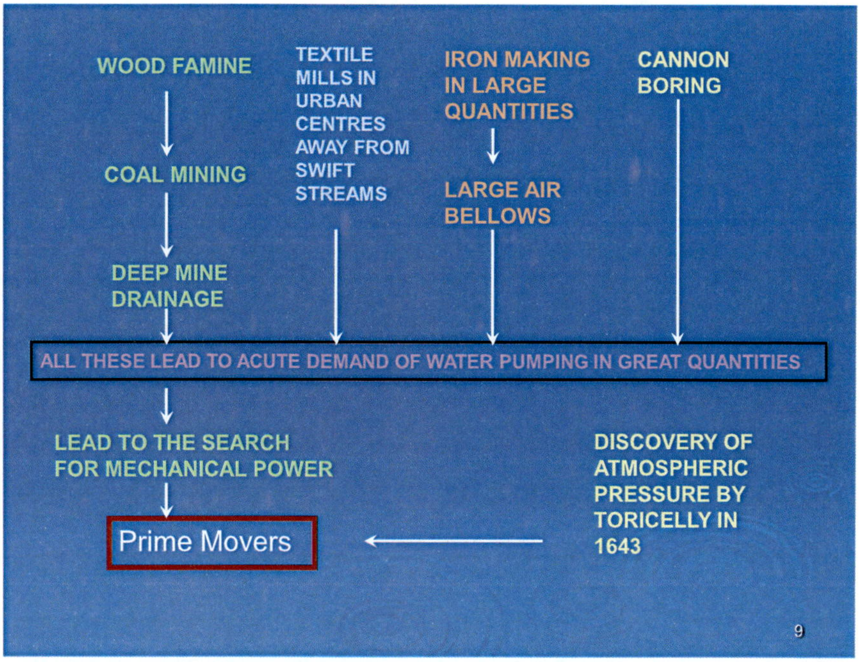

Fig. 3 Chain of events that led to the search of mechanical power and prime movers

water also became a major task. All these led to the acute demand for water pumping and forced all to search for a 'prime mover' as depicted in Fig. 3.

This led the scientists to look for various possibilities. After Torricelli discovered the existence of atmospheric pressure and its 'enormous' capacity was demonstrated by a number of experiments, the attention of the researchers was focussed on the idea 'how to extract mechanical power out of the atmosphere'. The obvious thinking was to create vacuum in a controlled fashion without using pumps. Christian Huygens thought of using explosives to throw away air from a closed container creating vacuum. For obvious reasons, this idea was not a practicable one and did not materialize into anything meaningful.

4 The First Solution

It was one of Huygens's assistants, Denis Papin, who broke the barrier. He noticed that if the steam in a closed container can be condensed, a partial vacuum can be created in the vessel (By the way he was also the inventor of pressure cooker. But the world had to wait for two centuries by the time pressure cooker became a popular device for quick cooking!). With plenty of natural ice-cold water being available in Europe, condensing steam was not at all a problem. Figure 4 shows the basic scheme

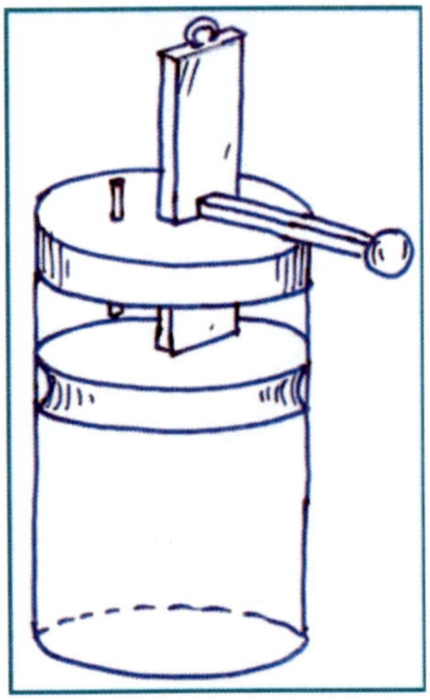

Fig. 4 Denis Papin's device for condensing steam

devised by Papin. It consists of a piston- and a cylinder-type combination. The lower part of the cylinder is filled with steam, when ice-cold water was sprinkled on the cylinder from outside the steam condensed and vacuum was developed causing the piston to be forced down by atmospheric pressure. Such a simple idea it was but that held the key to the Industrial Revolution that transformed the whole world. This idea of converting thermal energy (spent for converting water into steam) into mechanical force (and power) was the first step toward achieving something we may consider as the forerunner of 'Mechanical Engineering'. This article is not on the history of steam engines, but since the development of steam engine laid the foundation of Mechanical Engineering certain related key issues need to be touched upon in the next section.

5 The Early Prime Movers

It was not only the force caused by atmosphere against vacuum but the idea of using the expansive force of steam against atmosphere also came to people's mind, and some simple device to push out water by using steam was created. But in the initial stages, the atmospheric engines (which ran based upon atmospheric pressure working against created vacuum) played the dominant role for the whole of eighteenth century. To begin with the system did not have any mechanically moving members as indicated in Figs. 5 and 6. After Papin demonstrated his vacuum setup in 1690 (and left the further work on the idea), in 1698 Thomas Savery perfected a scheme working on the use of atmospheric pressure that became popular for pumping water out of deep mines. Thus the name given was 'Miners' Friend'. Figures 5 and 6 show the principle of operation and a drawing of a real-life version (primarily to demonstrate

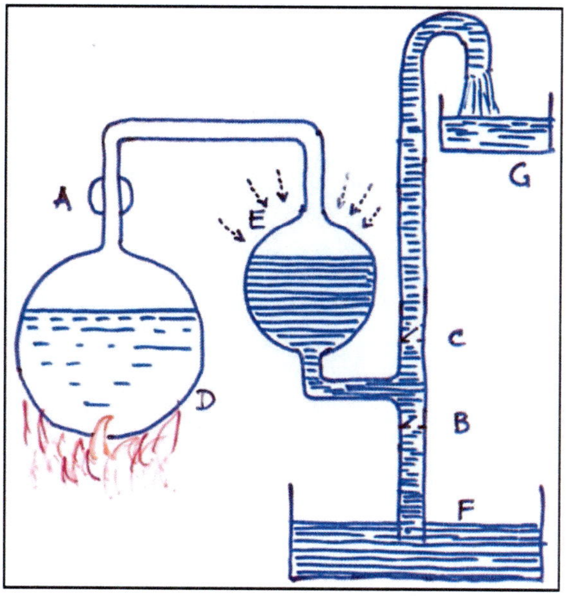

Fig. 5 Scheme of operation of Savery's 'Miners' Friend'

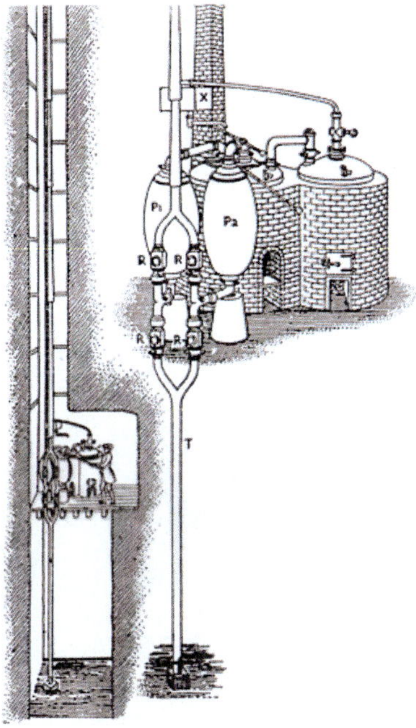

Fig. 6 A real version of Miners' Friend

the scale). In D water is heated and with the valve A open steam enters vessel E. The valve B gets automatically closed and water from E goes through valve C (that opens upward) to container G. In the next half of the pumping cycle, valve A is closed and ice-cold water is sprinkled on E causing condensation of the accumulated steam in E that results in a partial vacuum in E. Since valve C automatically closes valve B opens and water from tank F is sucked up by the vessel E due to the atmospheric pressure pressing the water in F. The next cycle starts with the opening of valve A, and water is continuously lifted up from tank F to container G. Except for the valves, there are no mechanical parts and no direct mechanical motion is generated. However, the water in the container G at a higher level could be used to run a water wheel if needed. Savery took a patent of his invention claiming overall right on a device for "raising water using fire." Although this was the first time heat energy was directly used, the device cannot be a true example of 'Mechanical Engineering'. Direct generation of mechanical motion was possible a few years later when Thomas Newcomen presented to the world the classic combination of a boiler, a cylinder, and a piston describing oscillatory motion.

Figure 7 schematically explains the principle of operation. The generated steam (with a little above the atmospheric pressure) pushes the piston in the cylinder upward causing the pump rod to descend. In the next half-cycle (the power stroke), the valve is closed and the valve connecting the cold water tank is opened causing cold water to be sprinkled inside the cylinder and condense the accumulated steam. This results in the creation of vacuum, and the atmospheric pressure pushes down the piston causing the pump rod to move up and work is done. The closing and opening of the valves were accomplished automatically with a system of strings and levers.

Figure 8 shows the engine erected by Savery and Newcomen at Dudley castle in 1712. The enormous size can be judged from the figure on the floor supplying coal! These engines got erected at many sites and became popular. However, their

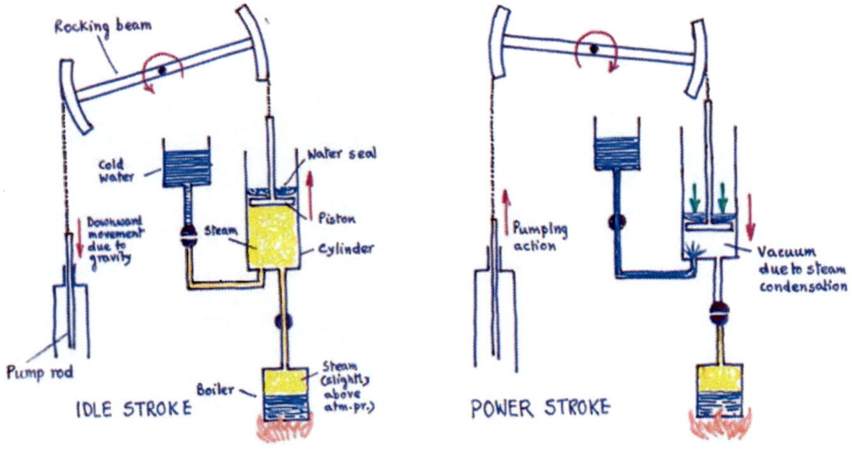

Fig. 7 Scheme of Newcomen atmospheric engine

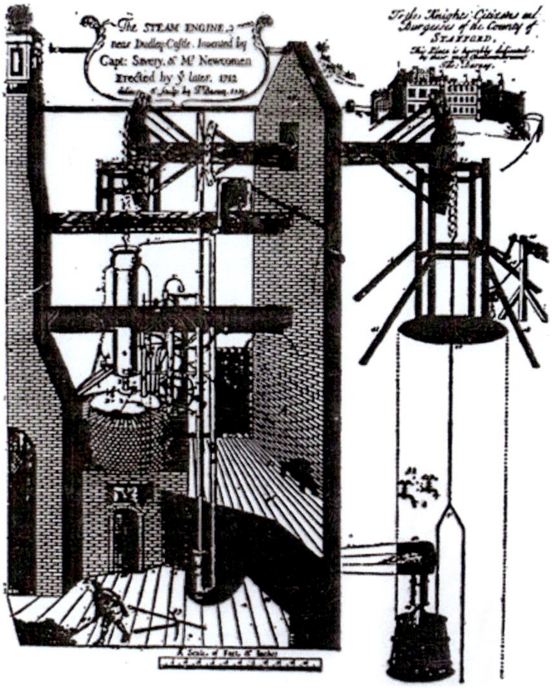

Fig. 8 Engine erected by Savery and Newcomen at Dudley castle in 1712

power was very little compared to the size and efficiency was below 0.5%! But this contrivance continued to serve unchallenged for six decades and made the wheel of Industrial Revolution to run and change the world. Even though the real inventor of these engines was Newcomen, the control of the whole affair was with Savery because of his all-encompassing patent! Newcomen gradually vanished into the oblivion—poor and wretched. But his creation continued, and the performance was improved by John Smeaton who was the first to study it scientifically. Figures 9 and 10 show an improved Newcomen engine toward the middle of the eighteenth century and a version of the engine erected by Smeaton in 1772. The efficiency became almost 1%!

It is important to mention at this stage that though Smeaton was a mechanical engineer according to the present-day definition he was the first to coin the term ‘Civil Engineering’. In the earlier times, most of the activities we call ‘engineering’ was intended for warfare. He identified those areas of engineering activities meant to improve the life of civilian population and used the term ‘civil engineering’. One can notice how the various elements of ‘Mechanical Engineering’ are gradually appearing in the Newcomen engines as the time progressed.

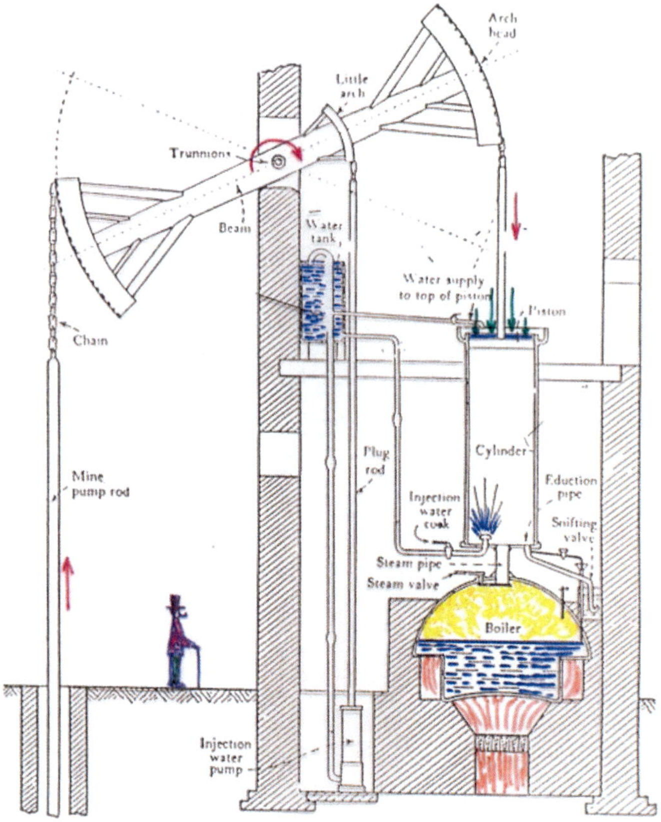

Fig. 9 A fully developed Newcomen engine (15 hp and 0.5% efficiency)

6 James Watt and Improved Atmospheric Engines

The popular belief is that James Watt invented steam engine. Although this is incorrect, it must be accepted that his improvement of Newcomen engine led to a quantum change in its performance. At the same time, most of the basic building blocks of Mechanical Engineering got developed along with the rotary engine developed by James Watt and his assistant William Murdoch. James Watt was born in 1736 and became an instrument maker by profession and joined Glasgow University. In 1764, he was asked to repair a Newcomen engine belonging to the philosophy department. In those days 'science' used to be called 'natural philosophy'. He quickly realized that repeated cooling of the cylinder was the main shortcoming of Newcomen engine. On a sunny May day in 1765 when he was walking across a park, the idea of a separate condenser came to him. He realized that if the condensation of the accumulated steam is done in a separate container (he called condenser) connected to the main cylinder, then this problem can be eliminated. When steam condenses in the condenser, the

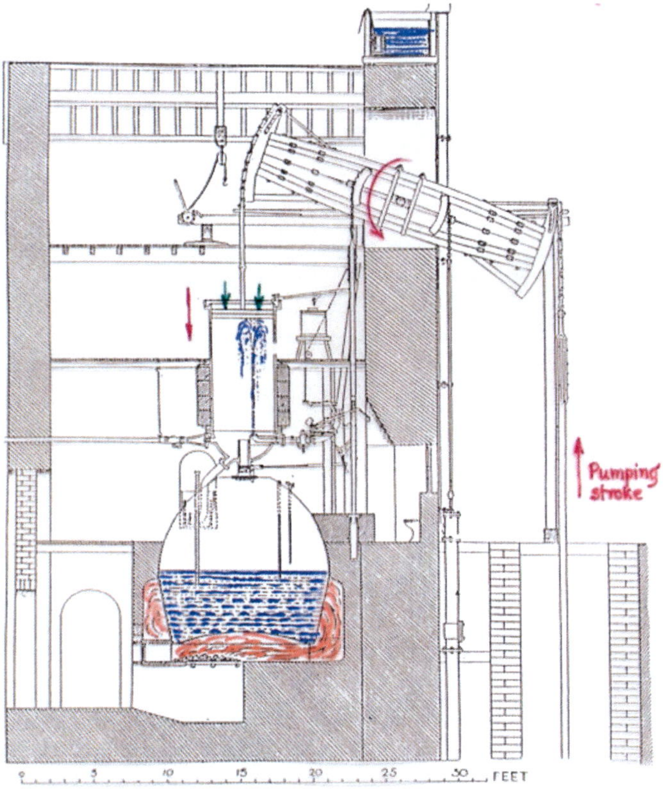

Fig. 10 Newcomen engine erected by John Smeaton in 1772 for Long Benton colliery

whole steam in the main cylinder flows to the condenser because of the vacuum created there and the main cylinder is not cooled down. In the original scheme of Newcomen engine after each cycle, it was necessary to heat up the cylinder before the steam could be collected there as the cylinder cooled while condensing the accumulated steam. Figure 11 shows the first experimental setup of Watt's condenser. It looks so insignificant a device but one must not forget that this helped the world we see today possible. The study of the phenomenon of condensation of steam and associated issues brought the subject 'thermodynamics' and 'heat transfer' into the realm of 'mechanical engineering'.

In 1769, he took a patent of an engine of his design in which the cylinder was surrounded by a steam jacket and condensation took place in a separate condenser. His financier was Dr. Roebuck and later Boulton. But they have been struggling for almost a decade without success. The reason was the lack in accuracy of cylinder boring keeping enough gap between the cylinder and the piston for the atmospheric air to pass through instead of moving the piston. In case of Newcomen engine, a leather

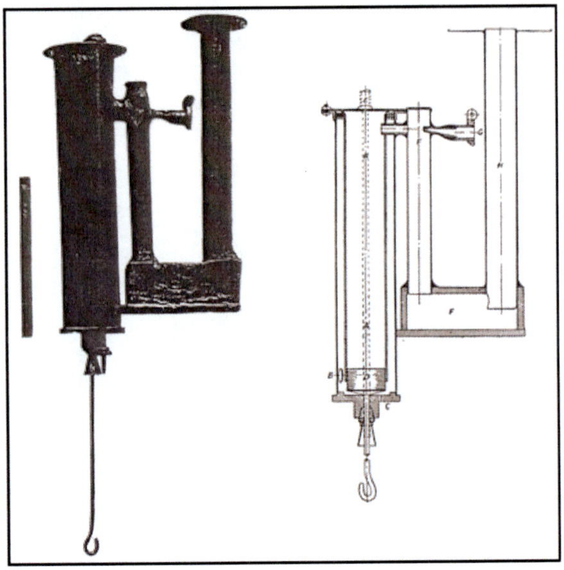

Fig. 11 James Watt's first experimental setup of a separate condenser

tray full of water above the piston kept the gap sealed. But such an arrangement using water seal was not allowed to avoid any contact of the cylinder wall with cold water.

John Wilkinson was the greatest ironmaster of his generation. He devised a new method of boring cylinders. Figure 12 shows the new version of boring mill developed

Fig. 12 John Wilkinson's new boring mill

by Wilkinson. This was the first step toward achieving accuracy in machining that was another founding stone of emerging discipline—'Mechanical Engineering'. It may be interesting for today's mechanical engineers to note the status of accuracy of machining in those days. In April 1775, Wilkinson delivered a cylinder which in Boulton's language "*— does not err the Thickness of an old shilling at any part —*". Later Wilkinson improved his technique and in Watt's language "*— a 72 inch cylinder being not further from absolute truth than the thickness of a new sixpence in the worst part —*". The thickness of a coin may be a monstrous error by today's standard BUT THE ENGINE RAN after a decade of unsuccessful continuous struggle.

At this stage, it is desirable to get a glimpse of the historical patent taken by Watt on January 29, 1769 for "Lessening the Consumption of Steam and Fuel in Fire Engines." It is given below:

> First, that vessel in which the powers of steam are to be employed to work the engine, which is called the cylinder in common fire engines, and which I call the Steam Vessel must, during the whole time the engine is at work, be kept as hot as the steam that enters it; first by enclosing it in a case of wood, or any other materials that transmit heat slowly; secondly, by surrounding it with steam or other heated bodies; and thirdly, by suffering neither water nor any other substance colder than the stream to enter or touch it during that time.
>
> Secondly, in Engines that are to be worked wholly or partially by condensation of steam, the steam is to be condensed in vessels distinct from the steam vessels or cylinders, although occasionally communication with them; these vessels I call Condensers; and; whilst the engines are working, these condensers ought at least to be kept as cold as the air in the neighborhood of engines, by application of water, or other cold bodies.
>
> Thirdly, whatever air, or other elastic vapor, is not condensed by the cold of the condenser and may impede the working of the engine, is to be drawn out of the steam vessels or condensers by means of pumps, wrought by the engines themselves, or other wise.
>
> Fourthly, I intend in many cases to employ the expansive force of steam to press on the pistons or whatever may be used instead of them, in the same manner as the pressure of the atmosphere is now employed in common fire engines; in cases where cold water cannot be had in plenty, the engines may be wrought by this force of steam only, by discharging the steam into open air after it has done its office.

It is to be noted that in his fourth claim Watt included the concept of using the expansive force of steam, instead of the atmospheric pressure working against vacuum. This prevented others in developing real 'steam engines' as Watt never allowed such attempts as in his opinion use of high-pressure steam was dangerous. This delayed the steam engine technology by almost 20 years. Although with a separate condenser the efficiency of the engine increased drastically, the basic concept was still the same as that of Newcomen's engines. The configuration also remained the same consisting of a typical rocking beam as shown in Fig. 13.

It was John Wilkinson who gave Watt the idea of a rotating steam engine instead of generating only a reciprocating motion. Such an engine can drive machineries and mills directly. Watt had to conceive of a mechanism and thought of a crank rocker mechanism for converting the reciprocating motion into a continuous unidirectional rotation. But his idea of a crank rocker mechanism was leaked by one of his assistants to some Mr. Pickard who took a patent. This became a stumbling block for

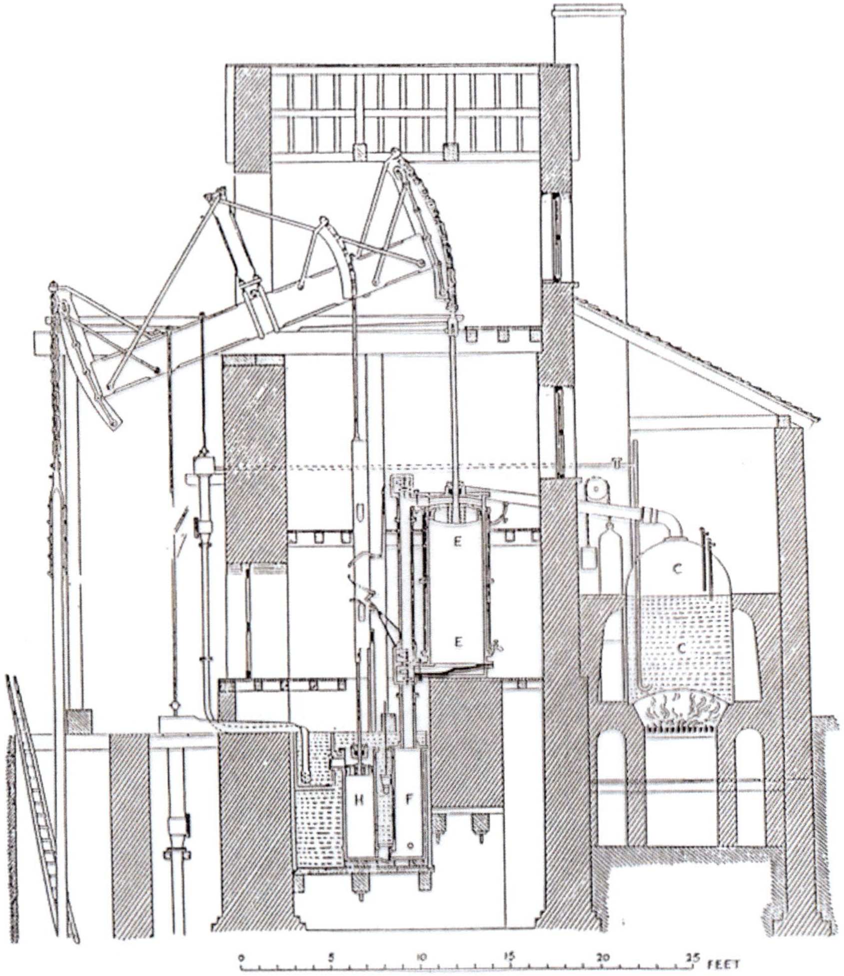

Fig. 13 Watt's single-acting engine for pumping water for draining mines in 1788. The boiler C is placed in an outhouse, and the steam passes to the cylinder E, which is maintained hot using a separate steam jacket. F is the separate condenser and H is an air pump

developing a rotating engine, and his assistant Murdoch suggested a way out without infringing upon the patent. It was the famous sun and planet gear system which has baffled many students of history of Mechanical Engineering. Figure 14 shows Watt's original crank rocker mechanism and the alternate arrangement using sun and planet gears. To automate various movements of different parts, mechanical means were being devised and gradually the subject 'kinematics and mechanisms' started taking shape as a part of mechanical engineering. Earlier it was in the domain of some mathematicians! Another problem related to mechanism science was solved by Watt. In

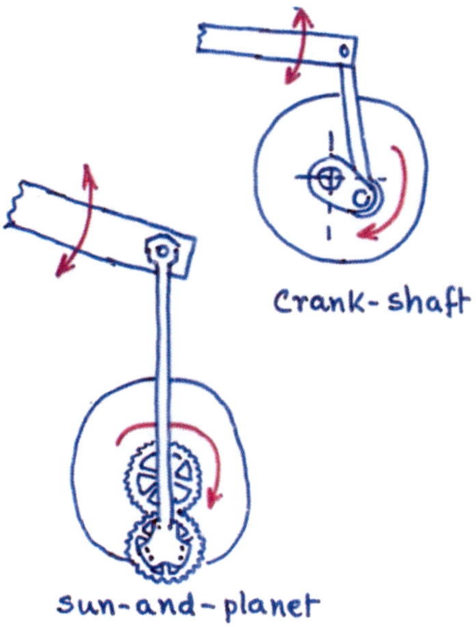

Fig. 14 Watt's original crank rocker design and the alternative sun and planet arrangement

constructing a double-acting engine, it was necessary to eliminate the circular sectors on the rocking beam and connecting it with the piston rod that required a rectilinear oscillating motion. He designed a straight-line generator with a four-bar mechanism that is still known as 'Watt's Straight-Line Generator' shown in Fig. 15. Generating a straight-line motion with the help of only hinge joints in a mechanism was a design challenge and Watt considered this mechanism to be his best work. Slowly the subject of mechanism theory started achieving maturity as the steam engines started taking modern shapes with the introduction of high-pressure steam engines by Richard Trevithick in 1801. For controlling the speed, a governor mechanism was also introduced by James Watt, and the centrifugal governor designed by him is still called 'Watt Governor'. Figure 16 shows Watt's design of the centrifugal governor he employed. Figure 17 shows a double-acting rotary engine constructed in 1788. It shows the introduction of the straight-line mechanism and the governor. Figure 18 shows the drawing of a Watt's double-acting rotary engine. Since the piston moved in a straight line, in this case, its connecting point driving the rocking beam needed rectilinear motion. The use of the straight-line mechanism and a Watt governor is clearly visible. The sun and planet mechanism to convert oscillatory motion to unidirectional continuous rotary motion is also conspicuous in the drawing. It is to be noted how gradually the steam engine started looking as a product of mature 'Mechanical Engineering' as time progressed.

As mentioned earlier, the real steam engine employing the expansive force of high-pressure steam could not be developed since Watt was against it, and his claim no. 4 in the patent use of high-pressure steam was mentioned. Only after the expiry of the patent period, Richard Trevithick developed the first high-pressure steam engine

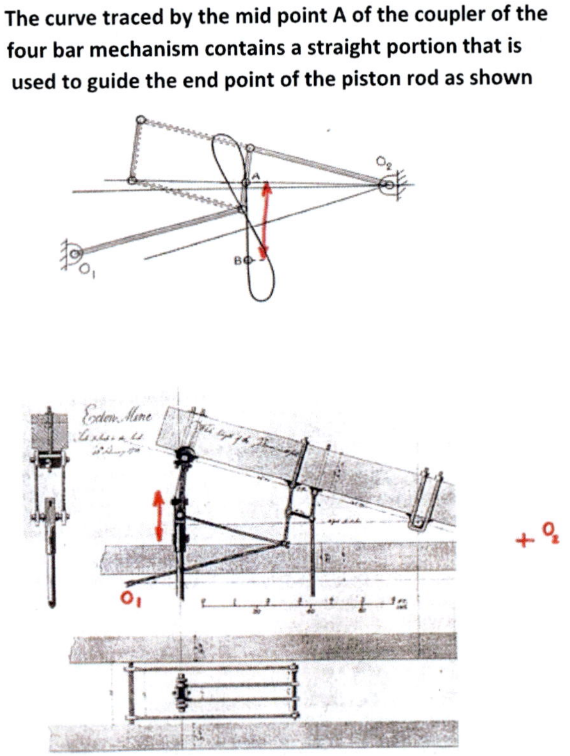

Fig. 15 Watt's design of a four-bar straight-line mechanism

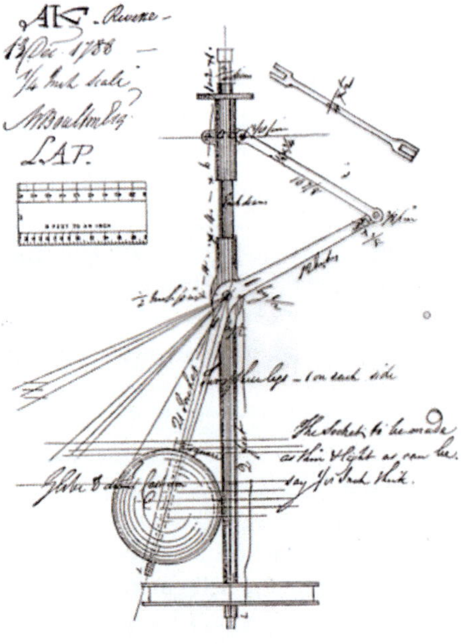

Fig. 16 Watt's design of centrifugal governor

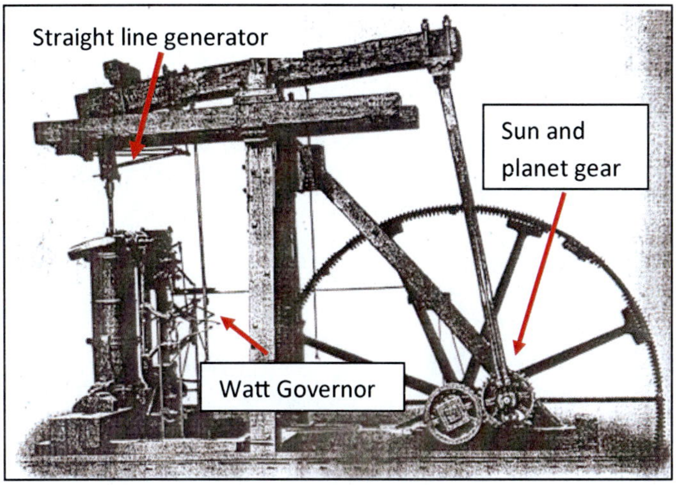

Fig. 17 A double-acting engine developed by Watt in 1788

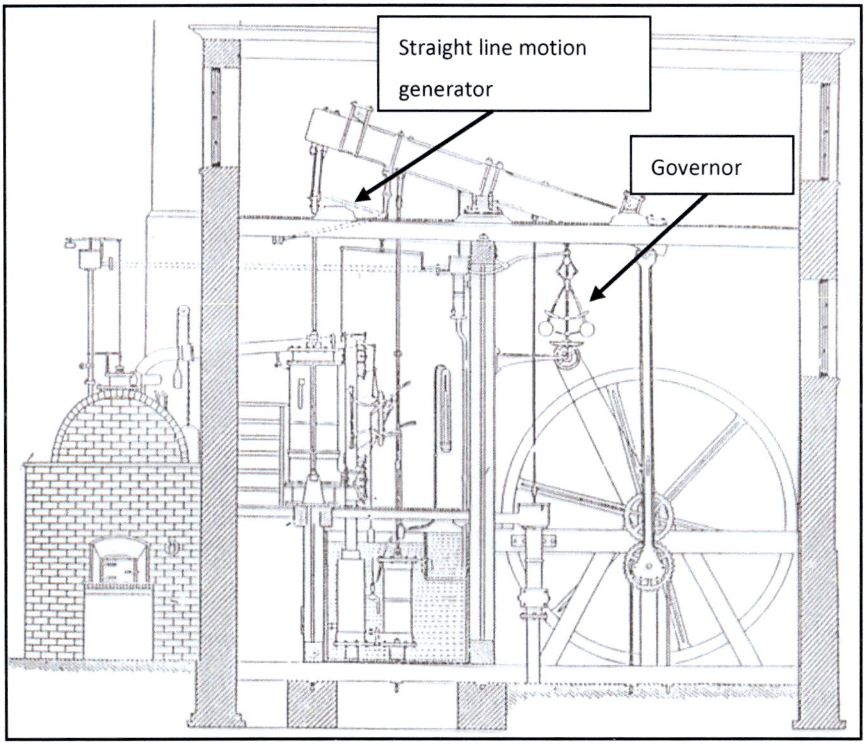

Fig. 18 Drawing of a double-acting rotary engine designed by Watt

and the era of 'atmospheric engines' finally was over. Trevithick's engine had the modern look, and the traditional rocking beam was eliminated and the 'direct-acting engine mechanism' was introduced. He developed both locomotives and stationary engines. Figures 19 and 20 show Trevithick's locomotive and engine.

It is not commonly remembered that the father of steam locomotive is not Stephenson as popularly believed but Richard Trevithick is the true inventor. Unfortunately, he was away from England when the first rail line was proposed, and George

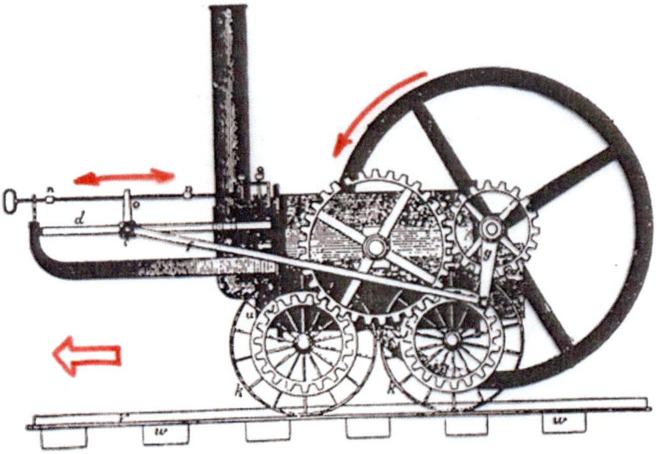

Fig. 19 Steam locomotive developed by Trevithick in 1801

Fig. 20 Stationary steam engine developed by Trevithick in 1805

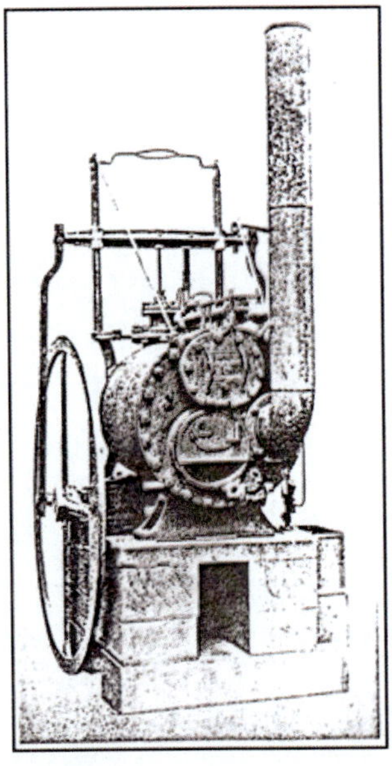

Fig. 21 The Rocket

and Robert Stephenson (father and son) won the contest in the Rain Hill in 1829 as Stephenson's 'Rocket' engine gave the best performance. They got the contract for the first railway. Figure 21 shows their engine–Rocket.

The above account shows that 'Mechanical Engineering', as a branch of engineering consisting of its basic foundational subjects evolved along with the evolution of the first prime mover and the period was about a century. Thus, using the term 'birth' in connection with 'Mechanical Engineering' is justified. The railway, steamships, and the new production industry and mills grew at a fast rate because of the demand of extensive colonization of a major part of the world by England. Once the IR started, the demand for machines, engines, etc. increased manifold demanding faster rate of production and easy repair. Henry Maudslay was the first to bring the era of precision engineering. He developed the first micrometer and accurate screw-cutting lathes so that interchangeable nuts, bolts, and parts could be produced. Whitworth further perfected these principles. Repair and production both became easy and economical. Eli Whitney, around the same time, came with the idea of mass production, and the cost of machines, engines, etc. become affordable making them popular with a great demand. George Stevenson's development for the rail transport brought revolution in many areas of administration and economy.

Proliferation of engines, machines, mills, etc. made it necessary for the practicing engineers to specialize in particular fields. Those who specialized in 'dynamic' problems such as power generation, machinery, manufacturing, and engines were called 'MECHANICAL ENGINEERS', while those concerned themselves with 'static' problems, such as land measurement, dams, bridges, buildings were still called 'CIVIL ENGINEERS'.

In 1818, the Institution of Civil Engineers (the first professional society) was founded in Great Britain with Thomas Telford as its first President. In 1847, a group of railway engineers, who felt that the Institution of Civil Engineers was uninterested in the new breed of engineers resulting from the development of railways, formed the Institution of Mechanical Engineers with George Stephenson as the first President.

7 The Trends of Development and the Future Scenario

An attempt to guess the nature of the subject 'Mechanical Engineering' in the future is an interesting exercise. Before taking up the matter it may be desirable to have a look at the manner in which the engineering disciplines have evolved. Figure 22 shows this evolutionary process in a graphical fashion.

This evolutionary process depended on the advancement of science. As new horizons of scientific knowledge continued to expand, newer areas of specialized applications emerged and the engineering branches continued to proliferate based upon the specialized needs of the society. The specific dependence of the engineering specializations upon different branches of science and mathematics is depicted in Fig. 23. The primary engineering activities included mining, metallurgy, and civil engineering. Civil engineering was dependent primarily on the knowledge of mechanics

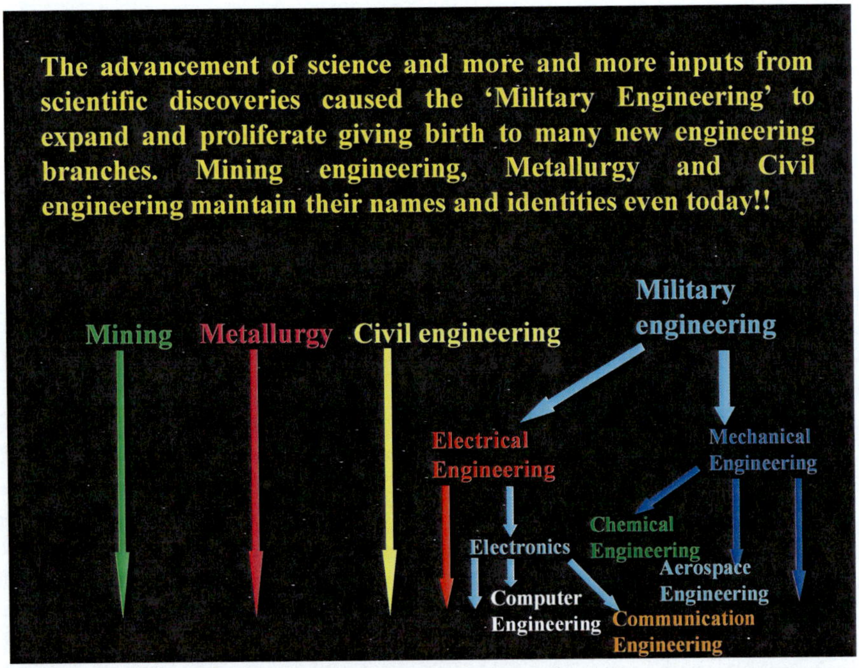

Fig. 22 The evolution of the engineering disciplines with time and advancement in science

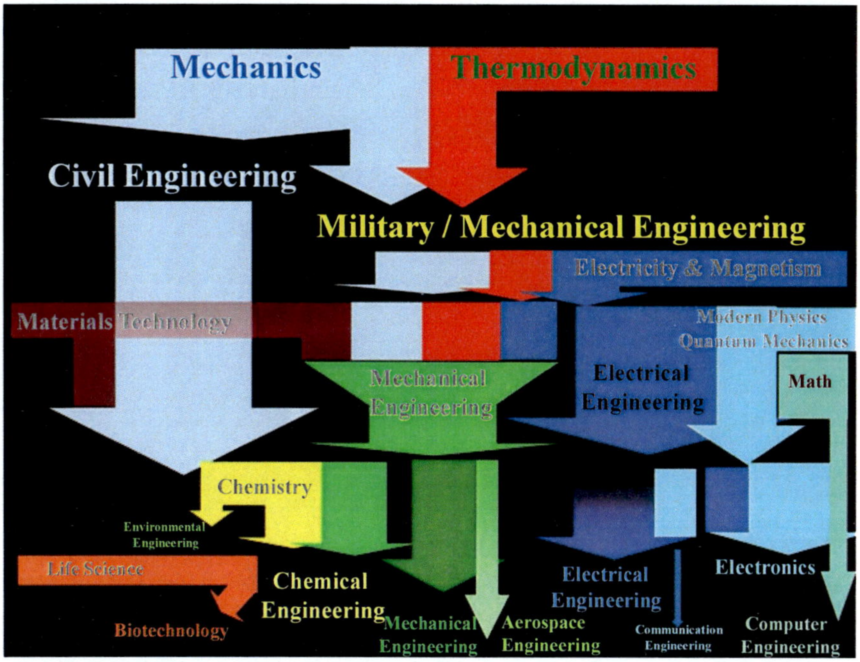

Fig. 23 Advancement in science and the corresponding evolution of engineering disciplines

and materials. Once thermodynamics got combined with mechanics, it gave rise to a specialization called 'Mechanical Engineering'. The next development in science was electricity and magnetism, and mechanical engineering was split into two important core branches—mechanical engineering and electrical engineering. Once atomic physics developed, electronics became an important subject and its further development led to computer engineering. In certain areas of mechanical engineering, chemistry started playing a major role and this was the impetus for the development of chemical engineering. Aircrafts and space devices coming into the picture in a major way caused a further split of mechanical engineering, and the result was the emergence of aerospace engineering.

It has been mentioned earlier that the invention of prime movers led to an Industrial Revolution—called the First Industrial Revolution. If production of iron is taken as a measure of the degree of industrialization, then Fig. 24 shows the characteristic change in the growth rate of Iron production. The sudden change in the slope indicates a paradigm change and the Industrial Revolution. The Second Industrial Revolution (IR) is considered to be caused by the emergence of computer technology and the semiconductor industry. This can be identified with the help of the economic growth characteristics shown in Fig. 25. The primary impetus to the Second Industrial Revolution came from the miniaturization of electronic devices. Figure 25 shows mainly two points where the slope changes and, so, the author feels that there have been really two revolutionary changes in the industry. Some historians have made finer

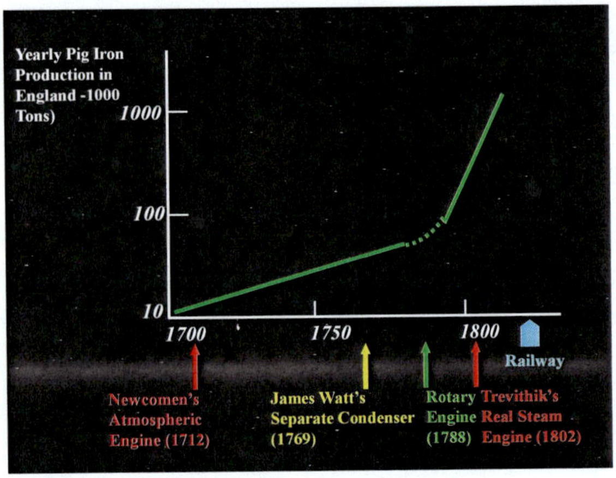

Fig. 24 The sudden change in the growth rate of iron production

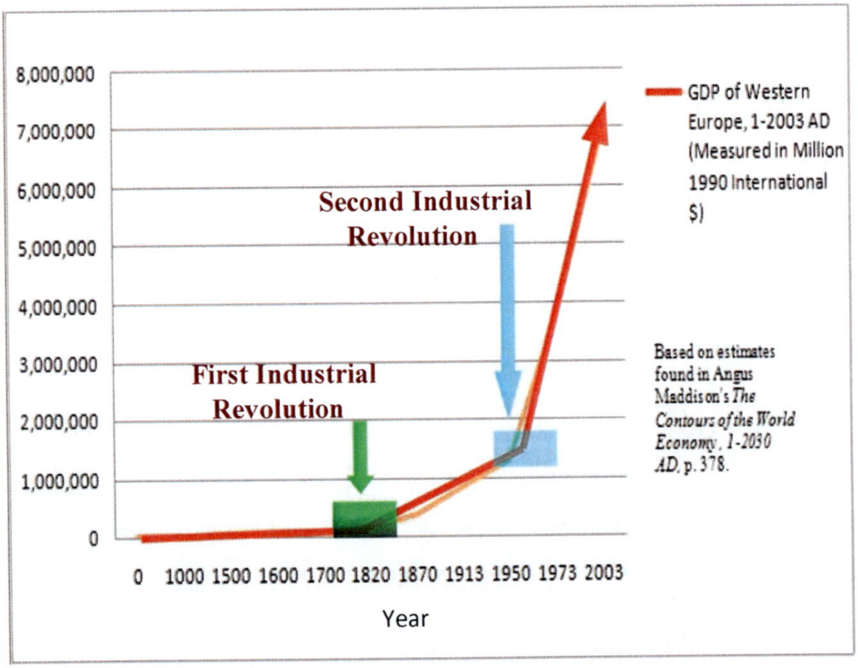

Fig. 25 The Industrial Revolutions indicated by the changes in slope of the GDP growth rate

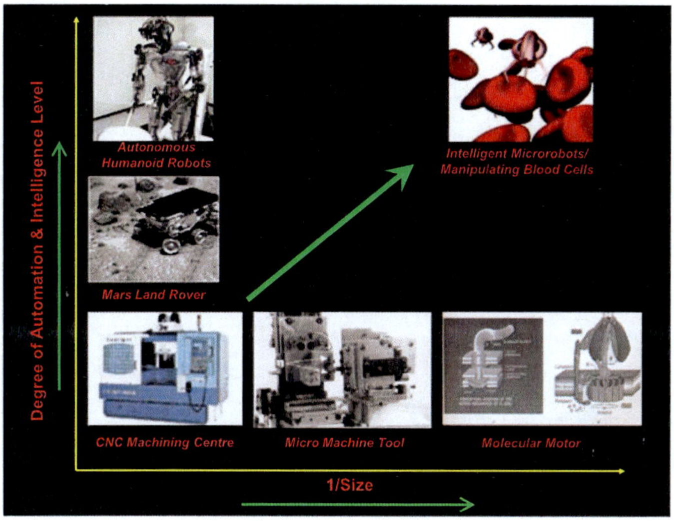

Fig. 26 Trend of engineering in intelligence and size

divisions and suggested more numbers of IR. The trend of miniaturization in engineering activities can be observed in the trend of development shown in Fig. 26. There are a number of reasons behind this trend of miniaturization. Ability to incorporate much higher level of intelligence in the machines and devices is one of the reasons. The other important reason is the opportunity to utilize the advantages derived from the scaling laws. Miniaturization will become increasingly more important because of the following reasons:

(i) Lower cost and higher packing density make it possible to incorporate a very large number of variety of sensors which is essential to make a machine intelligent and autonomous. (ii) Non-invasive health care can be possible through miniaturization of instruments. (iii) Low energy and low material consumption will lead to better conservation of resources. (iv) Response of devices will be faster and disposal will be easier, and (v) Using new laws of physics and chemistry for small scale has started becoming necessary for many applications.

The Second Industrial Revolution was made possible because of two-dimensional miniaturization of electronic circuits. They empowered the systems for manipulation of information. However, many experts believe that the Third Industrial Revolution will be made possible through three-dimensional miniaturization of devices, and the objective will be miniaturized manipulation of material and physical work. It should not be forgotten that miniaturization will be unavoidable for making not only small things but for developing systems of macroscopic sizes also miniaturization will be a very important activity. The major advantages will be compact packing of very large number of sensors, making actuators relatively more powerful and in reducing the cost for necessary social impact. Intelligent systems' major characteristics depend on the massively parallel arrangement of miniaturized devices like all living organisms.

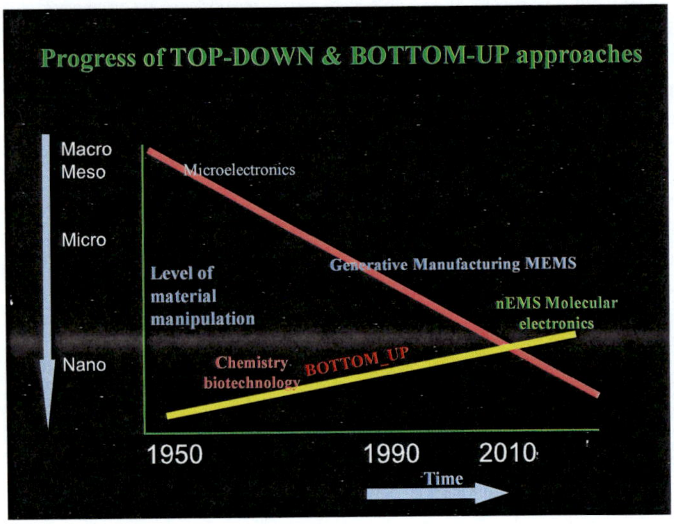

Fig. 27 Progress of miniaturization capabilities with time

This is not only to just increase the information collection ability but also to use the advantages of scaling laws in many cases as indicated before. Figure 27 shows how the capability of miniaturization has progressed over the years both through 'bottom-up' and 'top-down' approaches. The processes and their respective capabilities for miniaturization are shown in Fig. 27.

In top-down approach of making things, the desired shape is achieved by taking a piece of material, and shaping is either by subtraction of excess volumes (viz., machining) or by deforming the given material (viz., casting, forming). In making miniaturized shapes with complex geometries, the desired shape is generated by adding material in small quantities. Present-day rapid prototyping and 3-D printing processes belong to this class of shape generation technology. One major difficulty is that these processes are basically slow and may not be suitable for mass production. However, the emerging technology based upon self-assembly of material will remove that hurdle in the not-too-distant future.

The demand for miniaturization will have major impact on various aspects of mechanical engineering as mentioned below:

Configuration and Design
Material
Actuation and motion generation
Sensing and Control
Fabrication and Manufacturing
Energy source

Intelligent systems' major characteristics depend upon the massively parallel arrangement of miniaturized devices like all living organisms. This is not only to just increase the information collection ability but to use the advantages of scaling laws. A quick glimpse of the future trends in mechanical engineering is presented in Fig. 28. In fact the nature of 'Mechanical Engineering' in future may be very different and, even, beyond recognition.

For developing miniaturized mechanical systems and devices, it will be necessary to design those as monolithic entities since assembly of miniaturized parts is next to impossible. As an example, the monolithic miniaturized version of a four-bar straight-line mechanism is shown in Fig. 29. To achieve this, all hinge joints are replaced by flexure joints which simulate the relative motion among links as obtained through hinges.

Some major conceptual paradigm shifts are expected because of the tremendous advances in material science, computer science, and electronic devices. Bottom-up approach instead of top-down approach will play a predominant role in shaping objects eliminating the needs for tools, dies, and assembly operations. Manufacturing of most micro- and nano-sized machines and devices will be accomplished through coded self-assembly of material. Figure 30 shows some very elementary examples of self-assembly. By controlling the hydrophobic character of various surfaces, many different patterns can be generated by self-assembly.

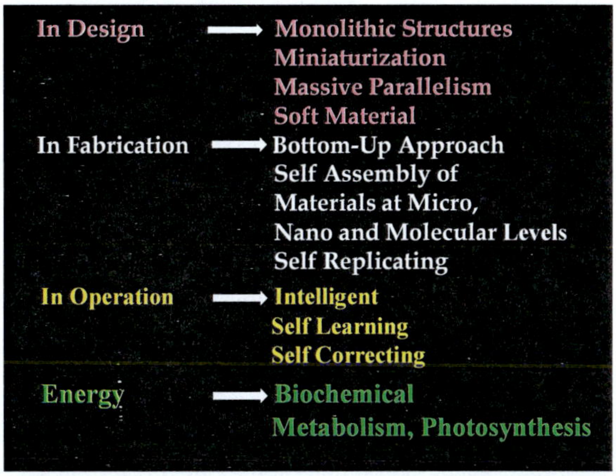

Fig. 28 Trends of change in various aspects of mechanical engineering

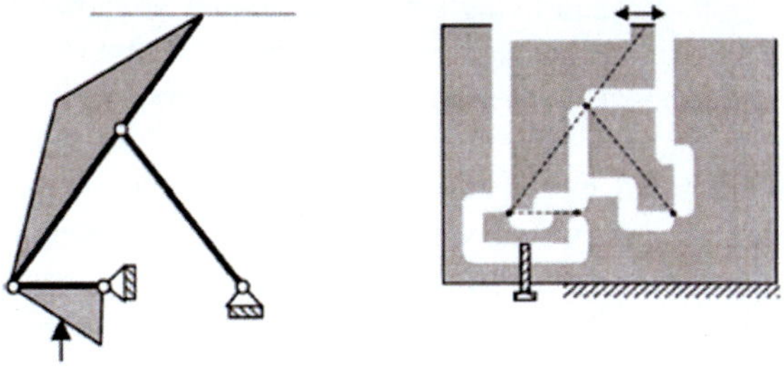

Fig. 29 Flexure-based complaint mechanism derived from a cognate of the Chebyshev's approximate straight-line mechanism

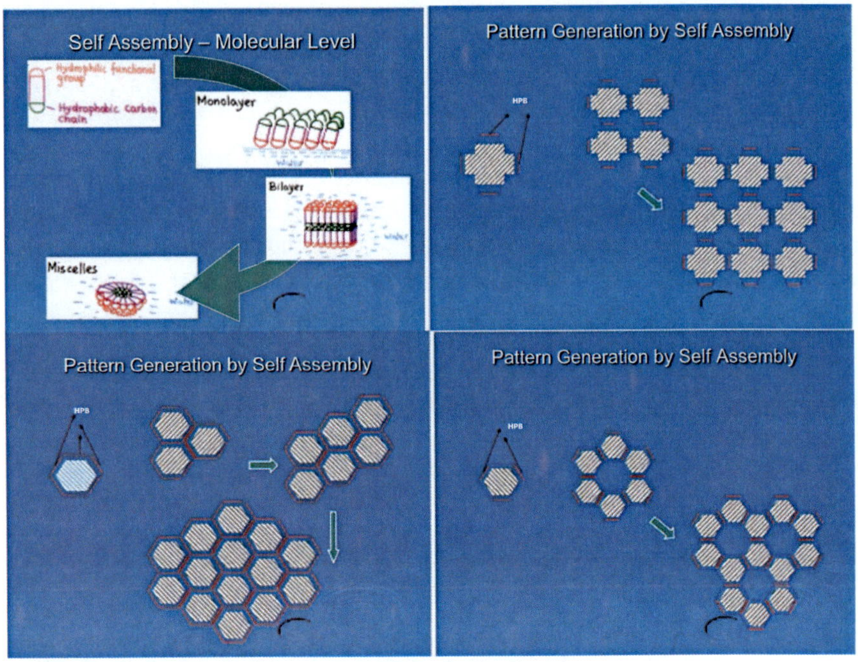

Fig. 30 Some examples of elementary self-assembly process for pattern generation

So far as material is concerned, more and more non-metallic soft materials will replace the present-day hard metals and alloys. In many cases, this will be a natural outcome of the self-generation process. New sources of energy including biochemical energy as in case of living objects will evolve. Static and direct conversion in many cases along with wireless transmission of large power will be developed. Space will be a prominent source of energy. Revolutionary concepts in storage and retrieval of energy will emerge. For micro-level actuation, microscopic actuators and even molecular motors will be used. Artificial muscle-like materials will be developed. The preliminary work has already started, and some electro-active polymeric materials have been developed which generates movement with electrical stimulation. 'ATP Synthase'-type molecular motors will emerge which will be driven by flow of ions as in the case of living organisms. Figure 31 shows the model of an ATP Synthase. ATP may act as a major energy currency of future artificial devices.

The above discussion indicates that many of the new-era machines and devices, particularly the micro-sized ones, will be of monolithic nature. The joints will be replaced by structured and localized compliance. Ionic polymer-based artificial muscles will play the role of actuators. Actuation can be distributed. Motors and transmission systems will be eliminated in such cases, and the use of smart materials will be common. Sensors will be miniaturized beyond recognition. Multiplicity and redundancy of sensed signals will help in achieving intelligent machines. Sophisticated distributed control will make parallel systems more common.

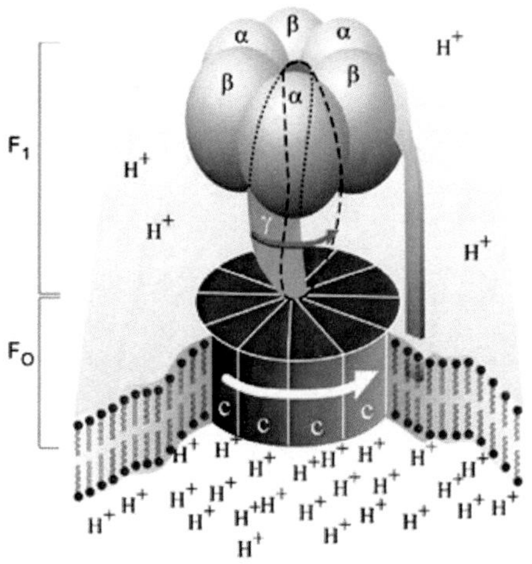

Fig. 31 Model of ATP Synthase

Apart from the trend of miniaturization, the shrinking of time involved is also a feature of the emerging technologies. So now engineers are handling time durations of picoseconds and femtoseconds. Another extremely important aspect of modern engineering is simulation and modeling. This requires sound understanding of the science behind a phenomenon and good expertise in the mathematical and computational tools. With the capability to peer down at the molecular level, the mysteries of life are unraveling many new concepts and information. The engineers have started realizing that the nature being the best designer it is worthwhile to imitate nature and many natural processes related to living objects. The ability to miniaturize and manipulate at the molecular level has opened up new avenues for achieving artificial intelligence and creating truly intelligent devices. This, however, requires the modern engineers to be far more knowledgeable in modern science including life science. An intensive integration of material physics/chemistry, material science with Mechanical Design and Manufacturing will be necessary.

Many areas of electrical and computer science will merge with the mechanical design of machines and devices. As machines will have much higher level of resemblance to living objects, many aspects of life science will become integral parts of design and manufacturing. Micro-system technology will play a predominant role. Thus, many new concepts in design and fabrication will take the center stage.

A new subject 'Synthetic Biology' which will lead to the design and development of machines and devices following the principles of biology is emerging. Such systems will be, obviously, inanimate objects but will behave like living objects in many respects.

A mechanical engineer of the future will have to acquire competence in many new areas as indicated below:

Molecular Engineering
Synthetic Biology
Smart Macromolecules and

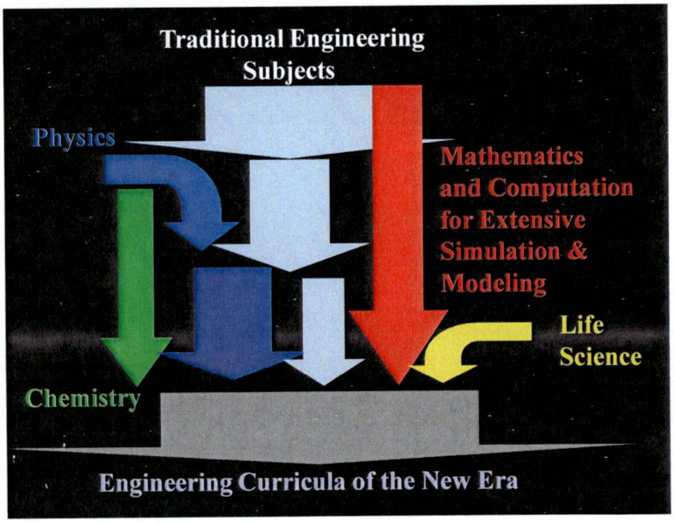

Fig. 32 The evolution of curriculum for future mechanical engineering

Intelligent Materials
Manufacturing by Self-Assembly of Materials
Artificial Intelligence, Self-Learning,
Self-Correction
Self-Replication.

This will require the curricula for the future generation 'mechanical engineers' to undergo major revision as indicated in Fig. 32. The future mechanical engineering curriculum is going to be far more interdisciplinary in nature and heavily dependent on basic sciences. A reasonable dose of life science will be also the essential part of all branches of engineering. This is mainly because the impending 3rd IR is going to be based upon the engineering of miniaturization and 'synthetic biology'.

All these revolutionary changes will result in the 3rd Industrial Revolution that will far surpass the 2nd IR in its impact on our civilization and the mankind. 'Mechanical Engineering' will once again take the center stage.

Bibliography

1. Dixit, U.S., Hazarika, M., Davim, J.P.: A Brief History of Mechanical Engineering. Springer (2017)
2. Dickonson, H.W.: A Short History of the Steam Engine. Cambridge University Press (2010)
3. Pelesco, J.A.: Self Assembly: The Science of Things That Put Themselves Together. Chapman and Hall/CRC (2007)
4. Baldwin, J., Bayer, T., Dickinson, R., Ellis, T., Freemont, P.S., Kitney, R.T., Polizzi, K., Guy-Bart, S.: Synthetic Biology: A Primer (Revised Edition). Imperial College Press (2012)

The Art and Signs of a Few Good Mechanical Designs in MEMS

G. K. Ananthasuresh

1 Introduction

Mechanical design is a triple helix intertwining geometry, material, and manufacturability. Geometry mostly decides the functionality of a device; materials dominantly determine the performance; and manufacturability arguably dictates the economic viability. While this is true of any field, it is particularly relevant where movable elements are involved. *Microelectromechanical Systems* (MEMS) field is unique in this respect. MEMS devices combine sensing, actuation, processing, control, communication, and power within a minute volume. It is important to note that miniaturized mechanical elements have made this possible. Some MEMS devices have nearly a million moving parts that are individually controllable (e.g., tilting mirrors of Digital Light Processor used in projectors)—a feat unmatched by any engineered mechanical systems in the macro world.

Mechanical design of MEMS has not been easy because MEMS designs, at the inception of the field, had to be made using the same unit processes that were available for making electronic circuits, and using the same materials. This continued along the same lines while a few more processes and materials were added to the repertoire [1]. Thus, design and fabrication are tightly integrated to achieve the desired functionality within the constraints. There aren't many other instances of such integrated systems other than biological cells and unicellular organisms. It is therefore pertinent to reflect on the role of mechanical design in MEMS.

When a MEMS researcher was asked to list a few good mechanical designs, the reply was that commercially successful devices have good designs. Furthermore, it was noted that great MEMS designs are those that actually made it into useful products and concurrently diverged from concepts that are routinely used in meso- and macro-sized devices and systems. This is indeed true. Some mechanical designs

G. K. Ananthasuresh (✉)
Mechanical Engineering, Indian Institute of Science, Bengaluru, India
e-mail: suresh@iisc.ac.in

© Springer Nature Singapore Pte Ltd. 2021

U. S. Dixit and S. K. Dwivedy (eds.), *Mechanical Sciences*,
https://doi.org/10.1007/978-981-15-5712-5_2

in MEMS are so clever that there were no such designs before in other disciplines. Some, of course, were taken from other disciplines to fulfill a need in MEMS. In this chapter, we consider a few good designs of both kinds. As we examine them, it might reinforce the idea that design is mostly an art. But that is not entirely true. There are systematic methods to obtain novel designs that are elusive to human designers [2]. And there are logical extensions of elegant concepts that point to new designs. It is thus important to notice the characteristics of good designs. Therefore, several mechanical designs used in MEMS are presented with an eye toward nine indicative attributes of good designs:

Simplicity	*Ease of making*	*Generality*
The "wow" factor	*Optimality*	*Economy*
Ubiquity	*Modularity*	*Hierarchy*

The collage of a few MEMS devices is depicted in Fig. 1. When seen in their microfabricated and assembled form, the intrinsic merit of these is not easily apparent. So, they are broken down into components that have the essential quality that defines a device. This helps us to adapt and use the design concept in other applications.

Fig. 1 A collage of a few good mechanical designs used in MEMS. Starting from top-left, and moving left to right row-wise: a folded-beam suspension and an electrostatic comb drive actuator [3], Agere two-axis tilting mirror array used in optical cross-connects [4], a Displacement-amplifying Compliant Mechanism (DaCM) that helps simultaneously increase sensitivity and bandwidth of a capacitive accelerometer [5], a two-axis in-plane accelerometer that mechanically decouples and amplified motion in x- and y-directions [6], an ultra-sensitive ring gyroscope [7], an array of flat beams in a polychromator [8], a bent-beam thermal actuator array [9], a herringbone fluidic mixture [10], a cell stretcher [11], and an array of heatuators that enhance effective electrothermal expansion [12]

2 *Simplicity* of a Folded-Beam Suspension

Joints are essential components of any mechanical device with moving parts. MEMS field got its recognition at the inception of the field when researchers showed that hinges and sliders could be made at the micron scale with polysilicon by incorporating a sacrificial layer in microelectronic fabrication processes [13]. Electrostatic micromotors [14], which could spin about an axis perpendicular to the silicon wafer, had used cleverly-crafted revolute joints, i.e., hinges. Similarly, prismatic joints (i.e., sliders) too were realized using silicon. Sacrificial layer process was the key to those developments. Years later, Sandia laboratory made a miniature revolute joint with a seven-layered process called SUMMiT [15], Sandia Ultra-planar Multi-level MEMS Technology. The geometry of the Sandia micro-hinge can be argued to be more complex than that of a macro-scale bearing. The relative tolerances were not stringent in those miniature revolute joints even though absolute values of tolerances were only a few microns. If MEMS field limited itself to such kinematic joints that merely imitate macro-scale joints, the field would not have attained the sophistication we see today. The key was simplicity offered by compliant design [16].

In *compliant design*, elastic deformation is utilized instead of kinematic joints such as hinges and sliders, to achieve relative motion. A compliant hinge can be as simple as a short beam segment or a narrow flexure that enables a body attached to it to rotate about a point. But this has three drawbacks: (i) the effective center of rotation keeps shifting during the rotation; (ii) the range of motion is limited because of stress in a deforming body; and (iii) there is inevitable resistance to motion because some effort is needed to deform a body. So, a compliant hinge cannot come close to a revolute joint in terms of functionality and performance. The situation is better with a compliant slider.

A kinematic sliding joint needs a guideway in which a block moves to provide relative translation. Such a moving block has zero resistance, barring friction, in the direction of translation; it has infinite resistance in all other directions, namely, two other translational directions and all three rotational directions. Resistance in a compliant slider is due to elastic stiffness. Therefore, a compliant slider should have as low a stiffness as possible in the intended translating direction and as high a stiffness as possible in all other directions. A pair of beams achieves that to some extent as can be seen in Fig. 2a.

Note that a cantilever beam has high stiffness in the axial direction and low stiffness in the transverse directions. So, we can use a beam to translate a body. But a block attached at the free tip of a single cantilever beam would rotate. On the other hand, a block attached to a pair of beams does not rotate much. So, it becomes a compliant slider. But then, we notice in Fig. 2a that the moving block experiences slight motion in the direction perpendicular to the intended translatory motion. This is avoided by using a folded-beam suspension [17] shown in Fig. 2b. The transverse motion of one pair of beams is perfectly compensated by that of another pair because of the folded-beam configuration. Figures 2c, d show the solid model and deformed configuration of a folded-beam suspension, respectively.

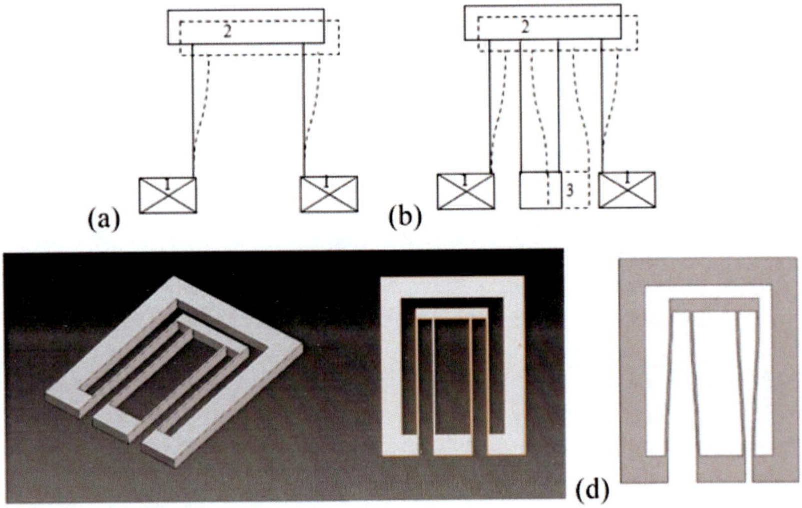

Fig. 2 The design principle of a compliant slider that uses a folded-beam suspension. **a** A parallel-beam suspension; while block 2 translates parallel to the anchored block 1, it also moves closer in the transverse direction; **b** the downward motion of block 2 is compensated by the upward motion of block 3 relative to block 2 because of the extra folded-beam pair; **c** a solid model of the folded-beam suspension in isometric and planar views; **d** the deformation of the compliant slider as obtained using finite element simulation

The folded-beam design for a compliant slider was borrowed by MEMS researchers from the precision mechanisms field that needed backlash-free precise motion not offered by kinematic joints.

A folded-beam suspension, which serves as a compliant sliding joint, has simple yet remarkable design. It enables one body to perfectly translate relative to another. The stiffness in the translating direction is much lower than that in the other directions. This is especially true when the height of the beams is larger than the in-plane width. Even more interesting is the fact that the stiffness in the translating direction is nearly constant over a long range of motion. Its range of motion is limited by the spacing of the beam when it is made in a single layer. The range can be extended when it is made of two or three layers.

In summary, the functionality of pure translation without any offset motion of a folded-beam suspension arises because of the flipping of one pair of beams. The performance parameters, such as the stiffness, maximum stress, range, etc., can be easily calculated using elementary beam theory. Finite element analysis can also be used if accurate nonlinear behavior is to be captured. The simplicity of this design enables it to be made using a single releasable layer in microfabrication. Its simplicity is also because the dimensions (length and cross section of the beams and their spacing) would not alter the characteristic behavior much. Furthermore, it uses distributed compliance [18] and thus keeping the stress low and hence increasing the range of motion. This design is commonly found in micromachined accelerometers, electrostatic comb drive actuators, and other MEMS devices.

3 *Ease of Making* of a Compliant Two-Axis Gimbal

When Micro-Opto-Electro-Mechanical Systems (MOEMS) started to emerge, researchers needed an array of tiny mirrors to reflect and steer light beams. Texas Instruments (TI) pioneered a tilting mirror, which could rotate about a single axis parallel to the silicon wafer. It used a short beam segment that twists to provide rotation. It was a simple design and served its purpose then. When it came to mirrors that needed to rotate about two orthogonal axes, more ingenuity was required. Such a need is depicted in Fig. 3. A light ray from any single input optical fiber in an input bundle needs to the directed to any fiber in an output bundle. For this, two arrays of mirrors were used by Agere Systems [4]. Each mirror was expected to tilt about two in-plane orthogonal axes. And the mirrors should almost fill the space in a plane, which means that the suspension of the mirror should be as compact as possible. The design question was this: How do we obtain a two-axis gimbal design that has a large range of rotation about both axes and one that can be compactly microfabricated in a single releasable layer? The answer can be seen in Fig. 4a–c.

A short beam of the kind used in the TI mirror does not give much rotation; it is severely limiting because the stress in a beam that twists a lot is very high. The alternative is to use bending using slender beams, which embody the concept of distributed compliance [18]. This concept is shown in Fig. 4d where a serpentine beam that is anchored at both the ends can be seen. When one end is fixed to a frame

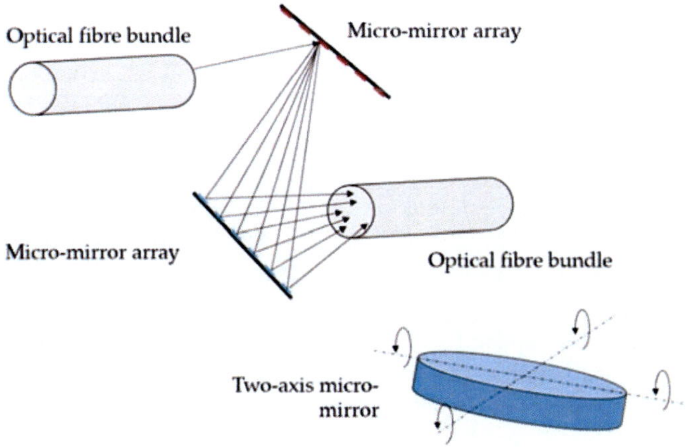

Fig. 3 The schematic of Optical Cross-Connect (OXC) developed at Agere Systems in early 2000s [4]. Each mirror in the two planar arrays should be tiltable about two axes in order to steer light from any fiber in the input bundle to any fiber in the output bundle

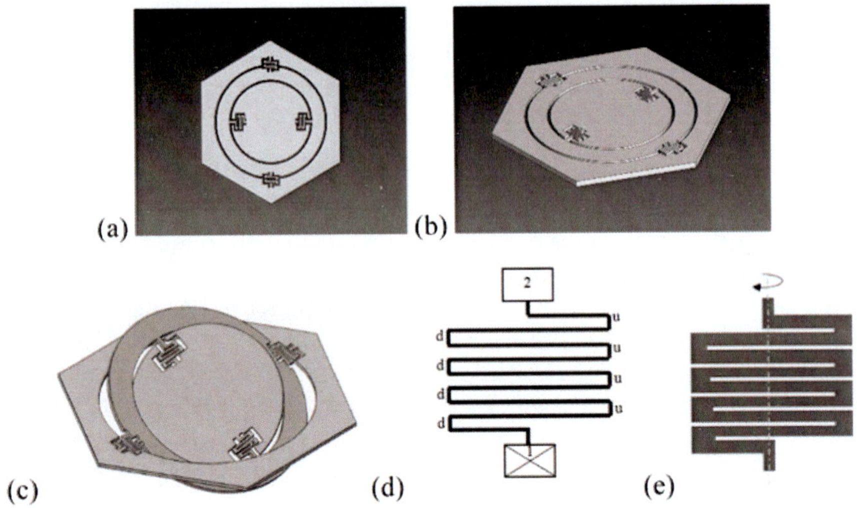

Fig. 4 A two-axis micro-mirror: **a**, **b** top and isometric views of a hexagonal block with two pairs of torsional hinges compactly arranged to give two-axis gimbal; **c** two-axis rotation of the mirror; **d** serpentine beam serving as a rotational hinge: when it twists, those marked with "u" move out of the plane and those with "d" move into the plane; and **e** a simple way to realize the serpentine beam rotational hinge using a few slits etched in a rectangle

and the other end is connected to a body that is required to rotate, the beams bend alternately up and down. This is indicated with letters "u" and "d" in Fig. 4d. When the serpentine beam *twists* by bending about the axis as shown in Fig. 4e, the beam segments marked with "u" move up and those with "d" move down. The longer beam segments deform accordingly, also up and down. A simple way to realize a serpentine beam in a single layer can be understood from Fig. 4e: We just need to etch a few narrow rectangular slits. As shown in Fig. 4a, b, such slits can be etched in a hexagon along with annular slots. The result is an amazing disk at the center with the capability to rotate about two in-plane orthogonal axes and by large angles. Once again, the distributed compliant design with bending of slender beams keeps the stress low. Furthermore, the rotational stiffness of the rotary joint in this design is quite low. The range of rotation and the rotational stiffness is limited by the microfabrication process. If lithography permits 1-μm-wide slits, many beams can be incorporated into the serpentine beam of a given size. However, even with a 5-μm-wide slit, substantial range of rotation with sufficiently low stiffness is possible with this design. Thus, ease of manufacture is a remarkable trait of this design that gives a rather complex functionality. This design is so simple and elegant that its ingenuity can be appreciated only if one contemplates alternative designs that surpass this.

4 *Generality* of an Electrostatic Comb Drive Actuator

Electrostatic force is a fundamental force. It exists between any pair of conductors that hold electric charge. It is also a large force as compared to another fundamental force, the gravitational pull. The magnitude of electrostatic force is large enough to cause sizeable motion at the micro-scale. It is inversely proportional to the gap between the conductors and directly proportional to the area of the conductor-surface patches that face each other. This is best understood when we consider two parallel plates shown in Fig. 5a.

When two parallel plates of overlapping area, $w \times l = wl$, of surfaces that face each other, are separated by a gap, g_0, and with a potential difference, V, the force in the gap direction between them is given by $\frac{\varepsilon_0 wlV^2}{2g_0^2}$, where ε_0 is the permittivity of free space. When the two plates come together with a reduced gap (Fig. 5b), the force increases a lot because of the inverse-square relationship. When the two plates displace along the width direction (Fig. 5c) or the length direction (Fig. 5d), the force is approximately constant. The force in the width direction is given by $\frac{\varepsilon_0 lV^2}{2g_0}$ and that in the length direction is $\frac{\varepsilon_0 wV^2}{2g_0}$. This simple principle is used for conceiving the comb drive actuator. Incidentally, if we use capacitance which has a similar relationship, it becomes the principle of a capacitive sensor.

The electrostatic comb drive uses the parallel-plate capacitor displaced in the length direction (Fig. 5d). As per the formula for the force, $\frac{\varepsilon_0 wV^2}{2g_0}$ is large when the gap is small. The gap is decided by the lithography limit of the microfabrication process used. If the process does not allow a small gap, a way to get around this problem is to use many pairs of parallel plates. This leads to the comb drive actuator shown in Fig. 6.

A comb in the comb drive actuator has many fingers. There are two combs, one is an anchored comb and another a moving comb. In Fig. 6, we see two anchored combs and two moving combs. The moving combs are attached to a shuttle mass. It is called a shuttle mass because it shuttles between two extremes limited by the suspension. The suspension here is the folded-beam suspension that acts like a compliant slider.

The electrostatic force between the moving shuttle mass and the fixed comb can be enhanced by packing many fingers into a given space. As said earlier, the narrower the gap between the fingers, the more the force. The range of displacement is limited by that of the compliant slider. It can be seen that the force of the comb drive actuator

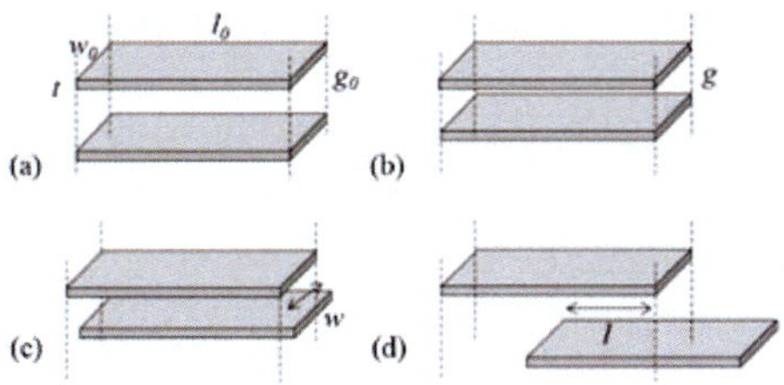

Fig. 5 Parallel-plate capacitor. **a** complete overlap in width and length separated by a gap; **b** reduced gap; **c** reduced width of overlapping surface; **d** reduced length of overlapping surface

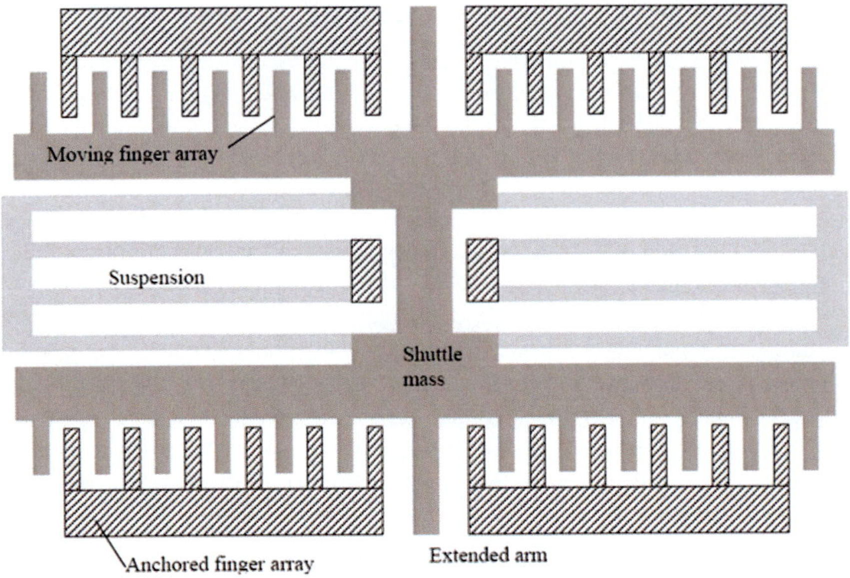

Fig. 6 Schematic of an electrostatic comb drive actuator. The hatched regions are anchored to the substrate, while the shaded regions are suspended above the substrate. The lighter shaded region indicates the folded-beam suspension. The moving comb fingers are attached to the shuttle mass that moves with the help of a compliant slider (i.e., the folded-beam suspension)

has to overcome the force required to deform the folded-beam suspension. So, the available output force tapers off with the stroke of the actuator. Most often, the force of a comb drive actuator is just sufficient to move the shuttle mass. So, it is useful in sensor applications (e.g., resonant sensors that need a mass to move back and forth). It is indeed a *prime mover* for microsensors and some micromechanisms. Sandia National Laboratories used a pair of comb drive actuators to turn a wheel continuously (not unlike a piston-crank mechanism) and called it a *microengine* [19].

Even though the force of a pair of fingers does not vary with their relative displacement in the length direction (to first order as the formula cited earlier neglects the fringing fields), the force is not constant due to the folded-beam suspension. This can be overcome by shaping the fingers. That is, one can vary the gap nonlinearly to increase the force with the displacement in the length direction [20]. However, this is not common because rarely MEMS applications need a constant force over a large distance.

As can be discerned from Fig. 6, the comb drive actuator needs a single releasable layer, which most microfabrication processes can provide. A small complication arises if the bond pads within the device are not possible (e.g., SOIMUMPs [21]). In such a case, the folded-beam suspension can be turned inside out to move the bond pads out to the periphery of the device.

The design of a comb drive actuator is simple, and it is easy to microfabricate. There is another feature in it that makes it general. Myosin motor, which is the fundamental building block for skeletal muscles, has the same interdigitated design [22]. The thick and thin myosin filaments have an arrangement similar to the comb drive design. It is hard to say, however, that myosin filaments inspired the comb drive. Nevertheless, the comb drive actuator is a general feature of many MEMS devices. It is the most widely used actuator in MEMS. This design is here to stay.

5 The *"Wow" Factor* of a Polychromator-Beam Mechanism

Sometimes a brilliant idea calls for a new design, and another brilliant idea is needed for practical realization of that design. An example of this is a polychromator [23], a MEMS device that is a miniaturized version of a correlation absorption spectrometer to identify a substance from a distance. The brilliant idea was to use diffraction of light to generate light of any wavelength from broadband white light, for the reference spectrum, using an array of long beams suspended above the substrate. The idea is illustrated in Fig. 7. Imagine a grating cell composed of an array of beams separated from one another in the top view and set at different heights above the substrate. If the gap between the beams and the substrate is of the same dimension as the wavelength of light, the gaps can be adjusted by applying electrostatic force using individually addressable drive electrodes. Then, the light rays reflecting from the top surfaces of the beams and the substrate underneath diffract to give out light of a particular wavelength. Different patterns of gaps give out diffracted light of desired spectral content.

The design question then is: How do we make the beams move up and down continuously for achieving any gap without distorting the beam? Since it is an optical

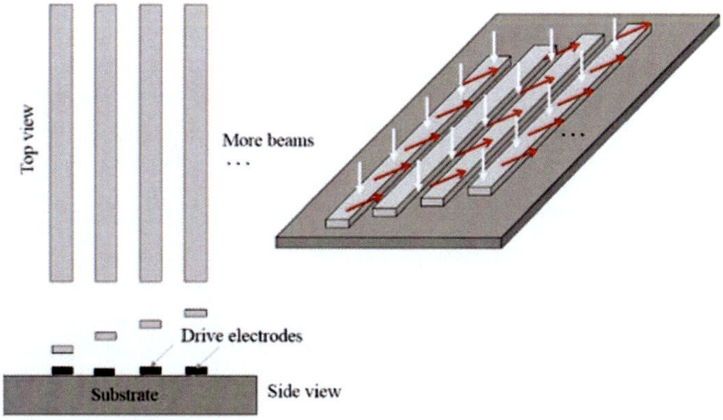

Fig. 7 An array of diffracting beams that can generate diffracted light of desired spectral content by adjusting the vertical gaps between beams and the substrate

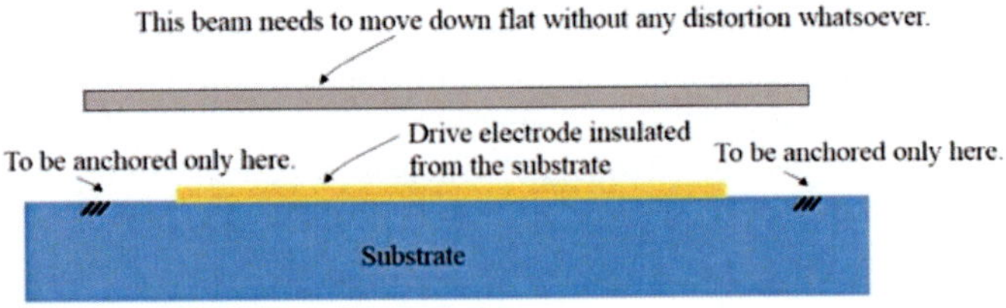

Fig. 8 An unusual requirement of an electrostatically actuated beam: How should the beam at the top be anchored to the substrate at the two points so that it moves down without bending, under the influence of the electrostatic force from the drive electrode?

application, the beam should not deform at all. But then, as in any microfabrication using photolithography, the beam has to be anchored to the substrate. The driving electrode ought to be underneath to pull the beam down. This situation is shown in Fig. 8. Those who know about elastic deformation of beams know that when a beam is anchored at the ends, it will deform. But here, we want the beam to not deform but still move down in the transverse direction. Much creative thinking is necessary to solve this unusual design problem. A reader is urged to spend at least 15 min before looking at the solution.

An ingenious solution to the problem posed in Fig. 8 was reported in [8]. As shown in Fig. 9, it uses a two-layered structure sandwiching a sacrificial layer. This design uses the well-known fact that a fixed–fixed beam under symmetric transverse load deforms with zero slope at the midpoint. Therefore, by attaching a short segment at the midpoint to the top beam, we get its downward motion without any bending whatsoever. This basic idea is further extended by concatenating two beams so that any tilting or vibration is also avoided. This is shown in Fig. 10, which shows a two-segment design in undeformed and deformed configurations. Figure 11 illustrates how an array of such designs can be realized to form a grating element, wherein each beam can be set at a different height, as desired by the intended wavelength of light.

The simple and elegant design of a polychromator serves its function really well. It maintains bending-free transverse motion of a long beam. It is the kind of a design that might occur quickly to a creative person in a flash or perhaps might occur (or not) to anyone after much thought. In any case, when we see it, we get the "wow" feeling. If it does not appear as an amazing design to a particular reader, he/she

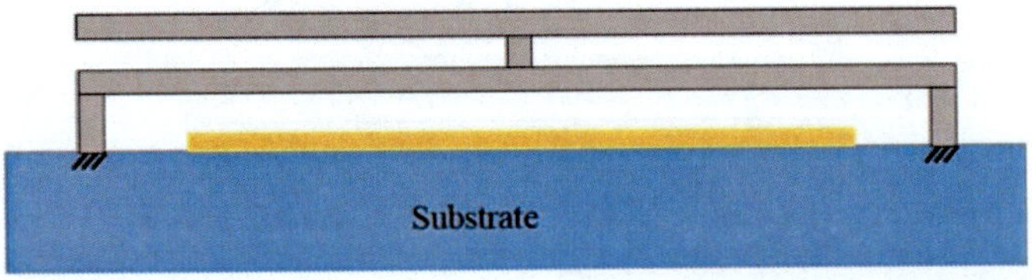

Fig. 9 An ingenious design that makes a long beam move down without bending when actuated from underneath and anchored at either end with an intermediate beam

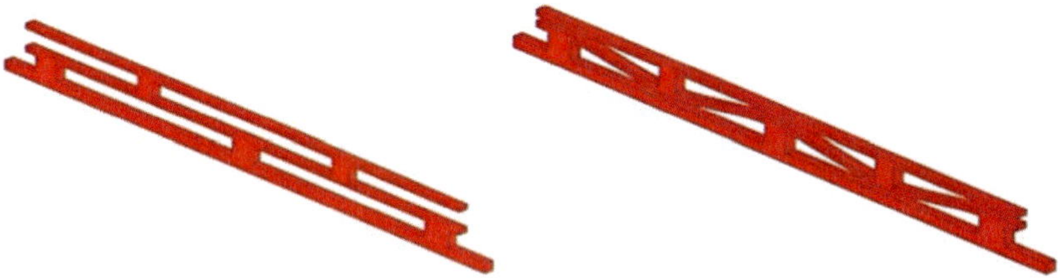

Fig. 10 A two-segment version of the design in Fig. 9 to make a longer beam stay flat. Undeformed and deformed configurations are shown

Fig. 11 A diffracting cell using an array of beams used in a polychromator

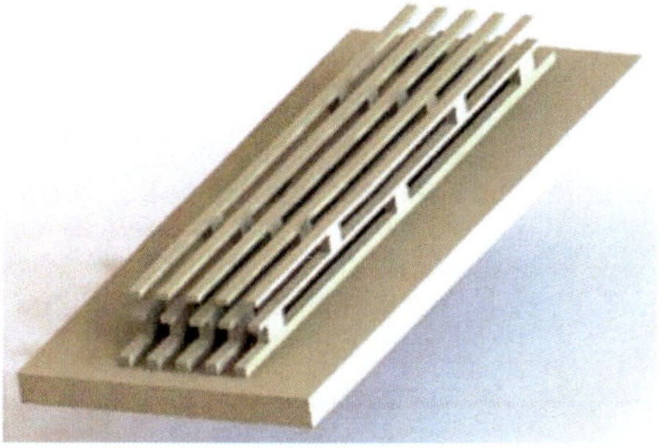

should contemplate a similar design in Sect. 7 wherein it is wrapped around a circle for even more amazing functionality.

6 *Optimality* of a Displacement-Amplifying Compliant Mechanism

The design discussed in the preceding section was most likely intuitively conceived. It was a remarkable example of human ingenuity. Those who use optimization techniques or those who develop them tend to think that algorithms too can give rise to clever designs. Designs generated by optimization algorithms can be counterintuitive. They are often beyond the grasp of human creativity. Thus, designs obtained using optimization algorithms are sometimes superior. Additionally, they are also optimal for the conditions set in formulating the optimization problem unlike intuitive designs that could be further tweaked and optimized. We illustrate this using an amplifying mechanism.

A lever, immortalized by Aristotle, comes to mind when we think of a mechanical amplifier. When a pin joint (i.e., a revolute joint) is not practically viable (as is the case with microfabrication), a simple lever is not a preferred choice. A flexure that replaces a pin joint is also not a good choice because of high stress and limited range of rotation. Furthermore, for large amplification ratio of output and input displacements, a lever needs a large space. The question, therefore, is how one can design a displacement-amplifying mechanism in a given compact space. A solution for this, given by a topology optimization algorithm [24], is shown in Fig. 12a. The squares in the top-left and top-right corners are anchors. When a force is applied vertically up on the flat segment at the bottom, the point in the middle of the top part of the compliant mechanism moves down by a large amount. This can be understood from Fig. 12b. First, the motion of this mechanism is counterintuitive: when we push a point up, another point of the mechanism comes down. Second, this design was optimal in the sense that algorithm that generated this satisfied the conditions of optimality set for it. Third, it is often difficult to argue which part of the mechanism or which beam segment is contributing to the amplifying behavior of this compliant mechanism.

Another Displacement-amplifying Compliant Mechanism (DaCM) is shown in Fig. 13. It is also obtained using a systematic design method and an optimization algorithm [6, 25], but it is more intuitive than the one shown in Fig. 12 but one that a human designer might not be able to conceive easily. Such mechanisms are

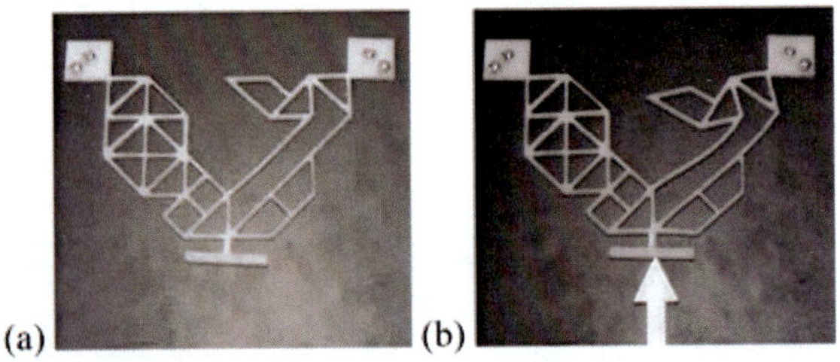

Fig. 12 A displacement-amplifying compliant mechanism designed using a topology optimization algorithm. **a** Undeformed, **b** deformed

Fig. 13 A displacement-amplifying compliant mechanism microfabricated with silicon [26]

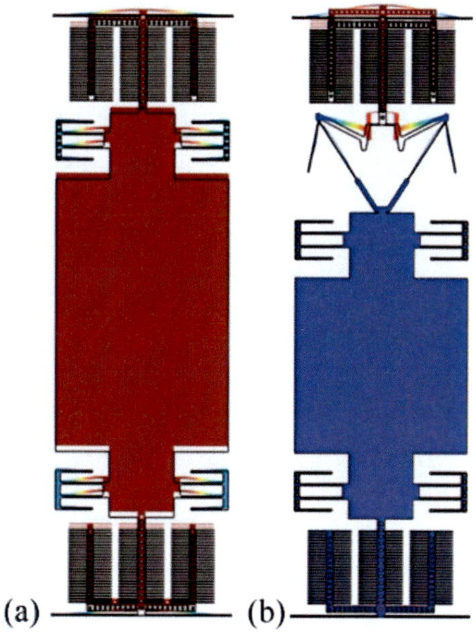

Fig. 14 Two designs of a micromachined accelerometer **a** without a DaCM, and **b** with a DaCM

designed with a purpose because an optimization problem with an objective function and some constraints would have been formulated. The purpose of the DaCM shown in Fig. 12 is shown in Fig. 14. Figure 14a, b shows the simulation of two capacitive micromachined accelerometers [6].

In Fig. 14a, one can see the familiar components of an accelerometer. It has a proof mass held by a compliant slider mechanism on the top and bottom. There are also electrostatic comb drives for the purpose of estimating the displacement of the proof mass indirectly through capacitance measurement. The design in Fig. 14b occupies the same footprint. For the same applied acceleration, the sensing combs at the top in Fig. 14b have larger displacement than that in Fig. 14a. This is because of the DaCM in the latter. Here, it can be seen that even with a smaller proof mass, we get larger displacement. It was also shown in [6] that the resonance frequency of the design with the DaCM is larger than that of the one without it.

Optimal designs, such as the DaCM just discussed, too have multiple uses as do intuitively conceived designs. Generality is indeed a common trait of good designs. The DaCM used in an accelerometer was also used in a micro-newton force sensor [26]. In this, the force applied at the input point of the DaCM results in amplified displacement at the output point. With optical measurement of the output displacement using a digital microscope, a DaCM becomes a force sensor. It is illustrated in Fig. 15. Thus, if a design is indeed optimal for a specific purpose (here, amplification of displacement) and could be used for other purposes, it falls into the category of good designs.

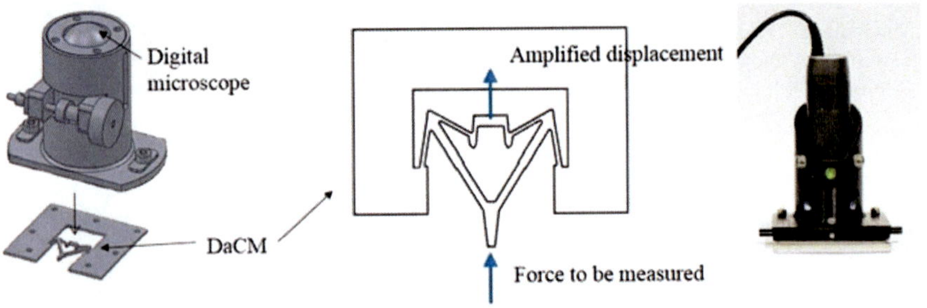

Fig. 15 A micro-newton force sensor using a DaCM and a digital microscope [26] and its product version (bendflex.in)

7 *Economy* of Material and Manufacturing

Benefits of minimalism are well known. It rings true for structural design as embodied in the adage: *the more you think, the less material you need*. Sometimes it also means *the less material you cut*. Serving a function well with a design that occupies a small footprint and minimal manufacturing effort is a goal in MEMS devices because the "real estate value" on a wafer is very high. And there are such MEMS designs that use almost all the material in a compact footprint and serve a useful function.

Shown in Fig. 16a is a rectangular layer patterned with a few slits indicated in white lines. What is left with these slits is a highly flexible spring. The width of the slit can be as small as a microfabrication process allows. The narrower the slit, the more flexible the spring is. This is because what we see here is an array of springs in series wherein each spring is a pair of two fixed-guided beams forming a rectangular box. It is indeed economical use of material and manufacture. Such a stack of springs has many uses.

Figure 16b shows two such springs (white and gray colors interchanged from Fig. 16a with white representing beams here) with a central ring. If the central ring is moved to one side, one spring expands and the other contracts. This structure was used as a cell stretcher [27]. As illustrated in Fig. 17a, b, when biological cells

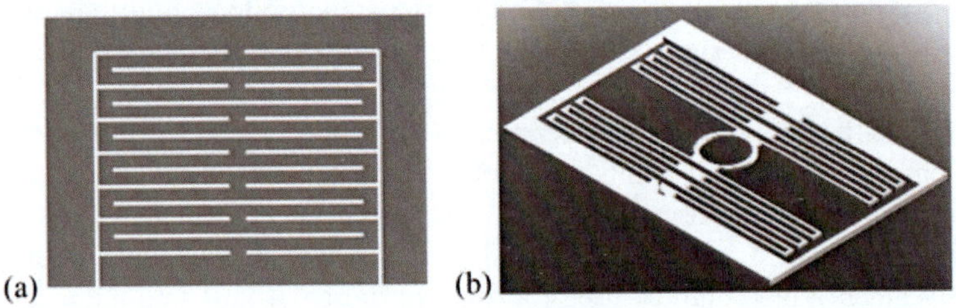

Fig. 16 An example of economy material and manufacture: **a** a very flexible spring is realized by cutting out a few slits (white lines); **b** a stack of two such springs on either side of the central ring

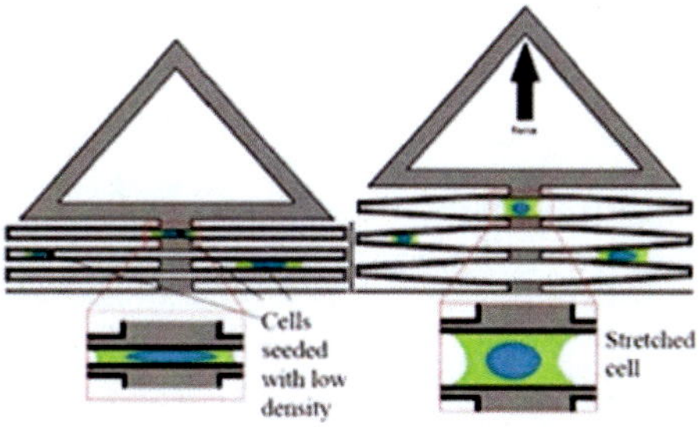

Fig. 17 A single-axis cell stretcher in unstretched and stretched configurations (adapted from [27]) using the spring shown in Fig. 16

are seeded on a patterned polymer layer, SU-8, in particular, cells are stretched by applying a force to stretch the spring. In fact, the stiffness of the spring is so low that cells themselves would stretch the spring, paving the way to measure forces applied by the cells [27].

There is another significant use for this kind of a spring. To see that, let us recall the concept of a micromachined ring gyroscope [28]. In a ring gyroscope, a ring is connected to a central post with spokes. In [28], semicircular spokes were used, see Fig. 18a. The ring is set into resonant motion by applying electrostatic force around it by using one set of electrodes on the periphery. The resonant mode shapes are two ellipses, as shown in Fig. 18b, c. The ring has degenerate mode shapes, which are inclined at 45° to each other. When the substrate to which the central post is attached rotates, then the ring starts to vibrate in the second elliptical mode shape. This motion is capacitively measured by the second set of electrodes on the periphery. This design, although worked well, could not compete with the resolution of the macromachined hemispherical-shell gyroscopes. One of the reasons for this is that the spokes connecting the ring to the central post distort the mode shapes. The second reason is that the actuation force and the change in capacitance were low.

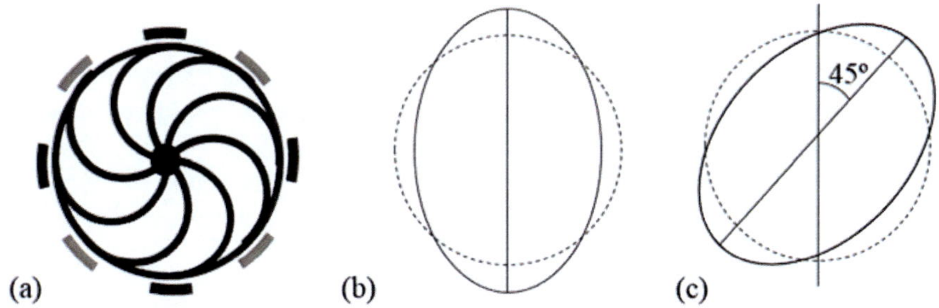

Fig. 18 Ring gyroscope concept. **a** A ring gyroscope with semicircular spokes attached to the central anchor with two sets of arc-shaped electrodes around, one to actuate the ring electrostatically and another to sense the change in capacitance as the ring deforms; **b** the first degenerate mode shape; and **c** the second degenerate mode shape

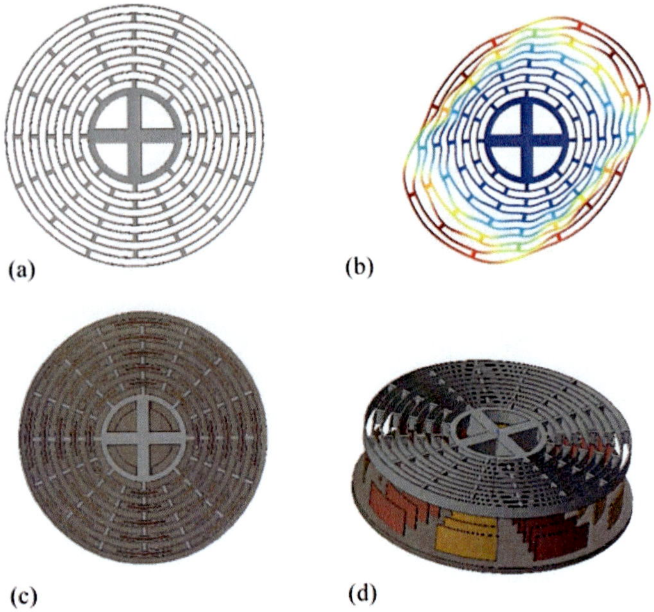
(a) (b) (c) (d)

Fig. 19 A multi-concentric ring gyroscope that combines the features of polychromator design in Fig. 10 and the compactness of the very flexible spring of Fig. 16a. It also embodies the principle of the ring gyroscope shown in Fig. 18 in terms of placing the electrodes in two groups, but with more concentric electrodes for resonance actuation and capacitive sensing

These two limitations were overcome to a large extent with a clever and economical design [7], as explained next.

Figure 19a shows concentric rings where adjacent ones are connected to each other in a staggered manner. Compare this with the designs in Figs. 10 and 16a. The design in Fig. 19a inherits the features of these two designs. Consequently, the stiffness of the concentric ring structure is made low, and the distortion at the midpoints of the arc beams is reduced (see Fig. 19b). In Fig. 19c, d, the electrodes for actuation and sensing are shown in top view and isometric views. This design of the micromachined gyroscope achieved resolution better than 0.01 °/h [29]. It is a great example of economical use of material with minimum etching of material resulting in superior performance.

8 *Ubiquity* of Electrothermal Microactuators

Electrostatic actuators, comb drive actuator being the prime example, are widely used in many MEMS sensors. But the force they generate is rather small—barely moving themselves and only occasionally moving something else. Electromagnet-based and electrothermal actuators, on the other hand, can generate much larger force.

Between these two, it is easier to design electrothermal actuators and much simpler to fabricate. They are increasingly used in MEMS devices where force is needed to move components other than themselves. The force can be an order of magnitude larger than that generated by electrostatic force. Hence, they are preferred actuators when large force is needed, and high speed of actuation is not very important. Common among them are two building blocks: bent-beam electrothermal actuator [9] and heatuator [29]. These two designs are ubiquitous in electrothermal actuators.

The extreme simplicity of a bent-beam actuator is apparent from Fig. 20a. It is simply a fixed–fixed beam with a slight kink—it is essentially a wide V-shaped beam. There are anchors, which also serve as bond pads for voltage application, at either end. As shown in Fig. 20a, upon application of a voltage difference between the two anchors, current flows through the bent beam. Consequently, Joule heating ensues. Heated beam expands. Since it is anchored at both the ends, it has no other way than push itself in the transverse direction. This is the basic principle of a bent-beam electrothermal actuator. The 3D view of a bent-beam actuator is shown in Fig. 20b. Just so that the beam does not move much in the out-of-plane direction, it is important to make it taller. Large thickness makes it strong as the area of cross section is large. Since the beam has to be sufficiently narrow in the in-plane direction in order to bend, height should be large.

A single bent-beam actuator cannot generate much force, but an array of them can. This is shown in Fig. 21. In such an array, individual bent-beam actuators are in parallel arrangement and hence their forces add up. This is also an example of economy of material and manufacture because very little space is left empty in the rectangular block. A microfabricated bent-beam actuator array is shown in Fig. 22.

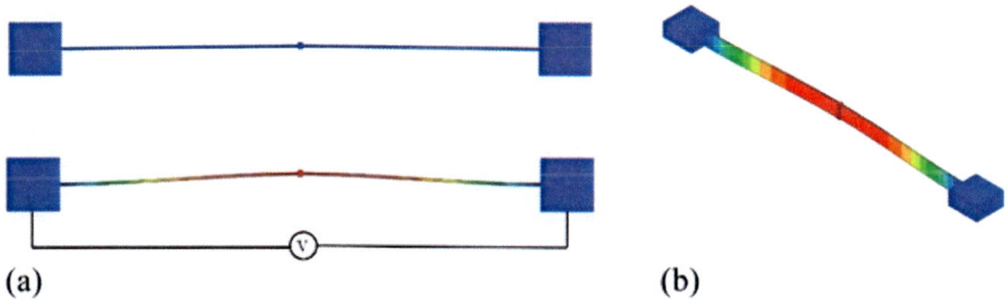

Fig. 20 A bent-beam electrothermal actuator: **a** in unactuated and actuated states; **b** 3D rendering of the same

Fig. 21 An array of bent-beam actuators that adds up the force

Fig. 22 A microfabricated array of bent-beam electrothermal actuators; the material is single-crystal silicon

The second building block of electrothermal actuators is called a *heatuator*, a portmanteau of heat and actuator [29]. Two versions of it are shown in Fig. 23 on the left and right sides. Both are ingenious designs. The one on the left side is the original heatuator conceived by Henry Guckel and his team [29]. It is a folded beam with one half much wider than the other. Two beams in the fold are connected with a short segment, and the other ends of the narrow and wide beams are anchored to bond pads. When a voltage difference is applied between the two anchor pads, current flows through the narrow and wide beams. Both heat up but the narrow beam gets hotter as the current density is higher in it. Consequently, it expands more than the wide beam. In order to achieve equilibrium under the ensuring thermal loads, the folded beam bends up as shown in the figure on the left side of Fig. 23. The principle of this actuator is similar to that of a bimorph [30] but there is a clever twist: a bimorph needs two materials with dissimilar coefficient of thermal expansion but here it is made of a single material. The disparity in the cross-sectional area is utilized here. Like the bent-beam actuator, it is also easy to realize in microfabrication as it needs a single releasable layer.

The figures on the right side of Fig. 23 show a variant of the original heatuator. Here, the wide and narrow beams are arranged electrically in parallel [12]. Therefore, the current flowing through them is different. The wide beam has more current than

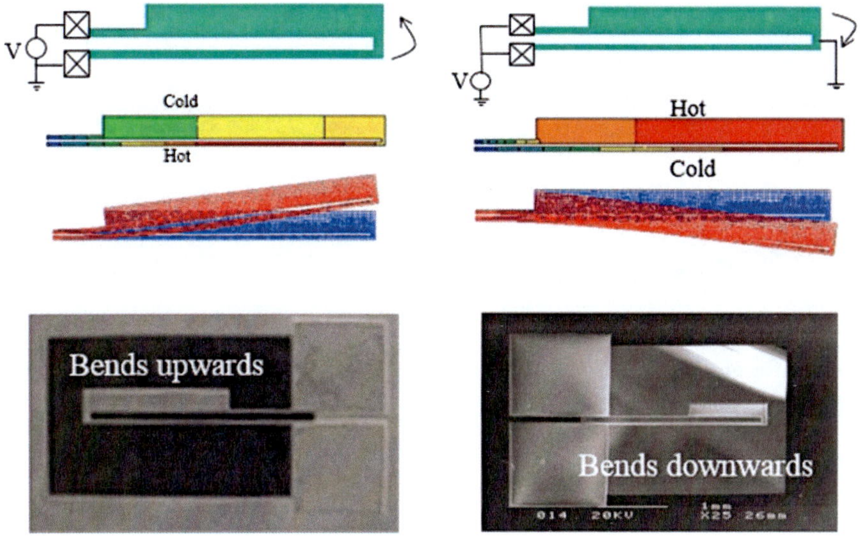

Fig. 23 Heatuators in series (left column) and parallel (right column) electrical connection

the narrow beam because the former has less electrical resistance than the latter. Therefore, now the wide beam gets hotter than the narrow beam. Of course, the length and cross-sectional areas have to be chosen suitably so that overall heating in the wide beam is more than that in the narrow beam. With the wide beam expanding more than the narrow beam, the same folded beam now bends downward as shown in the right-hand side of Fig. 23. As we will see later, there are many uses of this parallel electrical connection. But first, we should take note of the fact that both series and parallel versions have the same mechanical construction and that there is a short flexure that enables the wide beam to rotate. Also to be noticed is how parallel electrical connection could be given to a heatuator in the layout of microfabricated components.

As in the bent-beam actuator, many series heatuators can be made into an array to generate large force [12]. But we can do more with the parallel heatuators. This is shown in Fig. 24. In Fig. 24a, we see an expanding actuator that has four parallel heatuators. When voltage is applied between its two anchor pads at the bottom, they remain stationary but the movable pad at the top moves upward. Notice how the wide and narrow beams are arranged in order to achieve the upward or expanding motion, which can be seen in Fig. 24b. On the other hand, the narrow and wide beams are switched, and the heatuators on either side are kept at an angle, as shown in Fig. 24c. Now, we see a contracting actuator as can be discerned from Fig. 24d. The possibilities for design are almost unlimited with electrothermal actuation. It paves the way for topology optimization involving multiphysics simulation [31–33]. Analysis of electrothermal actuators is challenging because three simulations—electrical, thermal, and elastic—are to be performed in a sequence. Conduction,

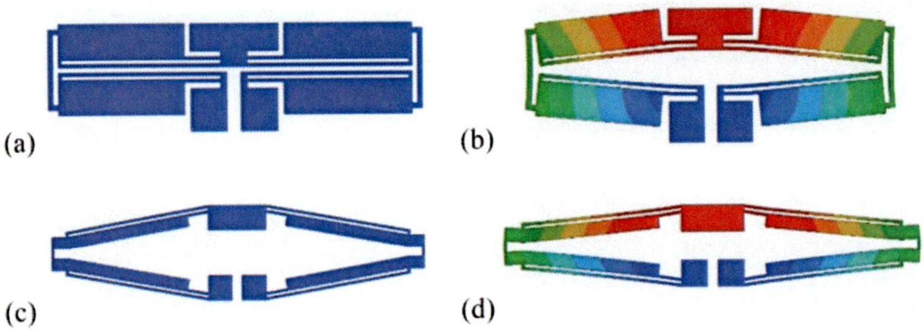

Fig. 24 Four parallel-connection heatuators forming expansion building block (top row) and contraction building blocks (bottom row)

convection, and radiation effects as well as temperature-dependent properties are to be accounted for in simulation and design [34]. Both intuition and systematic optimizations can lead to interesting designs. Shown in Fig. 25 is a planar robotic platform that uses three arrays of expanding parallel heatuators.

The bent-beam actuator and heatuator building blocks are used in arrays or in special arrangements. We see them in many actuators that use electrothermal principle. They are thus ubiquitous—the same principle used in many places.

Fig. 25 A microfabricated planar platform actuated with electrothermal expansion building blocks (adapted from [12]): by applying different voltages at the three anchor pads, the central problem can be positioned in the plane. The orientation is not independently controlled here as there are only two independent voltage differences among the three pads

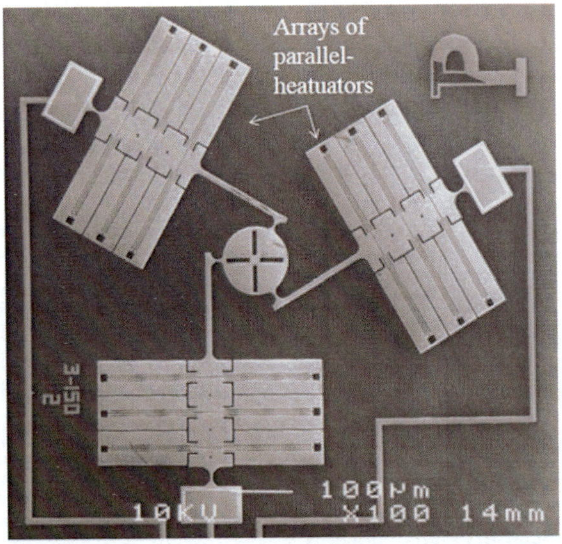

9 *Modularity* in a Compliant xy-Stage

Many designs illustrated so far have modular construction: a building block repeated in different ways in a specific arrangement to serve a function that is beyond what the building block does. The two-axis gimbal mechanism, comb drive actuator, polychromator, cell stretcher, concentric ring gyroscope, bent-beam actuator, and expanding and contracting heatuator blocks—all of them used building blocks. Modularity is thus a feature of a good design. It makes design and fabrication easy. The modules need not be identical. Even when they are identical, their interesting arrangements can lead to novel functionality. We illustrate novel arrangements of identical and non-identical building blocks, in this section.

Consider the compliant slider (or, the folded-beam suspension) once again. The slider and its schematic are shown in Fig. 26a, b. Shown in Fig. 26c is an arrangement of eight sliding joints connected in a specific arrangement to give the platform at the center two translational degrees of freedom. The platform of this kind cannot rotate. It is an *xy*-stage. By examining the motion of the platform in the *x*- and *y*-directions, one at a time, it can be seen that there is perfect decoupling of the two translatory motions in the plane. That is, when the platform moves in the *x*-direction, four sliders oriented in the *x*-direction move without disturbing the other four. The same is true

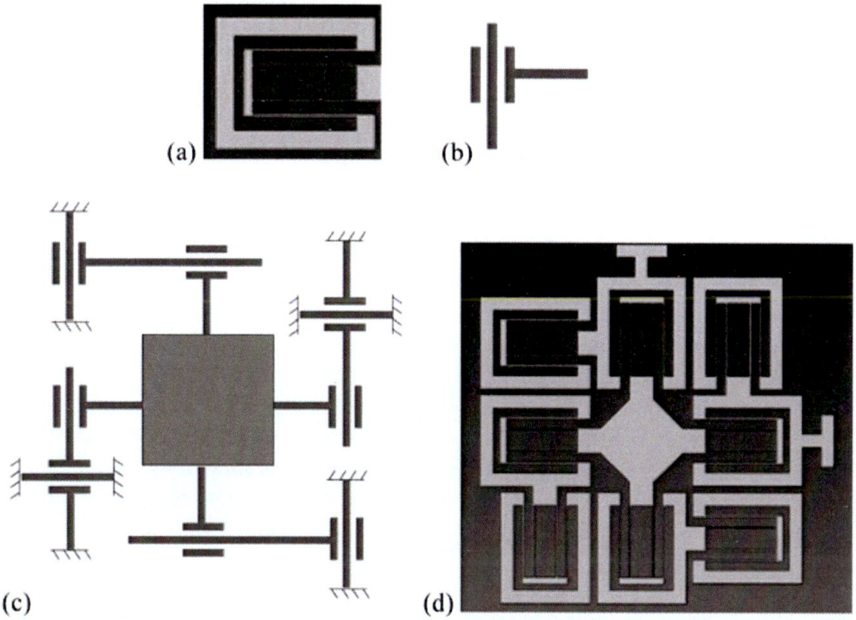

Fig. 26 A single-layer compliant stage with independent motion in *x*- and *y*-directions: **a** a compliant slider; **b** schematic of a compliant slider as a sliding joint; **c** an arrangement that illustrated how independent in *x*- and *y*-movements can be conceived; and **d** compliant version of the scheme in **c**

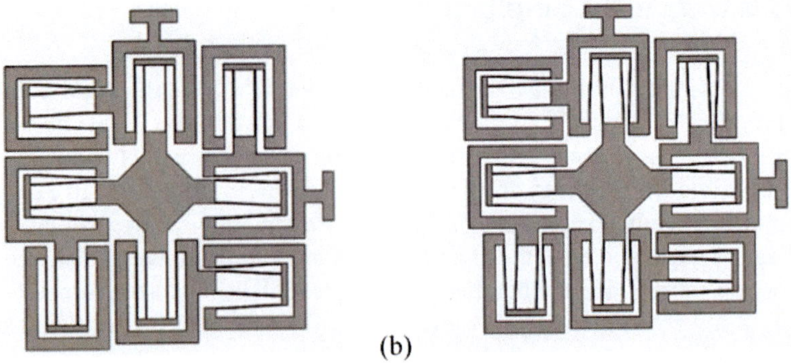

(a) (b)

Fig. 27 **a** Movement of the compliant xy-stage of Fig. 26d in the *y*-direction; notice that only four compliant sliders are deforming; **b** movement in *x*- and *y*-directions

for the motion in the *y*-direction. Now, consider the equivalent compliant mechanism where the eight sliding joints are replaced with compliant sliders, as shown in Fig. 26d [35]. Perfect decoupling is preserved in this conversion. This is thus a single-piece compliant mechanism that can be made in a single layer—a feature that makes it perfect for MEMS.

When the central platform in Fig. 26d is moved in the *y*-direction, only four sliders oriented in the *y*-direction deform as shown in Fig. 27a. Notice that there is no motion of the platform in the *x*-direction. Likewise, there is no motion in the *y*-direction for the T-shaped extension on the right-hand side of the mechanism. When the platform is moved in *x*- and *y*-directions simultaneously, the T-shaped extension at the top undergoes only the *y*-motion and that on the right only the *x*-motion. Thus, decoupling of the central platform can be clearly discerned in Fig. 27b. If we think of the combined motion of the platform in *x*- and *y*-directions as a combined signal, this mechanism can split that into two separate signals. Thus, it is mechanical signal amplifier. So, we see how a building block can be cleverly used to achieve a unique functionality.

Figures 28a, b show another arrangement comprising 12 compliant sliders and 12 DaCMs separated equally in two layers stacked one above the other. This *xy*-stage enhances the motion of the central platform when the rectangular extensions on the four sides are actuated [36]. This is because of the DaCMs. As compared to Fig. 26d, we need 12 building blocks here to achieve decoupling of the two axes and amplification along both the axes. This two-layered design can be used as an *xy*-stage with enhanced range of motion without sacrificing the dynamic characteristics. This is an example of two different kinds of building blocks used to create a unique new arrangement.

The design in Fig. 28 is a two-layered structure. Therefore, it is not easily amenable for microfabrication. A variant of this was used in [6] to make a single-layered mechanism without losing much in terms of functionality. Figure 29 shows such a design. It is a dual-axis, in-plane, capacitive accelerometer that has enhanced sensitivity

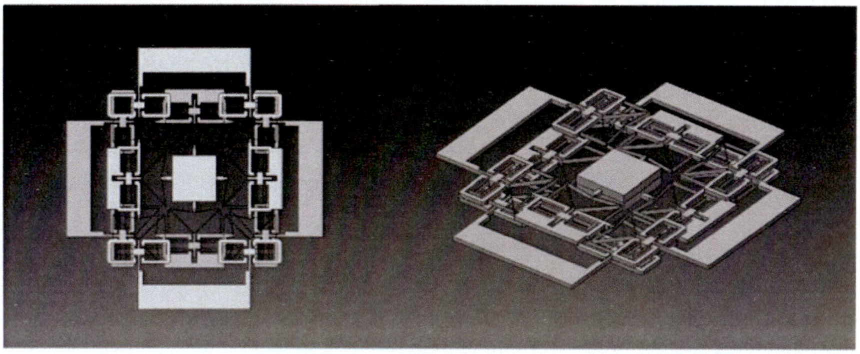

Fig. 28 Two-layered design of a compliant stage with DaCMs and compliant sliders. This design has decoupled as well as amplified motion of the stage in x- and y-directions

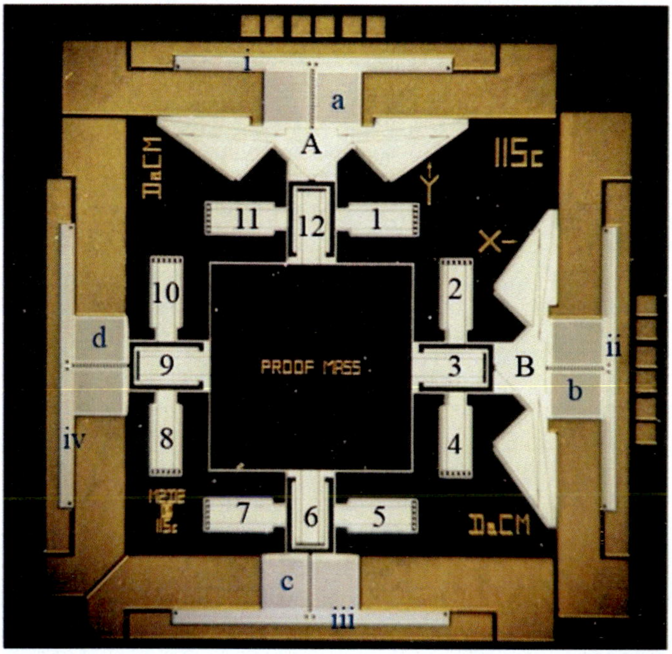

Fig. 29 A single-layer planar design of Fig. 28 with 12 compliant sliders, four compliant sliders, and two DaCMs. Here, the movement of the proof mass is amplified at the output points of DaCMs to enhance the sensitivity and bandwidth of the accelerometer

because of DaCMs. It does not use 12 DaCMs; it uses only two. To compensate for off-axis motion, it uses additional compliant sliders. As can be seen in Fig. 29, it has 12 compliant sliders of one kind (marked 1 to 12), four compliant sliders of another kind (marked i to iv), two DaCMs (marked A and B), and four electrostatic comb drives (marked a to d). Thus, it is modular design with four kinds of building blocks. Two DaCMs amplify the motion of the proof mass at the center. The purpose of 12 compliant sliders is to decouple the movements of the proof mass in x- and y-directions and transmit that to two sensing combs (a and b) though two DaCMs. Four more compliant sliders prevent the motion of the sensing combs in unintended directions. The other two combs (c and d) are used for self-test through actuation. Once we see the intent of each building block, the function of the entire device becomes clear and the interoperability of this modular design becomes clearer. It also makes it easy to arrive at the parameters of all the building blocks.

10 *Hierarchy* in Flow Paths and Solid Structures

What holds true at one size scale might change slightly or a lot at another size scale. Scaling arguments are popular in MEMS and nanotechnology [36, 37]. That line effects (e.g., surface tension) and surface effects (e.g., heat exchange) dominate over volume effects (e.g., inertia) at micro- and nano-scales is well known. Quantitatively speaking, length is the largest at the small scales as compared to area, and area looks larger than volume. This reversal of sequence as compared to the macro-size brings in interesting consequences in many physical and chemical phenomena [38]. There is much more to size effects within a given MEMS device. That has to do with hierarchy in a design.

Hierarchical design is most evident in Nature. The circulation paths of blood flow comprise a prime example. From the big aorta, the arteries multifurcate into small vessels and end up in small capillaries, only to unite later to veins and then vena cava. It is a tree structure. A big flow path divides into small, and then smaller flow paths. They combine in reverse. Andrian Bejan calls it *constructal law* [39]. A tree is another example. From the trunk up, branches emerge and then divide into sub-branches and twigs and then to leaves. Trunk down, the root system too has this hierarchy. For flowing from one source point to multiple points with least resistance requires a hierarchical tree structure [40]. Such a design also helps in mixing because microflows are known to have extremely low Reynolds number and hence are hard to mix otherwise. As we go down a tree structure, the cross-sectional area of the flow path decreases. This has to be calculated carefully; optimization helps and also points what is optimal under what conditions [40].

Hierarchy in structural design is much older and dates back to at least Gustave Eiffel [41, 42]. Eiffel exploited structural hierarchy by making his designs optimally *porous*. That is, he used beam segments crisscrossing with another rather than using a solid piece. If we examine the Eiffel tower, we see cross-beams with smaller cross-beams inside them. It resembles fractal-like design. While Eiffel took it to three levels of hierarchy in the famous tower in Paris, Nature has exploited it up to five levels [41]. Trabecular bone is made of collagen molecules making fibrils, fibrils combining to become fiber, fibers forming lamella, and lamella leading to cancellous bone. All five size scales have their own unique design unlike Eiffel's hierarchy where the same design exists at three different size scales, albeit with reduced cross-sectional sizes like in the flow paths. Solid structures at the macro-scale are replete with hierarchical designs (e.g., bridges and towers) but MEMS structures are yet to exploit it. Nevertheless, it is a trait of good design that leads to economic and efficient use of material.

11 Prospects for MEMS Design

In the last nine sections, we elaborated on nine traits of good MEMS designs with examples. As can be seen here, good designs share one or more traits. Simple designs are easy to make. Many of them are general in the sense that they can be used in other design situations.

A few simple designs serve as building blocks so that modular structures can be conceived. The "array" principle dominates MEMS designs. Just like a single transistor is not useful in itself unless millions of them combine to give a processor chip, the MEMS field uses arrays of single or multiple elements. Many examples presented in this chapter illustrate this. This is not as widely seen in macro-mechanical systems.

It is fair to conclude that MEMS designs depended on intuition and experience at the inception of the field. Later, optimization was used to aid intuition and somewhat automate the design process. There is much more to do in optimal design. It poses challenges as most MEMS devices operate in multiple energy domains [38, 43]. Therefore, multiphysics leading to coupled partial differential equations make design a challenging task. Several examples of optimizing in multiphysical domains were reported in [2].

Enhancing the quality factor of MEMS structures is a continuing challenge and a great opportunity to come up with innovative designs. Many modes of dissipation exist. Many clever designs that minimize dissipation are reported [44, 45]. Systematic study of the phenomena and how we can gain insights into their design are exciting avenues for further research on MEMS design.

Incorporating multiple materials in a single design and finding interesting combinations is also a future prospect in MEMS. Finally, many MEMS devices are focused on sensors. Actuators and mechanisms are far fewer. While we can argue that the demand drives designs (i.e., we needed microsensors more than microactuators), good designs can create demand. Upcoming areas of micro- and nano-robotics surely will venture into actuators and mechanisms, thus giving rise to new designs. Good designs in the future might have more than nine traits that are discussed in this chapter. Many consider design as an art. And there are surely some revealing signs to spot good designs also to conceive them either intuitively or systematically.

Finally, it is important to note that a design conceived in one situation for one application could be used by others in another situation. An example is that of a continuous-membrane mirror of [46] used in [8] for an array of beams that move down without distortion, as discussed in Sect. 5. Good designs are always recognized and used even if they are obscurely buried in the literature. Sometimes, reinvention of good designs happens and that is inevitable despite the widespread availability of research publications.

Acknowledgements The author is indebted to his students in the Multidisciplinary and Multiscale Design and Device (M2D2) laboratory in Mechanical Engineering at the Indian Institute of Science, Bengaluru. The synergy that exists among the group members plays a pivotal role in pursuing varied problems in the realms of compliant mechanisms, MEMS, topology optimization, and biomechanics of cells. Special thanks to Shamanth Hampali, R. Harisankar, Anoosha Pai, and Nitish Sathyamurthy who created the solid models of most of the designs presented in this paper.

References

1. Muller, R.S. (ed.): Microelectromechanical Systems: Advanced Materials and Fabrication Methods. NMAB-483, National Academy Press (1997)
2. Ananthasuresh, G.K. (ed.): Optimal Synthesis Methods for MEMS. Kluwer Academic Publishers (2003)
3. Tang, W.C., Nguyen, T.-C., Howe, R.T.: Laterally driven polysilicon resonant microstructures. Sens. Actuators **20**(1–2), 25–32 (1989)
4. Aksyuk, V.A., Pardo, F., Bolle, C.A., Arney, S., Giles, C.R., Bishop, D.J.: Lucent Microstar micromirror array technology for large optical crossconnects. In: Proceedings of the SPIE, MOEMS and Miniaturized Systems, vol. 4178 (2000). https://doi.org/10.1117/12.396503
5. Krishnan, G., Ananthasuresh, G.K.: Evaluation and design of compliant displacement amplifying mechanisms for sensor applications. J. Mech. Des. **130**(10), 102304, 1–9 (2008)
6. Khan, S., Ananthasuresh, G.K.: Improving the sensitivity and bandwidth of in-plane capacitive micro-accelerometers using compliant mechanical amplifiers. IEEE J. Microelectromech. Syst. **23**(4), 871–887 (2014)
7. Challoner, A.D., Ge, H.H., Liu, J.Y.: Boeing disc resonator gyroscope. In: IEEE/ION Position, Location, and Navigation Symposium—PLANS 2014 (2014)
8. Hung, E.S., Senturia, S.D.: Extending the travel range of analog-tuned electrostatic actuators. J. Microelectromech. Syst. **8**(4), 497–505 (1999)
9. Que, L., Park, J.-S., Gianchandani, Y.B.: Bent-beam electrothermal actuators-part i: single beam and cascaded devices. J. Microelectromech. Syst. **10**(2), 247–254 (2001)
10. Yang, Y.-J., Liao, H.-H., Huang, K.-H., Huang, Y.-Y., Lin, C.-W., Yang, L.-J., Jaw, F.-S.: Novel designs of herringbone chaotic mixers. In: Proceedings of the 1st IEEE International Conference on Nano/Micro Engineered and Molecular Systems (2006)
11. Kollimada, S., Balakrishnan, S., Malhi, C., Raju, S.R., Suma, M.S., Das, S., Ananthasuresh, G.K.: A micromechanical device for in situ stretching single cells cultured on it. J. Micro-Bio Robot. **13**, 27–37 (2018)
12. Moulton, T., Ananthasuresh, G.K.: Design and manufacture of electro-thermal-compliant micro devices. Sens. Actuators Phys. **90**, 38–48 (2001)
13. Fan, L.-S., Tai, Y.-C., Muller, R.S.: Intergrated movable micromechanical structures for sensors and actuators. IEEE Trans. Electron Devices **35**(6), 724–730 (1988)
14. Mehregany, M., Bart, S.F., Tavrow, L.S., Lang, J.H., Senturia, S.D., Schlecht, M.F.: A study of three microfabricated variable-capacitance motors. Sens. Actuators **A21**(22/23), 173–179 (1990)
15. Sandia Ultra-planar Multi-level MEMS Technology (2020). https://www.sandia.gov/mesa/_assets/documents/design_documents/SUMMiT_V_Dmanual.pdf
16. Kota, S., Ananthasuresh, G.K., Crary, S.B., Wise, K.D.: Design and fabrication of microelectromechanical systems. J. Mech. Des. **116**(4), 1081–1088 (1994)

17. Smith, S.T., Chetwynd, D.G.: Ultraprecision Mechanism Design. Gordon and Breach (1992)
18. Yin, L., Ananthasuresh, G.K.: Design of distributed compliant mechanisms. Mech. Based Des. Struct. Mach. **31**(2), 151–179 (2003)
19. Miller, S.L., Sniegowski, J.J., LaVigne, G., McWhorter, P.J.: Performance trade-offs for a surface micromachined microengine. In: Proceedings of the SPIE, Micromachined Devices and Components II, vol. 2882. (1996). https://doi.org/10.1117/12.250702
20. Ye, W., Mukherjee, S.: Design and fabrication of an electrostatic variable gap drive in micro-electro-mechanical systems. Comput. Model. Eng. Sci. **1**, 111–120 (2000)
21. Silicon-on-insulator Multi-user MEMS Processes (SOIMUMPs) (2020). http://www.memscap.com/products/mumps/soimumps
22. Howard, J.: Mechanics of Motor Proteins and the Cytoskeleton. Oxford University Press (2005)
23. McAllister, A., Smith, M., Zafirou, K., Day, D., Butler, M.: Apparatus and Method Providing a Hand-held Sprectrometer. US Patent US20070194239A1 (2007)
24. Saxena, A., Ananthasuresh, G.K.: On an optimal property of compliant topologies. Struct. Multidiscip. Optim. **19**(1), 36–49 (2000)
25. Kota, S., Rodgers, S.M., Hetrick, J.A.: Compliant Displacement-multiplying Apparatus for Microelectromechanical Systems. US Patent US6175170B1 (2001)
26. Baichapur, G.S., Gugale, H., Maheshwari, A., Bhargav, S.D.B., Ananthasuresh, G.K.: A Vision-based micro-Newton static force sensor using a displacement-amplifying compliant mechanism. Mech. Based Des. Struct. Mach. **42**(2), 193–210 (2014)
27. Kollimada, S., Khan, S., Balakrishnan, S., Raju. S.R., Suma, M.S., Ananthasuresh, G.K.: A micro-mechanical compliant device for individual cell-stretching, compression, and in-situ force-measurement. In: Proceedings of the International Conference on Manipulation, Automation, and Robotics at Small Scales. Montreal, Canada (2017)
28. Putty, M., Najafi, K.: A Micromachined vibratory ring gyroscope. In: Proceedings of the Hilton Head Workshop on Solid State Sensors and Actuators (1995)
29. Guckel, H., Klein, J., Christenson, T., Skrobis, K., Laudon, M., Lovell, E.G.: Thermomagnetic metal flexure actuators. In: Technical Digest of Solid-State Sensors and Actuators Workshop, Hilton Head Island, SC, 1992, p. 73 (1992)
30. Timoshenko, S.: Analysis of bimetal thermostats. J. Opt. Soc. Am. **11**, 233–255 (1925)
31. Yin, L., Ananthasuresh, G.K.: A novel topology design scheme for the multi-physics problems of electro-thermally actuated compliant micromechanisms. Sens. Actuators A **97–98**, 599–609 (2002)
32. Mankame, N., Ananthasuresh, G.K.: Topology synthesis of electro-thermal-compliant mechanisms using line elements. Struct. Multidiscip. Optim. **26**, 209–218 (2004)
33. Sardan, O., Petersen, D.H., Molhave, K., Sigmund, O., Boggild, P.: Topology optimized electrothermal polysilicon microgrippers. Microelectron. Eng. **85**(5–6), 1096–1099 (2008)
34. Mankame, N., Ananthasuresh, G.K.: Comprehensive thermal modeling and characterization of an electro-thermal-compliant microactuator. J. Micromech. Microeng. **11**(5), 452–462 (2001)
35. Awtar, S., Slocum, A.H.: Constraint-based design of parallel kinematic XY flexure mechanisms. J. Mech. Des. **129**(8), 816–830 (2006)
36. Dinesh, M., Ananthasuresh, G.K.: Micromechaical Stages with enhanced range. Int. J. Adv. Eng. Sci. Appl. Math. **2**(1), 35–43 (2010)
37. Baglio, S., Castorina, S., Fortuna, L., Savalli, N.: Scaling Issues and Design of MEMS. Wiley-Interscience (2008)
38. Ananthasuresh, G.K., Vinoy, K.J., Gopalakrishnan, S., Bhat, K.N., Aatre, V.K.: Chapter 9 in Micro and Smart Systems: Technology and Modeling. Wiley, New York (2012)
39. Bejan, A., Zane, J.P.: Design in Nature: How the Constructal Law Governs Evolution in Biology, Physics, Technology, and Social Organization. Anchor Reprint Edition (2013)
40. Yan, S., Wang, F., Sigmund, O.: On the non-optimality of tree structures for heat conduction. Int. J. Heat Mass Transf. **122**, 660–680 (2018)
41. Lakes, R.: Materials with structural hierarchy. Nature **361**, 511–515 (1993)
42. Sundaram, M., Ananthasuresh, G.K.: Gustave Eiffel and his optimal structures. Reson. Sci. Educ. J. **14**(8), 849–865 (2009)

43. Senturia, S.D.: (2001) Microsystem Design. Springer
44. Pratap, R., Mohite, S., Pandey, A.K.: Squeze film effects in MEMS devices. J. Indian Inst. Sci. **87**(1), 75–94 (2007)
45. Candler, R.N., Duwel, A., Varghese, M., Chandorkar, S.A., Hopcroft, M.A., Park, W.-T., Kim, B., Yama, G., Partridge, A., Lutz, M., Kenny, T.W.: J. Micromech. Syst. **15**(4), 927–934 (2006)
46. Bifano, T.G., Mali, R.K., Dorton, J.K., Perreault, J., Vandelli, N., Horenstein, M.N., Castanon, D.A.: Contunuous-membrane surface-micromachined silicon deformable mirror. Opt. Eng. **36**(5), 1354–1360 (1997)

Nonlinear Dynamics of Resonant Microelectromechanical System (MEMS): A Review

G. Chakraborty and Nikul Jani

1 Introduction

Miniaturized sensors, known as microsensors, are often actuated such that one or more component structures are driven to resonance. In macromechanical systems, resonance is generally avoided for safety of equipment. However, micromachined resonant sensors are used for high accuracy, long-term stability and quasidigital output. They are widely used in detection of chemical and biological substances [1], measurements of rheological properties of fluid [2], energy harvesting [3], measuring pressure [4] and in many other diverse fields [5]. In order to increase the sensitivity of the instrument, the dissipation is kept to a very small value. The accuracy of a well-designed resonator is good enough to use them as gyroscopes, timing references, frequency filters in cell phones and computers. In the field of metrology, resonators capable of detecting mass of 1 picogram have been fabricated.

One drawback associated with the resonant devices is that during large amplitude oscillation, when the system is excited to resonance, the effects of nonlinearities become pronounced. First, the resonant frequency of the flexible microstructure depends on the amplitude of excitation (amplitude-frequency effect or *A-f* effect). Second, nonlinearities make the interpretation of sensor data difficult. Finally, the noise characteristics of the device deteriorate as the participation of nonlinear term increases. Nevertheless, paradoxical though it seems, nonlinearities can sometimes be helpful. One example is that, the pull-in phenomenon resulted by the presence of nonlinear terms can be advantageously used in electrostatically driven MEMS switches [6]. Further, nonlinearities sometimes improve the noise behaviour of the device.

The study of nonlinear dynamics of resonant microsystems has emerged as an active field of research because of two reasons: (a) Since the nonlinear terms invari-

G. Chakraborty (✉) · N. Jani
Department of Mechanical Engineering, Indian Institute of Technology Kharagpur, Kharagpur, India
e-mail: goutam@mech.iitkgp.ac.in

© Springer Nature Singapore Pte Ltd. 2021

U. S. Dixit and S. K. Dwivedy (eds.), *Mechanical Sciences*,
https://doi.org/10.1007/978-981-15-5712-5_3

ably show themselves up in the system's response characteristics, an analysis based only on linear term does not predict correct results. (b) Nonlinear terms can be advantageously exploited to improve system's performance and also to design new kinds of devices. Thus, whether good or bad, nonlinearities cannot be ignored altogether.

The effects of nonlinearities on the dynamics of resonant microdevices have been reviewed by many authors. Notable among them are the ones by Lifshitz and Cross [7], Rhoads et al. [8], Tiwari and Candler [9]. These reviews have emphasized on different aspects of nonlinearity. For example, Rhoads et al. [8] have studied effects of different types of excitation on the nonlinear dynamics of MEMS resonators, while Tiwari and Candler [9] have studied both types of devices where nonlinearity is undesirable and where it is to be exploited to the advantage.

The present review aims to provide a comprehensive overview of nonlinear effects on the response of resonant MEMS devices. Along with the usual sources of nonlinearity in a resonant MEMS, discussion has been made also on different ways by which nonlinearities are tailored to improve the system's performance. The beneficial and undesirable effects of nonlinearity have been pointed out by means of simple models, which are valuable tools for getting insight into the design of such systems. The review is divided into several sections. The basic operating principal of resonant MEMS has been explained in Sect. 2 restricting to the linear theory. The effects of nonlinearity on differently excited resonators are discussed in Sect. 3. Only one kind of nonlinearity, namely, duffing-type nonlinearity has been treated in this section in order to highlight various roles that the nonlinearities can play in the dynamics of such systems. The real nature and origin of nonlinearities that appear in resonant MEMS devices are explained in the next section (Sect. 4). In Sect. 5, we compare the effects of nonlinearity and discuss the desirable and undesirable effects. In the final section (Sect. 6), various methods of modifying the nonlinearities of the device have been outlined.

2 Working Principle of Resonant MEMS Sensor

The working principal of a resonant device can be explained with the help of a forced single-degree-of-freedom (SDOF) oscillator, whose equation of motion can be written as

$$m\ddot{x} + c\dot{x} + kx = f(t) \tag{1}$$

where the symbols have usual significance. In a resonant microsystem, the oscillator is driven to resonance by means of harmonic excitation $f(t) = f_0 \cos \omega t$. The steady-state response amplitude

$$X = \frac{F_0}{k\sqrt{\left(1 - \frac{\omega^2}{\omega_n^2}\right)^2 + \left(2\zeta\frac{\omega}{\omega_n}\right)^2}}, \quad \text{where, } \omega_n = \sqrt{\frac{k}{m}} \text{ and } \zeta = \frac{c}{2\sqrt{mk}} \tag{2}$$

attains a maximum value $X_{\max} = \frac{F_0/k}{2\zeta\sqrt{1-\zeta^2}}$, when $\omega = \omega_n\sqrt{1-2\zeta^2}$. Usually for resonant device the damping factor is kept small, so that $X_{\max} \approx \frac{F_0/k}{2\zeta}$ for $\omega \approx \omega_n$. It is customary to specify the damping factor in form of quality factor Q which is approximately equal to $\frac{1}{2\zeta}$. For resonant MEMS, the Q factor is very high.

In sensing devices, one of the physical parameters of the oscillator is changed depending on the external stimuli that need to be estimated. For example, in chemical gas sensor, the mass is altered when the gas is present near the vibrating body, in a pressure sensor the stiffness is changed. This change in inertia or stiffness causes the natural frequency to be shifted. The conditions for resonance of an initially resonating oscillator now get changed, resulting in significant change in the response amplitude and phase difference $\phi = \tan^{-1}\left(\frac{2\zeta\omega/\omega_n}{1-\omega^2/\omega_n^2}\right)$. Measurement of this change can enable one to quantitatively estimate the change in physical parameter and hence the stimuli.

For example, change of the mass from m to $m + \Delta m$ causes the steady-state amplitude to decrease by

$$\Delta X = \frac{F_0}{c\sqrt{k/m}}\left(1 - \frac{1}{\sqrt{1+\left(\frac{1}{2\zeta}\frac{\Delta m}{m}\right)^2}}\right) \tag{3}$$

or the phase angle is increased by

$$\Delta\phi = \tan^{-1}\left(-\frac{c}{\Delta m\sqrt{k/m}}\right) - \frac{\pi}{2} = \cot^{-1}\left(\frac{2\zeta}{\Delta m/m}\right) \tag{4}$$

If the excitation frequency is tuned then it may be possible to bring the system again to resonance by changing the frequency from $\omega = \omega_n = \sqrt{\frac{k}{m}}$ to $\omega' = \sqrt{\frac{k}{m+\Delta m}}$. The change in the frequency $|\Delta\omega| = \sqrt{\frac{k}{m}}\left(1 - \frac{1}{\sqrt{1+\frac{\Delta m}{m}}}\right)$ can be measured to estimate Δm, provided the values of $\omega_n = \sqrt{\frac{k}{m}}$ and m are known. It may be noted that since $\Delta\omega \propto \omega_n$, the measurement becomes easy as the value of ω_n is increased. This explains why the microscale or nanoscale sensors are better suited for this purpose; the natural frequency of a system increases as the dimension is reduced.

The efficacy of the above-described sensing scheme depends on the ability of detecting change in the response amplitude or phase as the natural frequency is shifted, i.e. on the value of $\frac{\Delta X}{\Delta\omega_n}$, which should be sufficiently large. However, from (3) it may be easily established that

$$\begin{aligned}\frac{\Delta X}{\Delta m} &= \frac{F_0}{4\zeta^2 km}\frac{1}{\lambda}\left(1 - \frac{1}{\sqrt{1+\lambda^2}}\right), \quad \lambda = \frac{1}{2\zeta}\frac{\Delta m}{m} \\ &= \frac{F_0}{c^2}\frac{1}{\lambda}\left(1 - \frac{1}{\sqrt{1+\lambda^2}}\right)\end{aligned} \tag{5}$$

It is therefore required to have large excitation amplitude F_0 in order to increase the sensitivity for small value of Δm.

This limitation can be avoided if the oscillator is excited by parametric excitation rather than direct excitation. For the simplest situation, the following equation of motion is obtained:

$$m\ddot{x} + c\dot{x} + k\,(1 + 2\varepsilon \cos 2\omega t)\,x = 0 \tag{6}$$

The response in this case shows two distinct behaviour depending on the values of the parameters ε, ω/ω_n and ζ. The amplitude either increases exponentially if certain inequalities $\varepsilon > f\,(\omega/\omega_n,\ \zeta)$ are satisfied or gets reduced to zero when $\varepsilon < f\,(\omega/\omega_n,\ \zeta)$. Thus, the system can be conceived to be either in 0-state corresponding to low amplitude of oscillator or in the 1-state which corresponds to high amplitude of oscillation. When the natural frequency is changed, the system may be easily brought from one state to other. Next by proper adjustment of the excitation frequency ω, the system can be restored to its original state. The measured value of $\Delta\omega$ required for that purpose is used to obtain the change in the mass or stiffness of the oscillator.

For example, when the excitation frequency ω is nearly equal to the natural frequency ω_n, primary parametric resonance in an undamped system ($c = 0$) occurs, if $1 - \left(\frac{\omega}{\omega_n}\right)^2 < \varepsilon < 1 + \left(\frac{\omega}{\omega_n}\right)^2$, i.e. within a narrow frequency zone of width, $1 - \varepsilon < \left(\frac{\omega_n}{\omega}\right)^2 < 1 + \varepsilon$ or approximately $1 - \frac{\varepsilon}{2} < \left(\frac{\omega_n}{\omega}\right) < 1 + \frac{\varepsilon}{2}$. The state of the oscillator changes when ω_n is changed by amount $\frac{\varepsilon\omega}{2} \approx \frac{\varepsilon\omega_n}{2}$.

3 Nonlinearities in MEMS

The response amplitudes of resonant MEMS devices are generally large and hence the linear approximations are often insufficient to predict their correct behaviour. Also nonlinearities are sometimes introduced deliberately to gain some advantages. Presence of nonlinearities makes the analysis complicated because of two reasons, (a) the model becomes quite complex and (b) unlike linear systems there is no generic behaviour of system response. Every nonlinear system behaves differently depending on the type of nonlinearity, excitation or initial conditions. Thus, a complete discussion of nonlinear dynamics of MEMS devices will be too lengthy to be considered in a review of modest length.

Resonant MEMS devices can be broadly classified as shown in Fig. 1.

Nonlinear analysis of the system is different for different subsystem. Further each of the subsystem can be classified according to the mode of excitation in the following manner:

1. system with direct excitation,
2. system with parametric excitation,
3. system with combined direct and parametric excitations and
4. system with self-excitation.

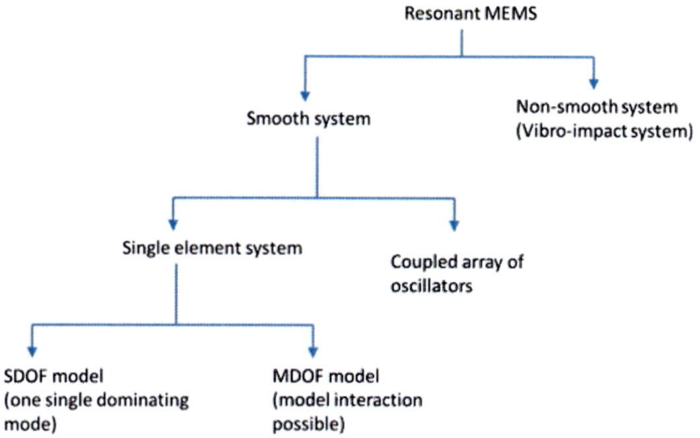

Fig. 1 A schematic explaining the classification of MEMS resonators

In the following pages, only one type of nonlinear system, namely, the one which can be modelled as a single-degree-of-freedom oscillator is considered for discussion. Further, out of numerous kinds of nonlinear terms that may be present in the system, only the cubic type of nonlinearity is taken up. Different cases of excitation are now considered separately.

3.1 System with Direct Excitation

The equation of motion of a harmonically excited system with cubic-type nonlinear term can be written as

$$m\ddot{x} + kx + c\dot{x} + \alpha x^3 = f_0 \cos \omega t \tag{7}$$

This quite simple looking system shows at times unexpected behaviour like sensitive dependence on initial conditions (chaos), although in most cases the behaviour is quite predictable. However, the behaviour depends on (i) the relationship between ω and ω_n, (ii) the magnitude of f_0, (iii) the sign of α, etc. Three cases can be distinguished.

Case 1: $\omega \approx \omega_n$. This is the case typically encountered in resonant MEMS [10]. The characteristic behaviour of the nonlinear system is mentioned here.

(a) The frequency response curve for steady-state amplitude bends towards right or left (see Fig. 2) according to the sign of α is positive(hardening-type nonlinearity) or negative (softening-type nonlinearity). In fact, the natural frequency of unforced system depends on amplitude of oscillation as $\omega_n^2 = \frac{k}{m} + \frac{3}{4}\alpha a^2$, where

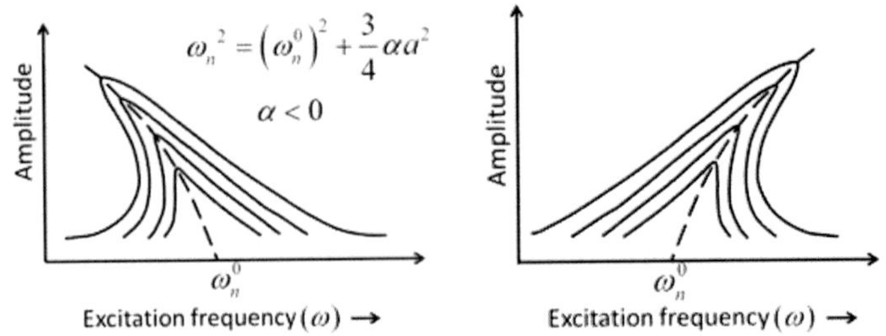

(a) Softening behaviour (b) Hardening behaviour

Fig. 2 Frequency–amplitude curves for forced excitation

a is the amplitude of oscillation. For a forced system, the response curve depends on the magnitude of f_0 and c.

(b) The response amplitude can take any of the three possible values depending on the initial conditions, when the excitation frequency is within a zone in between ω_1 and ω_2 (shown in Fig. 3).
The non-uniqueness in the amplitude gives rise to 'jump phenomena' when the frequency of excitation is quasistatically increased or decreased continuously. The direction of the jump (upward or downward) as well as its amount depends on the sign of α. For example, a hardening-type nonlinearity exhibits downward jump in the frequency sweep up operations while a softening-type nonlinear system shows upward jump. However, if the frequency is not changed quasistatically, then significant transient oscillation takes place before the amplitude sets down to a new value.

Case 2: $\omega \approx 3\omega_n$. Although the predicted response is small according to linear analysis, the presence of cubic nonlinear term can make it large. The steady-state response has two frequencies, namely, the excitation frequency ω and the natural frequency ω_n. The phenomena of large amplitude of oscillation is known as 'subharmonic resonance'.

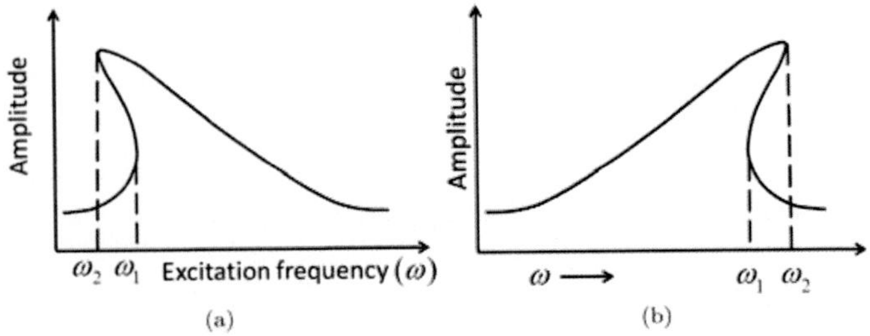

Fig. 3 Bifurcation frequency in case of softening and hardening nonlinearities

Case 3: $3\omega \approx \omega_n$. It is also possible to get large amplitude response because of 'superharmonic resonance'.
When the oscillator is excited by force with two or more different frequencies then in addition to above, large amplitude oscillation may take place due to 'combination resonance'.

3.2 *System with Parametric Excitation*

The effect of nonlinear terms in a parametrically excited system is to limit the oscillation amplitude to a finite value when the excitation causes instability to a linear system. For example, if the nonlinearity is of duffing type as in the following equation of motion:

$$m\ddot{x} + c\dot{x} + k(1 + 2\varepsilon \cos 2\omega t)x + \alpha x^3 = 0 \tag{8}$$

one or two limit cycles may exist depending on the value of $\frac{\omega}{\omega_n}$, $\zeta = \frac{c}{2\sqrt{mk}}$ and the sign of α. For a duffing-type nonlinearity, the limit cycle is stable if this is unique. For the system where two limit cycles exist, the one with higher amplitude of oscillation is stable. The instability region for a nonlinear system is shown in Fig. 4. The parametric space is divided into three regions. In the region I, both linear and nonlinear equations predict the same steady-state response which decays to insignificantly small value. In region II, the nonlinear equation predicts a stable limit cycle. However, the steady-state amplitude depends on initial condition in region III. For some initial conditions, the predictions of the nonlinear equation and the linear equations are identical, while for a different set of initial conditions a large amplitude of oscillation is predicted when the nonlinear term is present.

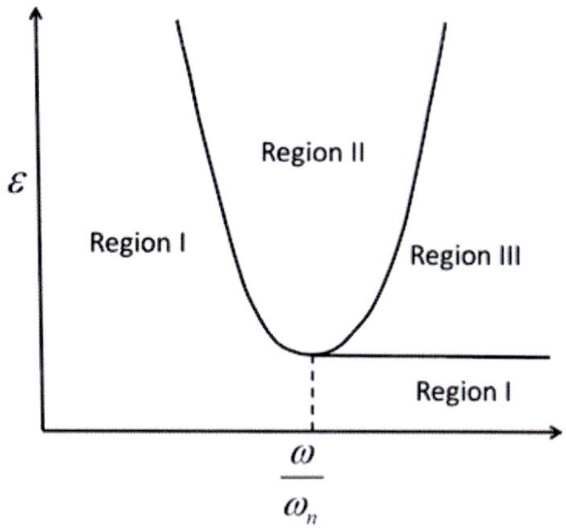

Fig. 4 Regimes for stable and unstable behaviours of parametrically excited system

3.3 System with Combined Direct and Parametric Resonances

When the stiffness term in a harmonically excited system is modulated periodically, the system shows often unexpected behaviour. Consider the following equation of motion [11]:

$$m\ddot{x} + c\dot{x} + k(1 + 2\varepsilon\cos 2\omega_p t)x = f_0\cos(\omega_d t + \phi) \tag{9}$$

If $\omega_d = \omega_p$, application of harmonic balance method gives the following steady-state response:

$$x = A\cos\omega_p t + B\sin\omega_p t \tag{10}$$

where A and B satisfy the following equation:

$$\begin{bmatrix} k(1+\varepsilon) - m\omega_p^2 & c\omega_p \\ -c\omega_p & k(1-\varepsilon) - m\omega_p^2 \end{bmatrix} \begin{Bmatrix} A \\ B \end{Bmatrix} = f_0 \begin{Bmatrix} \cos\phi \\ -\sin\phi \end{Bmatrix} \tag{11}$$

At resonance, i.e. $\omega_p = \omega_n = \sqrt{\frac{k}{m}}$, the response can be written in the following form provided the amplitude of parametric excitation remains below the threshold value:

$$x(t) = \frac{f_o\sqrt{4\zeta^2 + \varepsilon^2}}{k\left(4\zeta^2 - \varepsilon^2\right)}\left\{\sin\left(\phi - \tan^{-1}\frac{\varepsilon}{2\zeta}\right)\cos\omega_n t + \cos\left(\phi + \tan^{-1}\frac{\varepsilon}{2\zeta}\right)\sin\omega_n t\right\} \tag{12}$$

As $\varepsilon = 2\zeta$, the response amplitude becomes large leading to single amplification. Further, as $\tan^{-1}\left(\frac{\varepsilon}{2\zeta}\right) = \frac{\pi}{4}$ when $\varepsilon = 2\zeta$, the amplitude becomes small for $\phi = \frac{\pi}{4}$. Amplitude becomes very large when $\phi = \frac{-\pi}{4}$. Thus, amplification or quenching of the signal during resonance depends on the value of the phase difference between direct and parametric excitations.

Above the threshold level of excitation, the parametrically excited linear system becomes unstable. In this case, the response can be written as

$$x(t) = x_1(t) + x_2(t) \tag{13}$$

where $x_1(t)$ is obtained by the above analysis while $x_2(t)$ satisfies the damped Mathieu equation.

When $\omega_d \neq \omega_p$, the response can be written as

$$x(t) = A(t)\cos\omega_p t + B(t)\sin\omega_p t \tag{14}$$

where $A(t)$ and $B(t)$ satisfy the following differential equation:

$$\begin{bmatrix} c & -2m\omega_p \\ 2m\omega_p & c \end{bmatrix} \begin{Bmatrix} \frac{dA}{dt} \\ \frac{dB}{dt} \end{Bmatrix} + \begin{bmatrix} k(1+\varepsilon) - m\omega_p^2 & c\omega_p \\ -c\omega_p & k(1-\varepsilon) - m\omega_p^2 \end{bmatrix} \begin{Bmatrix} A \\ B \end{Bmatrix} = f_0 \begin{Bmatrix} \cos\{(\omega_d - \omega_p)t + \phi\} \\ -\sin\{(\omega_d - \omega_p)t + \phi\} \end{Bmatrix} \tag{15}$$

It is not difficult to see that if the solution is stable then the response consists of two harmonic terms, one with frequency ω_d and the other with frequency $2\omega_p - \omega_d$. When the nonlinear term is present, the response becomes generally quite complex. For a nonlinear oscillator driven by both parametric and external excitations (degenerate case), the equation of motion becomes [12]

$$m\ddot{x} + c\dot{x} + k(1 + 2\varepsilon\cos 2\omega_p t)x + \alpha x^3 = f_0\cos(\omega t + \phi) \tag{16}$$

The steady-state response can be obtained by assuming, as before

$$x(t) = A\cos\omega t + B\sin\omega t$$

where A and B are obtained using balancing the harmonics after substituting $x(t)$ into the governing equation of motion. When the system is driven below the parametric instability threshold, the system behaviour is the same as that of a duffing oscillator. When driven above the instability threshold, amplitude-frequency response shows five branches (unlike three in duffing oscillator) within certain frequency band, three of which are stable [12]. Two 'active' stable resonances are seen in this resonator. It is interesting to note that the maximum amplitude of the resonator does not depend on whether the parametric instability threshold is crossed or not. Thus, the parametric instability threshold is only of minor concern here, unlike in a linear resonator.

3.4 System with Self-Excitation

Some MEMS resonators are excited by itself through positive feedback mechanism [13]. Such systems are active and, of course, connected to an unlimited energy source. A simple mathematical model of a linear system with self-excitation is

$$m\ddot{x} - g\dot{x} + kx = 0 \tag{17}$$

whose response grows exponentially with time. Examples of self-excited MEM resonators are optically heated mechanical resonators [14]. Presence of nonlinearity limits the amplitude of oscillation to a limit cycle. For example, the following equation exhibits limit cycle:

$$m\ddot{x} - g\left(\dot{x} - \frac{1}{3}\dot{x}^3\right) + kx = 0 \tag{18}$$

This equation can be written in another form, associated with the name of Van der Pol.

$$m\ddot{y} - g\left(1 - y^2\right)\dot{y} + ky = 0 \tag{19}$$

by substituting $\dot{x} = y$.

If the oscillator is excited by harmonic excitation $f_0 \cos \omega t$, where $\omega \approx \omega_0 = \sqrt{\frac{k}{m}}$, then the steady-state motion becomes periodic with a frequency equal to that of excitation. Thus, the response gets synchronized at ω. An interesting fact is that a small change in excitation frequency can cause large change in response behaviour. This happens when the excitation frequency falls outside the band in which 'entrainment' takes place.

4 Sources of Nonlinearity

Any vibrating system is generally nonlinear. Only under special operating conditions the effect of the nonlinear terms can be ignored. For example, the system is adequately modelled as a linear system when the amplitude of vibration is small. In resonant devices, this is not the case since the amplitude of vibration is kept to a very high level in order to increase the sensitivity of the device. Furthermore, the MEMS device is made by interconnecting different subsystems which have their own dynamics [15]. Hence, modelling the nonlinear terms is a very difficult task which can be achieved only with a detailed knowledge of the individual subsystems. In what follows some common sources of nonlinearity are discussed.

The sources of nonlinearity can be broadly classified as following:

1. nonlinearity in mechanical structure,
2. nonlinearity in actuation system,
3. nonlinearity in sensing (measuring) devices and
4. nonlinearity in feedback and electrical circuit.

4.1 Nonlinearity in Mechanical Structure

The nonlinear terms appearing in the equation of motion of the structural component of the resonant device (for example, beam, plate, wire, etc.) are of following types:

(i) stiffness nonlinearity,
(ii) damping nonlinearity and
(iii) nonlinearity due to external forces.

4.1.1 Nonlinearity in Stiffness Terms

The stiffness or restoring force term in the equation of motion of the structure can be nonlinear because of (i) nonlinearity in strain–displacement relationship at large amplitude (geometric nonlinearity) [16], (ii) nonlinear constitutive relations of the material (material nonlinearity) and (iii) impact causing sudden change in the system behaviour [17].

In deriving the equation of motion of a vibrating structure, one needs the relationship between stress and strain components (constitutive equations) which can be mathematically expressed as

$$\sigma_{ij} = \sigma_{ij}(\epsilon_{kl})\,; \quad i, j, k, l = 1, 2, 3$$

Together with the relationship between strain and displacement gradient. In a linear system, both these are expressed as linear relations, namely,

$$\sigma_{ij} = \sum_k \sum_l C_{ijkl}\epsilon_{kl}, \quad \text{where } C_{ijkl} = C_{jikl} = C_{ijlk} = C_{klij}$$
$$\text{and } \epsilon_{ij} = \frac{1}{2}\left(\frac{\partial u_i}{\partial x_j} + \frac{\partial u_j}{\partial x_i}\right) \tag{20}$$

where $u_i (i = 2, 3)$ are the components of displacement field.

However, during large amplitude oscillation, significant deviations from the aforesaid relations occur. Even if the material nonlinearity can be ignored, the nonlinear relation between ϵ_{ij} and $\frac{\partial u_i}{\partial x_j}$ terms shows non-trivial effects. For example, the non-dimensional equation of motion of a MEMS beam fixed at both ends can be written as [16]

$$\frac{\partial^2 w}{\partial t^2} + \frac{\partial^4 w}{\partial x^4} - \left(\frac{Al^2}{2I}\right)\frac{\partial^2 w}{\partial x^2}\int_0^l \left(\frac{\partial w}{\partial x}\right)^2 \mathrm{d}x = 0, \quad 0 \le x \le l \tag{21}$$

Here, even if the constitutive relations are assumed to be linear, the nonlinear relationship between strain and displacement, i.e. $\epsilon_{xx} = \frac{\partial u}{\partial x} + \frac{1}{2}(\frac{\partial w}{\partial x})^2$, where $u(x, t)$ and $w(x, t)$ are axial and transverse displacements, respectively, couples the longitudinal and transverse motion of the beam. The net effect is stretching of the neutral axis of the beam during large deformation, a fact which is often ignored during small amplitude oscillation.

In a cantilever beam, the above-mentioned mid-plane stretching does not take place but the geometrical nonlinear terms exist. The equation of motion can be written as [18]

$$\frac{\partial^2 w}{\partial t^2} + \frac{\partial^4 w}{\partial s^4} + \frac{\partial}{\partial s}\left[\frac{\partial w}{\partial s}\frac{\partial}{\partial s}\left(\frac{\partial w}{\partial s}\frac{\partial^2 w}{\partial s^2}\right)\right] + \frac{\partial}{\partial s}\left[\frac{\partial w}{\partial s}\int_1^s \frac{\partial}{\partial s}\left(\int_0^s \frac{\partial w}{\partial s}\frac{\partial^2 w}{\partial s \partial t}\mathrm{d}s\right)\mathrm{d}s\right] = 0 \tag{22}$$

where s is the distance measured from the fixed end of the beam.

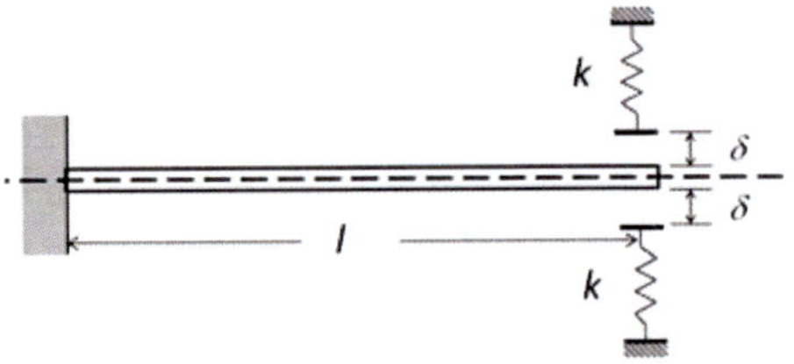

Fig. 5 Schematic of a compliant amplitude restraint

Material nonlinearity does not normally arise in MEMS devices unless special material is used. In systems where vibro-impact takes place, for example, the devices used in switching, positioning and tapping mode atomic force microscopy, the impacting process generates severe nonlinear behaviour. This kind of system is usually modelled as piecewise smooth (linear) system whose equations of motion are usually smooth. Specific rules, like Newtonian impact low, apply when the response crosses the regional boundaries. For example, for a cantilever beam with a compliant amplitude restraint (as shown in Fig. 5), the boundary conditions at the free end can be written as

$$(i)\frac{\partial^2 w}{\partial x^2} = 0 \text{ and } \frac{\partial^3 w}{\partial x^3} = 0 \text{ when } |w(x = l, t)| < \delta \tag{23}$$

$$(ii)\frac{\partial^2 w}{\partial x^2} = 0 \text{ and } \frac{\partial^3 w}{\partial x^3} = \frac{k}{EI}(w - \delta) \text{ when } w(x = l, t) \geq \delta$$

$$(iii)\frac{\partial^2 w}{\partial x^2} = 0 \text{ and } \frac{\partial^3 w}{\partial x^3} = \frac{k}{EI}(w + \delta) \text{ when } w(x = l, t) \leq -\delta$$

This system shows all the behaviour of a hard nonlinear system.

4.1.2 Nonlinear Damping Term

Different mechanisms are responsible for dissipation of energy from the vibrating structure in a resonant MEMS device. The sources of dissipation of energy can be broadly classified as follows:

(i) structural/internal damping,
(ii) support damping and
(iii) fluidic and acoustic damping.

Various mechanisms of energy dissipation have been identified within the structure (bulk dissipation) [19, 20] or on its surface (surface dissipation) [21]. Of the former, the thermoelastic dissipation is the most dominant. Usually the damping is treated as linear. A major loss of energy from a resonatory microstructure takes place in the fluidic medium surrounding the structure. In many situations, the vibrating body is placed near a static structure. The fluid entrapped between these structure causes what is known as squeeze-film damping. As the amplitude of response increases, the effect of nonlinearity also becomes significant. For example, the equation of motion of a pined-pined beam vibrating near a fixed plate is

$$\rho A \frac{\partial^2 w}{\partial t^2} + c\frac{\partial w}{\partial t} + EI\frac{\partial^4 w}{\partial x^4} - \frac{EA}{2L}\left[\int_0^l \left(\frac{\partial w}{\partial x}\right)^2 \mathrm{d}x\right]\frac{\partial^2 w}{\partial x^2} = \int_{-\frac{b}{2}}^{\frac{b}{2}} p\,\mathrm{d}y \tag{24}$$

where $p(x, y, t)$ is the pressure distribution on the beam surface which has width b. The pressure is calculated by solving the nonlinear version of Reynolds's equation

$$\nabla \cdot \left[w^3 p \nabla p\right] = \frac{12\mu}{1+6K}\frac{\partial (pw_0)}{\partial t} \tag{25}$$

where μ is the viscosity of the surrounding fluid and K is the Knudsen number (=mean free path of the gas/w). When the amplitude of vibration is small the damping is usually considered to be linear with damping coefficient, $C_{\text{squeeze}} = \frac{b^3}{d^3} l\mu$, where b, d, l and μ are, respectively, the beam width, gap between plates, length of the plates and air/fluid viscosity. When the amplitude of vibration increases, the nonlinearity of damping is modelled by considering the damping coefficient as [16, 22]

$$C_{\text{squeeze}} = \frac{\alpha}{(d-w)^3} \tag{26}$$

4.1.3 Nonlinear External Forces

As the dimensions of the structure are reduced, some kinds of interaction forces with the nearby objects become pronounced which can introduce nonlinearity. For example, in scanning probe microscopy, the tip of the microcantilever often experiences electrostatic interactions or attractive van der Waals forces from the surface under observation. Most of the interaction forces are highly nonlinear function of the separating distance between tip and sample [23]. A very general model for tip–sample interaction force is given by

$$F_{\text{tip}} = \begin{cases} -\frac{C}{d^\alpha}, & \text{for } d > a_0 \\ -\frac{C}{a_0^\alpha} + D(a_0 - d), & \text{for } d \le a_0 \end{cases} \tag{27}$$

Here C, D, d and a_0 are general attractive force parameters, tip–sample separation and intermolecular distance at which contact is considered to be initiated. The parameters C and D are functions of tip radius. The forces are dependent on material of tip and sample. They are attractive (repulsive) for long (short) range.

4.2 Nonlinearity in Actuating System

MEMS resonators are excited by different mechanisms. The most common of which are

(i) electrostatic actuation and
(ii) piezoelectric actuation.

The forces of actuation in an electrostatically actuated MEMS (also called capacitive MEMS resonator) is calculated by solving the field equations in an electrostatic problem. However, under certain assumptions the force can be modelled as

$$F_{\text{actuation}} = \frac{\epsilon A V^2}{2g^2}$$

where ϵ is the permittivity of the space, A is the area of the electrode, V is the voltage difference between the plates and g is the gap between the plates. For a capacitive actuation of the beam-type resonator, $g = d - w(t)$, where d is the initial gap and w is the transverse displacement of the beam. For a DC-driven system,

$$F_{\text{actuation}} = \frac{\epsilon A V_{DC}^2}{2(d-w)^2}$$

However, resonant MEMS are excited by time-varying voltage $V = V_{DC} + V_{AC} \cos \omega t$. In this case, the actuation force leads to both ordinary nonlinear terms and nonlinear parametric excitations [24]. In this case

$$F_{\text{actuation}} = \frac{\epsilon A}{2(d-w)^2} \left(\left(V_{DC}^2 + \frac{V_{AC}^2}{2} \right) + 2 V_{DC} V_{AC} \cos \omega t + \frac{V_{AC}^2}{2} \cos 2\omega t \right) \quad (28)$$

The nonlinearity is usually softening type since

$$\frac{1}{(d-w)^2} \approx \frac{1}{d^2} \left(1 + 2\frac{w}{d} + 3\left(\frac{w}{d}\right)^2 + 4\left(\frac{w}{d}\right)^3 + - - - \right)$$

Usually in the model, the fourth- and fifth-order nonlinear terms are generally neglected. However, for very high amplitude oscillation they become effective [25].

In a piezoelectrically actuated resonator, electrical voltage is applied to the piezoelectric material to induce deformation (strain actuation) within the structure. Linear assumption provides adequate accuracy to the model as long as the applied electric field and stress are low. They become, however, increasingly inaccurate as the stress level and electric field strength increase. Pronounced nonlinearity and hysteresis in the strain–field relationship are observed in such case. Usually the nonlinearity is softening type.

4.3 Nonlinearity in Sensing System

Different response parameters like amplitude, phase, etc. are used in various ways in a resonant MEMS device. This requires sensing of the pertinent quantity with the help of suitable sensing device. Often the resolution of the MEMS device gets limited by the limitation of the sensing device. Different types of sensors are used, like capacitive sensor, piezoresistive sensor and optical sensor. It is desirable that the sensor should be linear for wide range of input. In practice they often show nonlinearity.

For example, in capacitive-type sensing scheme, the change in capacitance is measured to detect the displacement of the movable plate of parallel-plate capacitor [26]. As the capacitance of such a capacitor is given by

$$C = \frac{\varepsilon A}{d - w}$$

one gets
$$\frac{dC}{dw} = \frac{\varepsilon A}{(d-w)^2} \approx \frac{\varepsilon A}{d^2}\left(1 + 2\frac{w}{d} + 3\left(\frac{w}{d}\right)^2 + \cdots\right). \tag{29}$$

Nonlinear terms in the aforementioned expression can be ignored if $w \ll d$. For large value of w, appreciable errors have been found out while using the linear relationship.

Apart from the sources of nonlinearity described above, the nonlinearities are deliberately introduced into the system by various mechanisms, like feedback, etc. Some of these methods are discussed later (see Sect. 6). In what follows the roles played by the nonlinearity are discussed. It will be seen that nonlinear terms sometimes enhance the performance measures of the device, while at other cases they play negative roles.

5 Effect of Nonlinearity on MEMS Performance

As mentioned earlier, nonlinearity can be the cause of performance degradation, or it can help to improve the device performance. Different effects of nonlinearities are discussed below.

5.1 Undesirable Effects of Nonlinearity

The following are the undesirable effects due to the presence of nonlinearities in the device:

a. *Frequency stability*
   The resonant frequency (the frequency at which the amplitude becomes highest) of a nonlinear oscillator depends on the amplitude of the excitation forcebreak [9,

27]. This dependance of resonant frequency or amplitude limits the performance of the device where they are required to resonate at a particular frequency, for example, time device, frequency filters, resonant accelerometers, resonant energy harvesters, gravimetric sensors, etc.

b. *Noise performance*
High amplitude of response is preferred in resonant devices for increasing the signal-to-noise (S/N) ratio. However, the presence of nonlinear term limits the amplitude of oscillation for a given excitation level.
To assess the noise performance of the oscillator, a block diagram of the open-loop system is used as shown in Fig. 6.
In Fig. 6 $N_1(t)$, $N_2(t)$ are the noise in excitation and transducer, respectively, and $y(t)$ is the measured output from the transducer. The maximum energy stored in the linear oscillator is

$$E_{\max} = \frac{1}{2} k_1 X_{\max}^2$$

While the maximum power that could be extracted from the oscillator is equal to

$$P_{\text{sig}} = \frac{\Delta E}{2\pi/\omega} = \frac{\omega E_{\max}}{Q}$$

The noise in $y(t)$ is generated by two main sources, namely, the noise generated from excitation and the thermal noise generated due to motion resistance. With a properly designed excitation mechanism the contribution of the first term can be minimized. The thermal noise power can be written approximately as [28]

$$p_{\text{noise}}^{\text{mech}} = 4k_b T \left(\frac{\omega_0}{2Q\Delta\omega} \right)^2$$

where ω_0 is the centre frequency and $\delta\omega$ is the offset from the former. The phase noise spectrum is obtained as

$$L(\Delta\omega) = \frac{p_{\text{noise}}^{\text{mech}} + p_{N_2}}{2p_{\text{sig}}} = \frac{2k_b T}{p_{\text{sig}}} \left(\frac{\omega_0}{2Q\Delta\omega} \right)^2 + \frac{p_{N_2}}{2p_{\text{sig}}} \tag{30}$$

where p_{N_2} is the noise in the transducer. In the simplified analysis, the effect of $1/f$ noise has not been considered.
It is seen from the above equation that a large value of $E_{\max}$ is required for improving the noise performance of the oscillator. However, the presence of the nonlinear term limits the value of $E_{\max}$ and hence decreases the S/N ratio. The limitation on amplitude of response due to nonlinearities puts a limit to the drive current that can be applied resulting in a far-from-carrier-phase noise. Also the frequency-amplitude dependence converts amplitude noise into phase noise [9, 29].

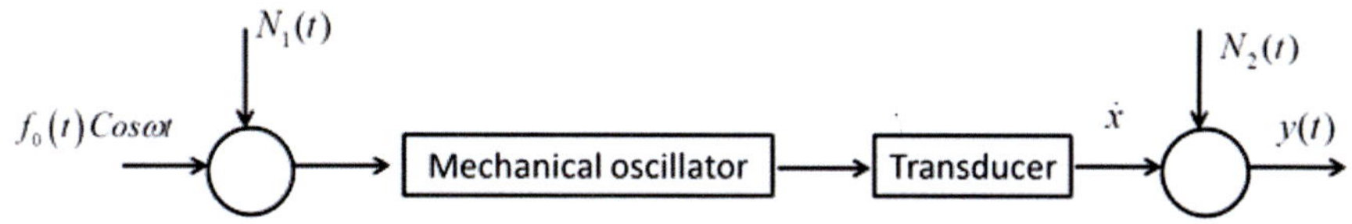

Fig. 6 Block diagram of a MEMS resonator considering the noise in excitation and detection

c. *Quality factor*
   Nonlinearity often decreases the quality factor of the resonator in MEMS device. This affects adversely the sensitivity of the instrument [15, 30]. Also amplitude of response during parametric excitation gets limited to a low value if nonlinear terms are significant.
d. *Sensing*
   Nonlinear terms introduce complexity in the sensing process [26, 31]. It is desirable to design and fabricate the resonator to drive in the linear regime. Nonlinear terms, for example, in capacitive sensing, affect the performance of the resonator. In optical sensing also the nonlinear effects are observed.
   Apart from the general effects of the nonlinearities present in almost all MEMS devices, some kinds of nonlinearities may degrade the quality of some special devices. For example, in dynamic atomic force microscopy, where near-resonant response of cantilever is measured to estimate topography of the sample, the nonlinear interaction strongly influences the resonant characteristics of the cantilever. Amplitude jump and hysteresis take place during forward and backward frequency sweeps. Transition between attractive-repulsive regime also leads to chaotic response. These effects have been found to affect the quality of image of the sample surface [32].

5.2 Desirable Effects of Nonlinearity

The desirable effects of nonlinearities are now listed below.

a. *Phase noise*
   In the previous section, it has been pointed out how the phase noise increases because of nonlinearities present in the system. However, it has been found that if nonlinearity is judiciously introduced then the phase noise can be reduced [33]. It has been found, contrary to the conventional phenomenological wisdom, that there exist a special region in the parameter space, lying above the nonlinear threshold, where the phase noise is reduced. By operating the oscillator in this region the signal level can be increased to large value without degrading the oscillator performance. However, to achieve this objective a feedback with a phase delay is required. By properly selecting the gain and phase delay the nonlinear frequency shift is made comparable to the linear resonance line width, but small

compared to the resonant frequency. This scheme is also known as phase feedback oscillator. Similar noise reduction has been experimentally demonstrated for a 2 DOF-coupled nonlinear oscillator [34].

b. *Frequency stability*
In timing devices, where stability of the output frequency is of greatest concern, the change in frequency, caused by increase in amplitude (i.e. amplitude-frequency, A-f effect), is a problem. However, the nonlinearity can help in achieving temperature compensation in silicon micromechanical resonator [35]. The main challenge in Si resonator is that Si has a temperature coefficient of frequency (TCF) near to 30 ppm/°C, which is large compared to that of a quartz oscillator (18 ppm/°C at 25 °C). With increasing temperature, both the resonant frequency and quality factor increase as follows:

$$\begin{aligned} f_r &= f_0 \left(1 + \mathrm{TCF}(T - T_0)\right), \\ Q_r &= Q_0 \left(1 + \mathrm{TCQ}(T - T_0)\right), \end{aligned} \tag{31}$$

where f_0 and Q_0 are the resonant frequency and quality factor at 12.5 °C and TCQ is the temperature coefficient of quality factor. In the linear regime, operating at resonance, the amplitude is proportional to the quality factor, i.e.

$$X = X_0 \left(1 + \mathrm{TCQ}(T - T_0)\right)$$

where X_0 is the amplitude of MEMS at T_0. From the above equations, we get

$$\begin{aligned} \Delta f_T &= \left(\frac{f_0}{X_0}\frac{\mathrm{TCF}}{\mathrm{TCQ}}\right)\Delta X \\ \text{or} \quad \frac{\Delta f_T}{f_0} &= \frac{\mathrm{TCF}}{\mathrm{TCQ}}\frac{\Delta X}{X_0} \end{aligned} \tag{32}$$

i.e. there exists a linear relationship between f_r and X. When nonlinearity is present the A-f dependence due to duffing-type nonlinearity can be written as

$$\frac{\Delta f_D}{f_0} = \frac{3\alpha/4(X_0 + \Delta X)^2}{2\pi f_0} \approx \frac{3\alpha}{4}\frac{X_0^2}{2\pi f_0} + \frac{3\alpha}{4\pi}\frac{X_0^2}{f_0}\frac{\Delta X}{X_0} \tag{33}$$

When both the effects are considered

$$\frac{\Delta f}{f_0} = \frac{\Delta f_T}{f_0} + \frac{\Delta f_D}{f_0} = \frac{3}{4}\frac{\alpha X_0^2}{2\pi f_0} + \frac{\Delta X}{X_0}\left(\frac{TCF}{TCQ} + \frac{3\alpha X_0^2}{4\pi f_0}\right). \tag{34}$$

The effect of temperature variation can be nullified if X_0 is selected for ($\alpha < 0$) in such a way that the last term gets cancelled out. Thus, the temperature dependence of frequency is minimized. Further, electrostatic tuning is also used to suppress the temperature-frequency drift [36].

c. *Sensing devices*
Nonlinearity can be useful in many sensing devices. Some examples are given below:

(i) The pull-in phenomena of electrostatically driven MEMS, which is due to nonlinear nature of the interacting force, is desirable in MEMS switches for reducing switching voltage [6].
(ii) Jump phenomena (Bifurcation) in a nonlinear system have been advantageously used in mass sensing. In a linear system, sweeping of excitation frequency is carried out to detect resonant frequency. Performance of the sensor depends on the quality factor or damping. Further, the minimum detectable mass is also limited by noise. The strategy of exploiting bifurcation phenomena has been found giving better results. It has been found to be applied in different ways.
    (a) One way is exciting microresonator at set point A′ (shown in Fig. 7a), close to the critical point, so that an addition of mass will cause a shift in the resonance frequency and also in the bifurcation frequency. If the change in bifurcation frequency is large enough, then it will cause saddle node bifurcation (jump to A) [10]. From the difference between set point and bifurcation frequency, minimum possible value of mass detected can be calculated. Such device has been used for switch-type operation, whether the minimum value of measurand is detected (1) or not (0).
    (b) Another way found in literature is bifurcation tracking through sweeping a parameter [37]. Parametrically excited systems contain stable and unstable regimes. As shown in Fig. 7b, excitation frequency is swept towards the separating boundary of these regions. At critical point, amplitude of oscillations goes up. On mass detection, boundary will shift towards lower frequency and finding the shift, amount of mass detected can be calculated. Noise will be a dominating factor for uncertainty in sensing. Rate of frequency sweeping is an important factor for getting results with maximum precision [38]. If the rate is too fast, then bifurcation will happen after critical point, and, if it is too low, then noise activated jump can occur. Even influenced by noise, sensor based on bifurcation tracking has been found giving better results than resonant sensing.
    (c) For continuous sensing, resonator has to be reset at low amplitudes, and again sensing can be carried out, which can be time-consuming. One technique was found very useful, which works on the phenomena of noise squeezing [39]. While approaching critical point, just before bifurcation, phase noise gets squeezed to a very small value. Considering it as sign of bifurcation gas sensing has been done with faster rate.
(iii) Nonlinear parametric resonance with broad driving range is useful in increasing the sensitivity of MEMS gyrosensor [40]. In a study, it has been demonstrated that by adjusting the electromagnetic and mechanical nonlin-

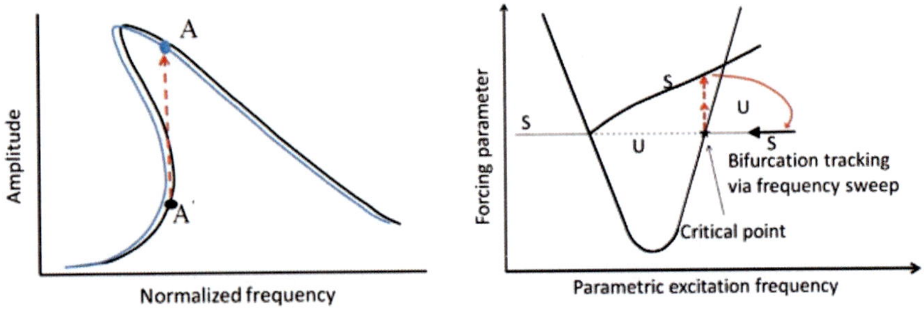

Fig. 7 **a** Utilizing jump phenomena for sensing purpose. **b** Tracking the bifurcation frequency in parametrically excited systems

earities, a reduction in angle random walk (ARW) and bias instability is possible.

(iv) Nonlinear coupling between two modes of a micromechanical resonatory disc gyroscope introduces the parametric amplification of the Coriolis force without the need of externally applied parametric pumping [41]. The amplification increases the rate sensitivity of vibrating gyroscope.

(v) As described in previous section, large displacement reduces the sensor performance and detecting the displacement also becomes difficult. To limit the displacement, nonlinearity is introduced in the excitation loop [42]. In some of the mass-sensing strategies, microcantilever is operated in self-excitation loop. Tip deflection of cantilever beam is detected and (as shown in Fig. 8) base displacement is given as cubic polynomial function of the sensor output [43]. Self-excitation strategy has also been utilized for creating parametric resonance [15]. Here, the introduced nonlinearity will not cause any jumps in amplitude while sweeping of the excitation frequency.

d. *Memory devices*

In memory elements, bistability is introduced into the system to generate two states in the response amplitude, namely, 0 and 1 states, which correspond to the low and high values of the amplitude, respectively. Such bistability, through which amplitude is switched from one state to the other, is possible only because of nonlinearities [44].

6 Tailoring Nonlinearity in MEMS Resonator

It is seen that adjustment of nonlinearity either by reducing it or enhancing may be required in MEMS resonator. Special arrangements are made to modify the nonlinearity in the existing microstructures. Some are listed below:

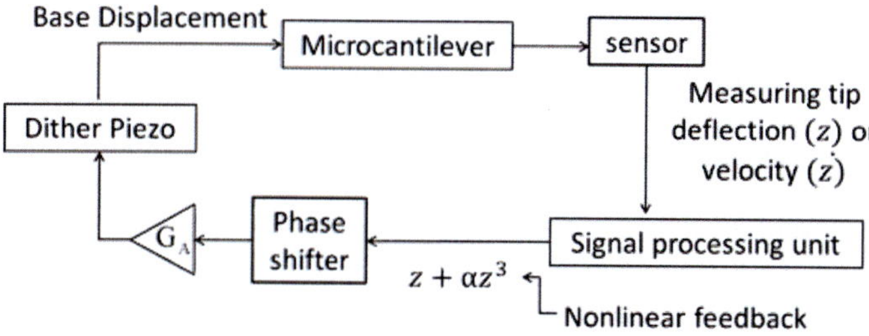

Fig. 8 Block diagram for self-excited MEMS resonator with induced nonlinearity in feedback

(i) Microbeams with initially curved shape have been fabricated to utilize the bistable and snap through behaviour of arch beam [45]. Initially, curved shape causes softening quadratic nonlinearity in the governing equation of motion (Fig. 9).

(ii) Nonlinearity may not be present due to geometric or actuation method, but for desirable results (as aforementioned in the previous section), it has been induced in artificial ways through applying nonlinear feedback [46]. Here, desired softening or hardening behaviour can be achieved via properly selecting feedback parameters.

(iii) Adding or removing material at location (as shown in Fig. 10), where slope in the mode shape of resonator is maximum, cubic nonlinearity due to geometric effect can be increased or decreased, respectively. The coefficient has been increased or decreased up to more than 2.5 and 3 times, respectively, via varying the thickness of microbeam [47]. Here γ is the coefficient for cubic nonlinearity. That influences the frequency-amplitude behaviour, by either broadening nonlinear resonance regime or making it nearly linear. Varying beam thickness, natural frequency also gets changed comparing to uniform beam.

(iv) Electrostatically actuated comb drives are widely used in MEMS. Electrostatic force is dependent on distance separating electrodes and electrode surface area. Via shaping comb fingers, coefficients for nonlinear terms in governing differential equation can be changed [48]. Linear electrostatic force–displacement behaviour can also be achieved.

(v) There is no mid-plane stretching in cantilever, but the stretching in the beam has been induced through a polymer attachment and similar effect is generated [49].

(vi) Electrostatic force is a highly nonlinear function of the distance between electrodes and this electrical nonlinearity has been used advantageously for cancelling mechanical nonlinearity or tuning the overall nonlinear behaviour [50, 51].

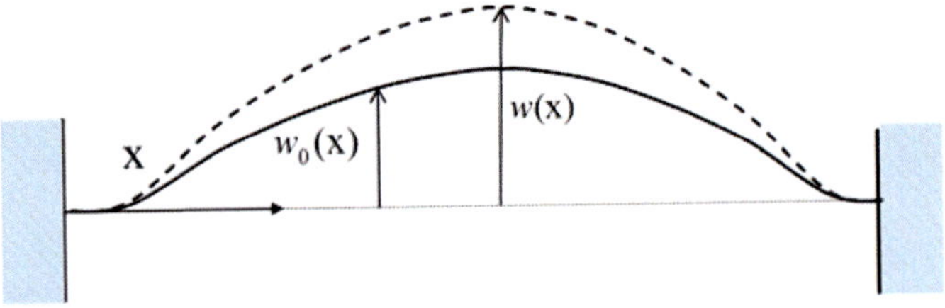

Fig. 9 Microbeam with initially curved shape (arch beam)

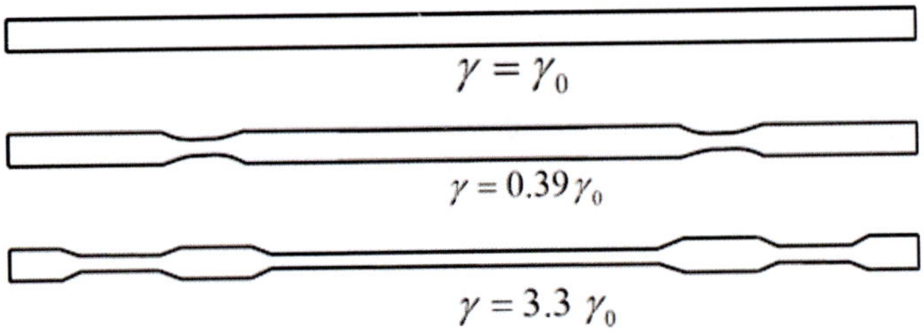

Fig. 10 Modifying the cubic nonlinearity coefficient through varying the thickness

7 Conclusion

Different aspects of nonlinearities on the overall dynamics of MEMS resonators have been reviewed. Both desirable and undesirable effects of nonlinearities are discussed with the help of simple mathematical models. Although many facts of nonlinearities are dealt with, an exhaustive study is impossible because, unlike linear system, there is no general solution of nonlinear equation of motion. Consequently, different excitations have different effects. Furthermore, only one type of system, namely, the one which can be modelled as single-degree-of-freedom system has been considered here. Coupled array of micromachined oscillators or the MEMS where multiple modes interact due to nonlinearity has not been discussed. Any meaningful discussion of the dynamic behaviour of such system will make the size of the article too large to be within the limit.

References

1. Lavrik, N.V., Sepaniak, M.J., Datskos, P.G.: Cantilever transducers as a platform for chemical and biological sensors. Rev. Sci. Instrum. **75**(7), 2229–2253 (2004)
2. Zhao, L., Hu, Y., Wang, T., Ding, J., Liu, X., Zhao, Y., Jiang, Z.: A mems resonant sensor to measure fluid density and viscosity under flexural and torsional vibrating modes. Sensors **16**(6), 830 (2016)
3. Todaro, M.T., Guido, F., Mastronardi, V., Desmaele, D., Epifani, G., Algieri, L., De Vittorio, M.: Piezoelectric mems vibrational energy harvesters: advances and outlook. Microelectron. Eng. **183**, 23–36 (2017)

4. Hasan, M.H., Alsaleem, F.M., Ouakad, H.M.: Novel threshold pressure sensors based on nonlinear dynamics of MEMS resonators. J. Micromech. Microeng. **28**(6), 065007 (2018)
5. Brand, O., Dufour, I., Heinrich, S., Heinrich, S.M., Josse, F., Fedder, G.K., Korvink, J.G., Hierold, C., Tabata, O.: Resonant MEMS: Fundamentals, Implementation, and Application. Wiley (2015)
6. Zhang, W.-M., Yan, H., Peng, Z.-K., Meng, G.: Electrostatic pull-in instability in MEMS/NEMS: a review. Sens. Actuators A: Phys. **214**, 187–218 (2014)
7. Lifshitz, R., Cross, M.: Nonlinear dynamics of nanomechanical and micromechanical resonators. Rev. Nonlinear Dyn. Complex. **1**, 1–52 (2008)
8. Rhoads, J.F., Shaw, S.W., Turner, K.L.: Nonlinear dynamics and its applications in micro-and nanoresonators. In: ASME 2008 Dynamic Systems and Control Conference, pp. 1509–1538. American Society of Mechanical Engineers Digital Collection (2009)
9. Tiwari, S., Candler, R.N.: Using flexural mems to study and exploit nonlinearities: a review. J. Micromech. Microeng. **29**(8), 083002 (2019)
10. Kumar, V., Yang, Y., Boley, J.W., Chiu, G.T.-C., Rhoads, J.F.: Modeling, analysis, and experimental validation of a bifurcation-based microsensor. J. Microelectromech. Syst. **21**(3), 549–558 (2012)
11. Mahboob, I., Yamaguchi, H.: Piezoelectrically pumped parametric amplification and Q enhancement in an electromechanical oscillator. Appl. Phys. Lett. **92**(17), 173109 (2008)
12. Rhoads, J.F., Shaw, S.W.: The impact of nonlinearity on degenerate parametric amplifiers. Appl. Phys. Lett. **96**(23), 234101 (2010)
13. Yabuno, H.: Self-excited oscillation for high-viscosity sensing and self-excited coupled oscillation for ultra-senseitive mass sensing. Procedia IUTAM **22**, 216–220 (2017)
14. Ramos, D., Mertens, J., Calleja, M., Tamayo, J.: Phototermal self-excitation of nanomechanical resonators in liquids. Appl. Phys. Lett. **92**(17), 173108 (2008)
15. Prakash, G., Raman, A., Rhoads, J., Reifenberger, R.G.: Parametric noise squeezing and parametric resonance of microcantilevers in air and liquid environments. Rev. Sci. Instrum. **83**(6), 065109 (2012)
16. Younis, M.I.: MEMS Linear and Nonlinear Statics and Dynamics, vol. 20. Springer Science & Business Media (2011)
17. Zhang, W., Zhang, W., Turner, K.L.: Nonlinear dynamics of micro impact oscillators in high frequency mems switch application. In: The 13th International Conference on Solid-State Sensors, Actuators and Microsystems, 2005. Digest of Technical Papers. TRANSDUCERS'05, vol. 1, pp. 768–771. IEEE (2005)
18. Delnavaz, A., Mahmoodi, S.N., Jalili, N., Mahdi Ahadian, M., Zohoor, H.: Nonlinear vibrations of microcantilevers subjected to tip-sample interactions: theory and experiment. J. Appl. Phys. **106**(11), 113510 (2009)
19. Park, Y.-H., Park, K.: High-fidelity modeling of mems resonators. Part I. Anchor loss mechanisms through substrate. J. Microelectromech. Syst. **13**(2), 238–247 (2004)
20. Lifshitz, R., Roukes, M.L.: Thermoelastic damping in micro-and nanomechanical systems. Phys. Rev. B **61**(8), 5600 (2000)
21. Sader, J.E.: Frequency response of cantilever beams immersed in viscous fluids with applications to the atomic force microscope. J. Appl. Phys. **84**(1), 64–76 (1998)
22. Younis, M.I., Alsaleem, F.M., Miles, R., Su, Q.: Characterization of the performance of capacitive switches activated by mechanical shock. J. Micromech. Microeng. **17**(7), 1360 (2007)
23. Hu, S., Raman, A.: Analytical formulas and scaling laws for peak interaction forces in dynamic atomic force microscopy. Appl. Phys. Lett. **91**(12), 123106 (2007)
24. Rhoads, J.F., Shaw, S.W., Turner, K.L.: The nonlinear response of resonant microbeam systems with purely-parametric electrostatic actuation. J. Micromech. Microeng. **16**(5), 890 (2006)
25. Sobreviela, G., Vidal-Álvarez, G., Riverola, M., Uranga, A., Torres, F., Barniol, N.: Suppression of the af-mediated noise at the top bifurcation point in a mems resonator with both hardening and softening hysteretic cycles. Sens. Actuators A: Phys. **256**, 59–65 (2017)
26. Trusov, A.A., Shkel, A.M.: Capacitive detection in resonant mems with arbitrary amplitude of motion. J. Micromech. Microeng. **17**(8), 1583 (2007)

27. Antonio, D., Zanette, D.H., López, D.: Frequency stabilization in nonlinear micromechanical oscillators. Nat. Commun. **3**, 806 (2012)
28. Kaajakari, V., Mattila, T., Oja, A., Seppa, H.: Nonlinear limits for single-crystal silicon microresonators. J. Microelectromech. Syst. **13**(5), 715–724 (2004)
29. Lee, S., Nguyen, C.T.-C.: Phase noise amplitude dependence in self limiting wine-glass disk oscillators. In: Solid State Sensor, Actuator, and Microsystems Workshop (2004)
30. Yie, Z., Miller, N.J., Shaw, S.W., Turner, K.L.: Parametric amplification in a resonant sensing array. J. Micromech. Microeng. **22**(3), 035004 (2012)
31. Thormann, E., Pettersson, T., Claesson, P.M.: How to measure forces with atomic force microscopy without significant influence from nonlinear optical lever sensitivity. Rev. Sci. Instrum. **80**(9), 093701 (2009)
32. Hu, S., Raman, A.: Chaos in atomic force microscopy. Phys. Rev. Lett. **96**(3), 036107 (2006)
33. Sobreviela, G., Zhao, C., Pandit, M., Do, C., Du, S., Zou, X., Seshia, A.: Parametric noise reduction in a high-order nonlinear mems resonator utilizing its bifurcation points. J. Microelectromech. Syst. **26**(6), 1189–1195 (2017)
34. Zhao, C., Sobreviela, G., Pandit, M., Du, S., Zou, X., Seshia, A.: Experimental observation of noise reduction in weakly coupled nonlinear MEMS resonators. J. Microelectromech. Syst. **26**(6), 1196–1203 (2017)
35. Defoort, M., Taheri-Tehrani, P., Horsley, D.: Exploiting nonlinear amplitude-frequency dependence for temperature compensation in silicon micromechanical resonators. Appl. Phys. Lett. **109**(15), 153502 (2016)
36. Chen, D., Wang, Y., Chen, X., Yang, L., Xie, J.: Temperature-frequency drift suppression via electrostatic stiffness softening in mems resonator with weakened duffing nonlinearity. Appl. Phys. Lett. **114**(2), 023502 (2019)
37. Turner, K.L., Burgner, C.B., Yie, Z., Holtoff, E.: Using nonlinearity to enhance micro/nanosensor performance. In: 2012 IEEE Sensors, pp. 1–4. IEEE (2012)
38. Burgner, C., Miller, N., Shaw, S., Turner, K.: Parameter sweep strategies for sensing using bifurcations in MEMS. In: Solid-State Sensor, Actuator, and Microsystems Workshop, Hilton Head Workshop (2010)
39. Li, L.L., Holthoff, E.L., Shaw, L.A., Burgner, C.B., Turner, K.L.: Noise squeezing controlled parametric bifurcation tracking of mip-coated microbeam mems sensor for tnt explosive gas sensing. J. Microelectromech. Syst. **23**(5), 1228–1236 (2014)
40. Oropeza-Ramos, L.A., Burgner, C.B., Turner, K.L.: Robust micro-rate sensor actuated by parametric resonance. Sens. Actuators A: Phys. **152**(1), 80–87 (2009)
41. Nitzan, S.H., Zega, V., Li, M., Ahn, C.H., Corigliano, A., Kenny, T.W., Horsley, D.A.: Self-induced parametric amplification arising from nonlinear elastic coupling in a micromechanical resonating disk gyroscope. Sci. Rep. **5**, 9036 (2015)
42. Endo, D., Yabuno, H., Yamamoto, Y., Matsumoto, S.: Mass sensing in a liquid environment using nonlinear self-excited coupled-microcantilevers. J. Microelectromech. Syst. **27**(5), 774–779 (2018)
43. Yabuno, H.: Review of applications of self-excited oscillations to highly sensitive vibrational sensors. ZAMM-J. Appl. Math. Mech./Z. für Angew. Math. und Mech. e201900009 (2019)
44. Hafiz, M., Kosuru, L., Ramini, A., Chappanda, K., Younis, M.: In-plane mems shallow arch beam for mechanical memory. Micromachines **7**(10), 191 (2016)
45. Younis, M.I., Ouakad, H.M., Alsaleem, F.M., Miles, R., Cui, W.: Nonlinear dynamics of mems arches under harmonic electrostatic actuation. J. Microelectromech. Syst. **19**(3), 647–656 (2010)
46. Bajaj, N., Sabater, A.B., Hickey, J.N., Chiu, G.T.-C., Rhoads, J.F.J.: Design and implementation of a tunable, duffing-like electronic resonator via nonlinear feedback. J. Microelectromech. Syst. **25**(1), 2–10 (2015)
47. Li, L.L., Polunin, P.M., Dou, S., Shoshani, O., Scott Strachan, B., Jensen, J.S., Shaw, S.W., Turner, K.L.: Tailoring the nonlinear response of mems resonators using shape optimization. Appl. Phys. Lett. **110**(8), 081902 (2017)

48. Jensen, B.D., Mutlu, S., Miller, S., Kurabayashi, K., Allen, J.J.: Shaped comb fingers for tailored electromechanical restoring force. J. Microelectromech. Syst. **12**(3), 373–383 (2003)
49. Asadi, K., Li, J., Peshin, S., Yeom, J., Cho, H.: Mechanism of geometric nonlinearity in a nonprismatic and heterogeneous microbeam resonator. Phys. Rev. B **96**(11), 115306 (2017)
50. Chen, D., Wang, Y., Guan, Y., Chen, X., Liu, X., Xie, J.: Methods for nonlinearities reduction in micromechanical beams resonators. J. Microelectromech. Syst. **27**(5), 764–773 (2018)
51. Agarwal, M., Chandorkar, S.A., Candler, R.N., Kim, B., Hopcroft, M.A., Melamud, R., Jha, C.M., Kenny, T.W., Murmann, B.: Optimal drive condition for nonlinearity reduction in electrostatic microresonators. Appl. Phys. Lett. **89**(21), 214105 (2006)

Automated Upgraded Generalized Full-Discretization Method: Application to the Stability Study of a Thin-Walled Milling Process

Check for updates

Chigbogu Ozoegwu and Peter Eberhard

1 Introduction

Lumped delayed models are usually used to describe the regenerative vibration of a flexible milling tool when cutting a rigid workpiece. In this case, the tool is the compliant body, see Fig. 1a. A few of the elaborate references on modeling and stability analysis of such systems include [1–4]. On the other hand, the tool can act as a rigid body when milling a thin-walled workpiece with high aspect ratio of length or width to thickness. This case is illustrated in Fig. 1b. Milling of thin-walled workpieces is encountered in the processing of monolithic parts for the aerospace industry where high strength-to-weight ratio is a major requirement. In such a milling process, much of the blank material is removed, and this has to be done at the maximum possible material removal rate for economic viability. Knowledge-based choices of the process parameters that guarantee optimal productivity and surface quality derive from modeling and analyzing the structural and regenerative dynamics of the cutting process. Since a thin-walled workpiece is a distributed elastic system, its regenerative machining is governed by a delayed continuum dynamics which is infinite dimensional in both spatial and temporal space. The infinite-dimensional response, which is dependent on the location of tool–workpiece contact, can only be made computationally feasible through spatial–temporal discretization and reduction.

Based on the finite element (FE) method and experimental modal analysis, a model for three-dimensional prediction of surface finish and displacements of a flexible T-shaped plate at the tool–workpiece contact zone and cutting forces was presented [5].

C. Ozoegwu
Department of Mechanical Engineering, University of Nigeria, Nsukka, Nsukka, Nigeria
e-mail: chigbogu.ozoegwu@unn.edu.ng

P. Eberhard (✉)
Institute of Engineering and Computational Mechanics, University of Stuttgart, Stuttgart, Germany
e-mail: peter.eberhard@itm.uni-stuttgart.de

© Springer Nature Singapore Pte Ltd. 2021

U. S. Dixit and S. K. Dwivedy (eds.), *Mechanical Sciences*,
https://doi.org/10.1007/978-981-15-5712-5_4

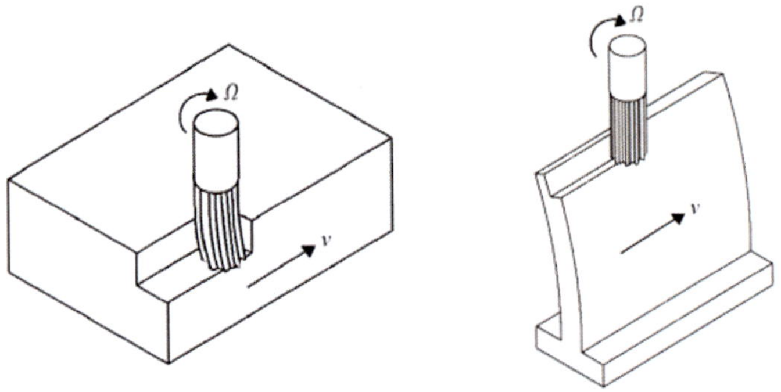

(a) A flexible tool milling a rigid workpiece

(b) A rigid tool milling a flexible workpiece

Fig. 1 Flexible tool and flexible workpiece scenarios

Regenerative chatter stability was not considered in the work. Subsequently, regenerative chatter stability was considered using frequency- and time-domain methods where the former predated the latter in application to the stability analysis of milling of thin-walled workpieces. In addition to considering the non-regenerative dynamics problems, a variant of the frequency-domain approach was used for the stability analysis of the relative regenerative vibration between a flexible T-shaped plate and a flexible milling tool [6, 7]. This allowed the avoidance of statics and dynamics problems, local and global structure deformations problems, and surface quality problems. The frequency-domain approach was used in [8, 9] to identify the three-dimensional stability boundaries of a thin-walled plate fixed at the base and at one side. The stability diagrams highlighted the productive stable cutting depths as a function of spindle speed and tool location. The stability limits of the three-dimensional dynamics of milling of thin-walled workpiece were constructed using the frequency-domain approach considering a nonlinear dependency of cutting force coefficients on axial depth of cut [10]. Considering the in-process structural variation due to material removal and time-dependent tool location on the basis of FE analyses of the workpiece, three-dimensional stability lobes were identified for the optimization of stable milling of flexible workpieces [11]. In [12], the cutting pass was divided into regular zones and the frequency response function was sequentially updated using the Sherman–Morrison–Woodbury formula such that it was not necessary, like in the earlier works, to re-build and re-mesh the FE model at each zone. This resulted in a computationally efficient stability identification in the frequency domain. Also, during a frequency-domain stability analysis of a curved thin-walled workpiece, the initial FE model was modified in-process to characterize the effect of material removal on the workpiece dynamics, thus pre-empting the need for FE model re-building with change of tool position [13]. Motivated by the problem of the computationally inefficient need of re-building FE models at each new position of the tool, a finite strip (FS) modeling was presented [14] for the structural dynamics analysis and regenerative chatter stability analysis in the frequency domain of milling thin-walled structures. Methods based on thin plate theory, mode superposition principle, and updated Rayleigh–Ritz solutions were used to prepare the regenerative dynamics of milling of thin-walled workpiece for stability analysis in the frequency domain [15, 16].

Application of time-domain methods to the stability analysis of milling of flexible workpieces is more recent and less extensively investigated than the frequency-domain methods. The simultaneous effect of tool position and material removal on the regenerative chatter stability of milling of thin-walled workpiece was studied with the semi-discretization method [17, 18]. A semi-discrete time-domain method based on Simpson Rules was applied in the stability analysis of milling of flexible workpiece considering process damping and non-uniform pitch effects [19]. This method of stability analysis was recently applied in identifying the enhanced stability boundaries of milling of thin-walled workpieces connected to appropriate additional masses [20]. Using the full-discretization method (FDM) based on the third-order Newton interpolation of the milling current state and first-order interpolation of the delayed state and considering process damping and helix angle, three-dimensional stability lobes which give the boundary axial depths of cut as a function of spindle speed and tool location were constructed [21]. The FDM based on numerical integration was used in [22] to construct the stability boundaries of milling of a thin-walled workpiece while submerged in a viscous fluid. The first-order semi-discretization was used in the stability analysis of a reduced order model of milling of a thin-walled workpiece [23]. The parametric model order reduction, which was based on modal truncation of location-dependent modal matrices and cubic spline interpolation of the individual reduced modes was utilized. An efficient decomposition-condensation method for chatter stability analysis of large-scale thin-walled structures has been proposed [24] by condensing and coupling the FE models of the machined workpiece and the initially removed material using the component mode synthesis method and subsequently integrating the reduced system to the discrete time-domain chatter prediction method [19]. Considering tool helix and mode coupling, the semi-discretization method is reconstructed based on the Shannon interpolation function for stability analysis of thin-walled workpieces [25]. A second-order polynomial tensor has been used to approximate both the current and delayed regenerative states for the stability analysis of a flexible workpiece milling [26].

Few works used a lumped parameter model of the flexible workpiece system. The flexible workpiece was considered as a lumped single-degree-of-freedom (1DOF) oscillator in [27] for maximization of milling material removal rate. The three-dimensional stability lobe diagrams, calculated using the frequency-domain approach, were presented on axial depth, radial depth, and spindle speed axes. In [28], the stability lobes of a $120 \times 100 \times 6$ mm thin-walled workpiece, partitioned into four zones, were determined from MATLAB time-domain simulations and statistical error analyses. The zone-dependent modal parameters, which determine the dynamics of the 2DOF lumped models of flexible workpiece–tool system in [28], were extracted from impact tests in each zone. A procedure based on operational modal analysis, which uses the cutting force model for scaling modal residues, and a frequency-domain approach was used to construct stability lobe diagrams that agree well with those based on impact tests [29].

One problem is clear from studying the above-reviewed works which are based on time-domain methods. The very complicated modeling and stability analysis of regenerative thin-walled workpiece milling is mostly done manually. After time-domain discretization, the three major stages for stability analysis of regenerative chatter of either a milling tool or a milled thin-walled workpiece are chatter state interpolation, monodromy matrix construction, and stability identification/eigenvalue analysis. Traditionally, the first two stages are done manually on a case-by-case basis while only the last stage is computerized, and when the degree of interpolation changes, the steps are repeated needing the last stage to be re-computerized. This work proposes to upgrade the fully general and programmed method in [30] by, among other things, incorporating modal reduction of elastic thin-walled workpieces so as to avoid all the stages of case-by-case manual symbolical and numerical analyses in chatter stability identification. A major motivation is to create a fully computerized framework for the first systematic study of the parametric effect of chatter states interpolation order on the computational precision of stability lobes identification. This aim can hardly be achieved on the basis of a partly computerized case-by-case approach because of the astronomically rising difficulty of symbolic analysis with rising interpolation order. Another major motivation is to improve the expediency of practical application since any ordinary industrial operator of thin-walled workpiece milling could make a simulation for precise selection of productive chatter-free process parameters.

Following this introductory section, Sect. 2 contains the theoretical model and the stability criteria of the studied system. In Sect. 3, a generalized monodromy matrix of reduced system is described. The general system is then computed for automatic study of the parametric effects of interpolation order on computational efficiency and for automatic simulations of stability diagrams in Sect. 4. Conclusions are presented in Sect. 5.

2 Model and Stability Condition

The force that excites the regenerative response of flexible tool/rigid workpiece model have been typically given on the basis of the nonlinear cutting force model as

$$\mathbf{F}(t) = \mathbf{H}(t)(\mathbf{z}(t) - \mathbf{z}(t-\tau)), \tag{1}$$

where $\mathbf{z}(t) \in \mathbb{R}^{n_d}$ is the state of the milling tool, $\mathbf{H}(t) \in \mathbb{R}^{n_d \times n_d}$ is the specific force variation matrix, and n_d is the number of directions of regenerative response of the tool. Assuming a non-helix tool, the axial component of the cutting force is ignored; therefore, the regenerative responses of the tool occur in the feed and feed-normal directions. This means that $n_d = 2$. The matrix-valued function $\mathbf{H}(t)$ is typically given, see [31], by

$$\mathbf{H}(t) = -w \begin{bmatrix} h_{xx}(t) & h_{xy}(t) \\ h_{yx}(t) & h_{yy}(t) \end{bmatrix}, \tag{2}$$

where

$$h_{xx}(t) = C_F \sum_{j=1}^{N} g_j(t)(\sin\theta_j(t))^{\gamma}(\chi \sin\theta_j(t) + \cos\theta_j(t)), \tag{3}$$

$$h_{xy}(t) = C_F \sum_{j=1}^{N} g_j(t)(\sin\theta_j(t))^{\gamma-1}\cos\theta_j(t)(\chi \sin\theta_j(t) + \cos\theta_j(t)), \tag{4}$$

$$h_{yx}(t) = C_F \sum_{j=1}^{N} g_j(t)(\sin\theta_j(t))^{\gamma}(\chi \cos\theta_j(t) - \sin\theta_j(t)), \tag{5}$$

$$h_{yy}(t) = C_F \sum_{j=1}^{N} g_j(t)(\sin\theta_j(t))^{\gamma-1}\cos\theta_j(t)(\chi \cos\theta_j(t) - \sin\theta_j(t)), \tag{6}$$

and w is the axial depth of cut. The value $\theta_j(t) = \frac{\pi\Omega}{30}t + \frac{2\pi}{N}(j-1)$ is the angular displacement of the j-th cutting edge for $j = 1, 2, \ldots, N$ and $g_j(t) = 0.5\left(1 + \text{sign}\{\sin(\theta_j(t) - \arctan\wp) - \sin(\theta_\text{s} - \arctan\wp)\}\right)$ is the screening function where $\wp = (\sin\theta_\text{s} - \sin\theta_\text{e})/(\cos\theta_\text{s} - \cos\theta_\text{e})$. When the j-th cutting edge is active, $g_j(t) = 1$ but when idle, $g_j(t) = 0$. The angles θ_s and θ_e are the start and end angles of the cutting interval, χ is the ratio of thrust to tangential cutting force and $C_F = C_t\gamma(v\tau)^{\gamma-1}$ is a constant where C_t is the tangential cutting coefficient, and γ is the feed exponent in the cutting force law. The start and end angles, θ_s and θ_e shown in Fig. 2, are given as

$$\theta_\text{s} = 0,\ \theta_\text{e} = \arccos(1 - 2\rho), \qquad \text{for up-milling}, \tag{7}$$

$$\theta_\text{s} = \arccos(2\rho - 1),\ \theta_\text{e} = \pi, \qquad \text{for down-milling}, \tag{8}$$

where $\rho = \frac{B}{D}$ is the radial immersion, B is the radial depth of cut, and D is the tool diameter.

The thin-walled workpiece being an elastic continuum with a large DOF second-order delayed model, as can be derived from FE analysis, is needed to capture the regenerative dynamics when excited by a milling tool. Since the tool–workpiece contact obeys Newton's third law of motion, the FE model for the regenerative dynamics of a thin-walled workpiece can be presented in the form

$$\mathbf{M}(p_y)\ddot{\mathbf{z}}(t) + \mathbf{D}(p_y)\dot{\mathbf{z}}(t) + \mathbf{K}(p_y)\mathbf{z}(t) = \mathbf{B}_H(p_y)\mathbf{H}(t)\,\mathbf{C}_H(p_y)(\mathbf{z}(t) - \mathbf{z}(t-\tau)), \tag{9}$$

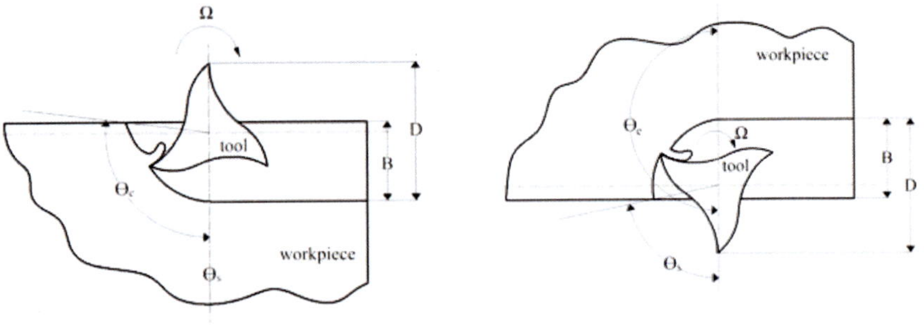

(a) Up-milling (b) Down-milling

Fig. 2 Start angle θ_s and end angle θ_e of milling

where $\mathbf{z}(t) \in \mathbb{R}^{d_F}$ is the state, $\mathbf{M}(p_y) \in \mathbb{R}^{d_F \times d_F}$ is the mass matrix, $\mathbf{D}(p_y) \in \mathbb{R}^{d_F \times d_F}$ is the damping matrix, and $\mathbf{K}(p_y) \in \mathbb{R}^{d_F \times d_F}$ is the stiffness matrix. The matrices $\mathbf{B}_H(p_y) \in \mathbb{R}^{d_F \times n_d}$ and $\mathbf{C}_H(p_y) \in \mathbb{R}^{n_d \times d_F}$ are used to project the n_d-dimensional tool excitation $\mathbf{H}(t)$ to the d_F-dimensional workpiece response. The normalized parameter p_y is indicative of the dependence of the modal matrices on tool location along the feed direction (y) where $p_y = 0$ at the beginning of tool pass and $p_y = 1$ at the end of the pass. Such high-fidelity FE model as given in Eq. (9) has many DOF and is thus computationally cumbersome. Based on modal truncation, Eq. (9) reduces, according to [23], to

$$\tilde{\mathbf{M}}(p_y)\ddot{\tilde{\mathbf{z}}}(t) + \tilde{\mathbf{D}}(p_y)\dot{\tilde{\mathbf{z}}}(t) + \tilde{\mathbf{K}}(p_y)\tilde{\mathbf{z}}(t) = \tilde{\mathbf{B}}_H(p_y)\mathbf{H}(t)\tilde{\mathbf{C}}_H(p_y)(\tilde{\mathbf{z}}(t) - \tilde{\mathbf{z}}(t-\tau)), \tag{10}$$

where $\tilde{\mathbf{z}}(t) \in \mathbb{R}^{d_R}$ is the reduced state, $\tilde{\mathbf{M}}(p_y) = \mathbf{V}^T(p_y)\mathbf{M}(p_y)\mathbf{V}(p_y) \in \mathbb{R}^{d_R \times d_R}$ is the reduced mass matrix, $\tilde{\mathbf{D}}(p_y) = \mathbf{V}^T(p_y)\mathbf{D}(p_y)\mathbf{V}(p_y) \in \mathbb{R}^{d_R \times d_R}$ is the reduced damping matrix, $\tilde{\mathbf{K}}(p_y) = \mathbf{V}^T(p_y)\mathbf{K}(p_y)\mathbf{V}(p_y) \in \mathbb{R}^{d_R \times d_R}$ is the reduced stiffness matrix, $\tilde{\mathbf{B}}_H(p_y) = \mathbf{V}^T\mathbf{B}_H(p_y) \in \mathbb{R}^{d_R \times n_d}$, $\tilde{\mathbf{C}}_H(p_y) = \mathbf{C}_H(p_y)\mathbf{V} \in \mathbb{R}^{n_d \times d_R}$, and $\mathbf{V}(p_y) \in \mathbb{R}^{d_F \times d_R}$ is the projection matrix. The projection matrix $\mathbf{V}(p_y)$ is a concatenation of the first d_R eigenvectors where $d_R \ll d_F$. In [23], each p_y-dependent matrix is derived from cubic spline interpolation of the values at a finite number of tool location zones under a consistent system of coordinates $\tilde{\mathbf{z}}(t)$. Making the substitution $\mathbf{x}_1(t) = \tilde{\mathbf{z}}(t)$ and $\mathbf{x}_2(t) = \dot{\tilde{\mathbf{z}}}(t)$, the first-order form of the reduced model for stability analysis using the FDM reads

$$\dot{\mathbf{x}}(t) = \mathbf{A}\mathbf{x}(t) + \mathbf{B}(t)\mathbf{x}(t) - \mathbf{B}(t)\mathbf{x}(t-\tau), \tag{11}$$

where $\mathbf{A} = \begin{bmatrix} \mathbf{0} & \mathbf{I} \\ -(\tilde{\mathbf{M}}(p_y))^{-1}\tilde{\mathbf{K}}(p_y) & -(\tilde{\mathbf{M}}(p_y))^{-1}\tilde{\mathbf{C}}(p_y) \end{bmatrix}$, and $\mathbf{B}(t) = \begin{bmatrix} \mathbf{0} & \mathbf{0} \\ (\tilde{\mathbf{M}}(p_y))^{-1}\mathbf{B}_H(p_y)\mathbf{H}(t)\mathbf{C}_H(p_y) & \mathbf{0} \end{bmatrix}$. Due to the time delay in Eq. (11), the spatially reduced dynamics is still infinite dimensional on the temporal axis. The solution of the reduced model according to the extended Floquet theory is generally

$$\mathbf{w}(t) = \Upsilon(t)\mathbf{w}(0) = \mathbf{P}(t)e^{\mathbf{B}_f t}\mathbf{w}(0), \tag{12}$$

where $\Upsilon : \mathbb{R}_{\geq 0} \rightarrow \mathbb{R}^{\infty\times\infty}$ is the transition matrix, $\mathbf{B}_f \in \mathbb{R}^{\infty\times\infty}$ is the fundamental matrix, and $\mathbf{P} : \mathbb{R}_{\geq 0} \rightarrow \mathbb{R}^{\infty\times\infty}$ is T-periodic with the initial condition being an identity matrix, that is, $\mathbf{P}(0) = \mathbf{I}$. Making use of similarity transformation and matrix exponential function gives

$$\Upsilon(t) = \mathbf{P}(t)e^{(\mathbf{V}_A\mathbf{D}\mathbf{V}_A^{-1})t} = \mathbf{P}(t)\mathbf{V}_A e^{\mathbf{D}t}\mathbf{V}_A^{-1}, \tag{13}$$

where $\mathbf{V}_A \in \mathbb{R}^{\infty\times\infty}$ is the matrix of all the eigenvectors of $\mathbf{B}_f$ and $\mathbf{D} \in \mathbb{R}^{\infty\times\infty}$ is a diagonal matrix with all the eigenvalues of $\mathbf{B}_f$, called characteristic exponents $\lambda_i = \sigma_i + j\omega_i$, as the diagonal elements. It is obvious from Eqs. (12) and (13) that the condition for asymptotic stability is that all λ_i have negative real parts, that is, $\sigma_i < 0$. From Eq. (13), the monodromy matrix becomes

$$\Upsilon(T) = \mathbf{V}_A e^{\mathbf{D}T}\mathbf{V}_A^{-1}. \tag{14}$$

The eigenvectors of $\Upsilon(T)$ are identical with those of $\mathbf{B}_f$, and therefore the similarity transformation gives

$$\mathbf{M}_T = e^{\mathbf{D}T}, \tag{15}$$

where $\mathbf{M}_T \in \mathbb{R}^{\infty\times\infty}$ is a diagonal matrix with eigenvalues of $\Upsilon(T)$ as the diagonal elements. The eigenvalues of $\Upsilon(T)$, called the characteristic multipliers μ_i, are seen from Eq. (15) to be given as $\mu_i = e^{\lambda_i T}$. Therefore, the condition for asymptotic stability is that all characteristic multipliers μ_i have modulus less than one, that is, $|\mu_i| = e^{\sigma_i T} < 1\ \forall_i$. The basis for identifying the stability boundary is $\max|\mu_i| = 1$. Stability analysis of the reduced order model based on Eq. (12) is not feasible; therefore, finite-dimensional approximations of the rather infinite-dimensional monodromy matrices are needed for computational purposes. It can then be said that in addition to spatial reduction, temporal reduction is also necessary for stability analysis of the system.

3 Generalized Discrete Map

The discrete time delay τ is also the period of the model for the studied case of uniform pitch milling. The period of the system is divided into k equal discrete time intervals $[t_i, t_{i+1}]$ where $i = 0,\ 1,\ 2,\ \ldots,\ (k-1)$ and $t_i = i\frac{T}{k} = i\Delta t = i(t_{i+1} - t_i)$. Equation (11) is solved in each discrete interval to give

$$\mathbf{x}_{i+1} = e^{\mathbf{A}\Delta t}\mathbf{x}_i + \int_{t_i}^{t_{i+1}} e^{\mathbf{A}(t_{i+1}-t)}(\mathbf{B}(t)\mathbf{x}(t) - \mathbf{B}(t)\mathbf{x}(t-\tau))dt. \tag{16}$$

Since the milling states do not have an exact analytical expression, the integration problem can only be solved if the states $\mathbf{x}(t)$ and $\mathbf{x}(t-\tau)$ are interpolated/approximated accurately enough. This is the underlying problem of the time-domain methods. In what follows, a generalized tensor-based analysis of the problem is adapted from [30]. The monodromy matrix, which is a finite-dimensional approximation of $\Upsilon(T)$, is

$$\Psi(p_c, p_d) = \prod_{i=0}^{k-1} \mathbf{M}_i(p_c, p_d), \tag{17}$$

where p_c and p_d are the degrees of the respective polynomials interpolating the states $\mathbf{x}(t)$ and $\mathbf{x}(t-\tau)$ and the discrete time transition matrices $\mathbf{M}_i(p_c, p_d)$ are given by

$$\mathbf{M}_i(p_c, p_d) = \begin{bmatrix} \mathbf{M}_{1,1}^{(i,p_c)} & \mathbf{M}_{1,2}^{(i,p_c)} & \cdots & \mathbf{M}_{1,p_c-1}^{(i,p_c)} & \mathbf{M}_{1,p_c}^{(i,p_c)} & \cdots & \mathbf{N}_{1,k+1-p_d}^{(i,p_d)} & \mathbf{N}_{1,k+2-p_d}^{(i,p_d)} & \cdots & \mathbf{N}_{1,k}^{(i,p_d)} & \mathbf{N}_{1,k+1}^{(i,p_d)} \\ \mathbf{I} & \mathbf{0} & \cdots & \mathbf{0} & \mathbf{0} & \cdots & \mathbf{0} & \mathbf{0} & \cdots & \mathbf{0} & \mathbf{0} \\ \mathbf{0} & \mathbf{I} & \cdots & \mathbf{0} & \mathbf{0} & \cdots & \mathbf{0} & \mathbf{0} & \cdots & \mathbf{0} & \mathbf{0} \\ \vdots & \vdots & \vdots & \vdots & \vdots & \vdots & \vdots & \vdots & \vdots & \vdots & \vdots \\ \mathbf{0} & \mathbf{0} & \cdots & \mathbf{0} & \mathbf{0} & \cdots & \mathbf{0} & \mathbf{0} & \cdots & \mathbf{I} & \mathbf{0} \end{bmatrix} \tag{18}$$

for $i = k-1,\ k-2,\ \ldots,\ 1,\ 0$ and

$$\mathbf{M}_{1,1}^{(i,p_c)} = \mathbf{P}_i^{(p_c)}(\mathbf{F}_0 + \mathbf{G}_{p_c,p_c}\mathbf{B}_i + \mathbf{G}_{p_c,2p_c+1}\mathbf{B}_{i+1}), \tag{19}$$

$$\mathbf{M}_{1,1-q}^{(i,p_c)} = \mathbf{P}_i^{(p_c)}(\mathbf{G}_{p_c,p_c+q}\mathbf{B}_i + \mathbf{G}_{p_c,2p_c+1+q}\mathbf{B}_{i+1}), \qquad \text{for } q = -1, -2, \ldots, 1-p_c, \tag{20}$$

$$\mathbf{N}_{1,k+1-q}^{(i,p_d)} = -\mathbf{P}_i^{(p_c)}(\mathbf{D}_{p_d,1+q}\mathbf{B}_i + \mathbf{D}_{p_d,p_d+2+q}\mathbf{B}_{i+1}), \qquad \text{for } q = p_d, p_d-1, \ldots, 0, \tag{21}$$

$$\mathbf{P}_i^{(p_c)} = [\mathbf{I} - \mathbf{G}_{p_c,p_c+1}\mathbf{B}_i - \mathbf{G}_{p_c,2p_c+2}\mathbf{B}_{i+1}]^{-1}. \tag{22}$$

The $\mathbf{G} \in \mathbb{R}^{n_d(d_R \times d_R)}$ matrices have the general forms

$$\mathbf{G}_{p_c,2p_c+1+q} = \mathbf{p}_L \odot \mathbf{F}_L = p_{L,i}\mathbf{F}_{L,i}, \tag{23}$$

$$\mathbf{G}_{p_c,p_c+q} = \mathbf{p}_H \odot \mathbf{F}_H = p_{H,i}\mathbf{F}_{H,i}, \tag{24}$$

where $\mathbf{p}_L = \{p_{L,1}\ p_{L,2}\ \cdots\}^T$ and $\mathbf{p}_H = \{p_{H,1}\ p_{H,2}\ \cdots\}^T$ are vectors and $\mathbf{F}_L = \{\mathbf{F}_{L,1}\ \mathbf{F}_{L,2}\ \cdots\}^T$ and $\mathbf{F}_H = \{\mathbf{F}_{H,1}\ \mathbf{F}_{H,2}\ \cdots\}^T$ are matrix concatenations. The vectors $\mathbf{p}_L$ and $\mathbf{p}_H$, which are the carriers of the effects of powers of Δt for inclusion in the post-integration results, are given as

$$\mathbf{p}_L = \frac{1}{(-q+1)!(p_c+q-1)!(\Delta t)^{p_c+1}}\mathbf{rep}(P_L(\Delta t), '+', ''), \tag{25}$$

$$\mathbf{p}_H = \frac{1}{(-q+1)!(p_c+q-1)!(\Delta t)^{p_c+1}}\mathbf{rep}(P_H(\Delta t), '+', ''), \tag{26}$$

and $\mathbf{rep}(P(t), '+', '')$ is a vector-valued function which produces a vector by replacing any plus sign in the coefficients of a polynomial $P(t)$ with a single void space. For example, $\mathbf{rep}(-t^4+t^2-t, '+', '') = \{-t^4\ 0t^3\ t^2\ \ -t\}^T$. The operator $\odot$ is a special form of dot product operator that multiplies every element of a vector with the correspondingly subscripted sub-matrix of a concatenation and sums the products. The polynomials in Δt, represented as $P_L(\Delta t)$ and $P_H(\Delta t)$, are given as

$$P_L(\Delta t) = \prod_{j=1-p_c}^{0}(-j\Delta t+1), \qquad \text{for } q=1, \tag{27}$$

$$P_L(\Delta t) = (-1)^{-q}(\Delta t-1)\prod_{j=1-p_c,\, j\neq q}^{0}(-j\Delta t+1), \qquad \text{for } q=0,\,-1,\,-2,\,\ldots,\,1-p_c, \tag{28}$$

$$P_H(\Delta t) = (\Delta t-1)P_L(\Delta t), \qquad \text{for } q=1,\,0,\,-1,\,-2,\,\ldots,\,1-p_c. \tag{29}$$

The terms of $P_L(\Delta t)$ and $P_H(\Delta t)$ are arranged from the lowest to the highest powers of Δt from left to right. Since the products $\mathbf{B}(t)\mathbf{x}(t)$ and $\mathbf{B}(t)\mathbf{x}(t-\tau)$ in the integrand of Eq. (16) are power polynomials in t, Eq. (16) simply requires the execution of the items $\int_{t_i}^{t_{i+1}} t^0 e^{\mathbf{A}(t_{i+1}-t)}dt$, $\int_{t_i}^{t_{i+1}} t^1 e^{\mathbf{A}(t_{i+1}-t)}dt$, $\int_{t_i}^{t_{i+1}} t^2 e^{\mathbf{A}(t_{i+1}-t)}dt$,..., $\int_{t_i}^{t_{i+1}} t^{p_c+1} e^{\mathbf{A}(t_{i+1}-t)}dt$ for $\mathbf{B}(t)\mathbf{x}(t)$ and the execution of the items $\int_{t_i}^{t_{i+1}} t^0 e^{\mathbf{A}(t_{i+1}-t)}dt$, $\int_{t_i}^{t_{i+1}} t^1 e^{\mathbf{A}(t_{i+1}-t)}dt$, $\int_{t_i}^{t_{i+1}} t^2 e^{\mathbf{A}(t_{i+1}-t)}dt$, ..., $\int_{t_i}^{t_{i+1}} t^{p_d+1} e^{\mathbf{A}(t_{i+1}-t)}dt$ for $\mathbf{B}(t)\mathbf{x}(t-\tau)$. The integration terms associated with $\mathbf{B}(t)\mathbf{x}(t)$ are the $\mathbf{F}$ matrices which are concatenated in $\mathbf{F}_L$ and $\mathbf{F}_H$ as

$$\mathbf{F}_L = \{\mathbf{F}_{p_c+2}\ \mathbf{F}_{p_c+1}\ \cdots \mathbf{F}_{p_c+3-\text{size}(\mathbf{F}_L)}\}^T \tag{30}$$

and

$$\mathbf{F}_H = \{\mathbf{F}_{p_c+2}\ \mathbf{F}_{p_c+1}\ \cdots \mathbf{F}_{p_c+3-\text{size}(\mathbf{F}_H)}\}^T, \tag{31}$$

where

$$\mathbf{F}_1 = (\mathbf{F}_0 - \mathbf{I})\mathbf{A}^{-1}, \tag{32}$$

$$\mathbf{F}_{p_c+2-l} = [(p_c+1-l)\mathbf{F}_{p_c+1-l} - (\Delta t)^{p_c+1-l}\mathbf{I}]\mathbf{A}^{-1}, \qquad \text{for } p_c,\, p_c-1,\,\ldots,\,0, \tag{33}$$

and $\mathbf{F}_0$ is the matrix exponential function evaluated at Δt with

$$\mathbf{F}_0 = e^{\mathbf{A}\Delta t}. \tag{34}$$

The $\mathbf{D} \in \mathbb{R}^{n_d(d_R \times d_R)}$ matrices have a similar general form

$$\mathbf{D}_{p_d, p_d+2+q} = \mathbf{p}_L \odot \mathbf{F}_L = p_{L,i}\mathbf{F}_{L,i}, \tag{35}$$

$$\mathbf{D}_{p_d, 1+q} = \mathbf{p}_H \odot \mathbf{F}_H = p_{H,i}\mathbf{F}_{H,i}, \tag{36}$$

$$\mathbf{p}_L = \frac{1}{(p_d - q)!q!(\Delta t)^{p_d+1}}\mathbf{rep}(P_L(\Delta t), '+', ''), \tag{37}$$

$$\mathbf{p}_H = \frac{1}{(p_d - q)!q!(\Delta t)^{p_d+1}}\mathbf{rep}(P_H(\Delta t), '+', ''), \tag{38}$$

$$P_L(\Delta t) = \prod_{j=1}^{p_d}(j\Delta t - 1), \qquad \text{for } q = 0, \tag{39}$$

$$P_L(\Delta t) = (-1)^{1+q} \prod_{j=1, j\neq q}^{p_d} (j\Delta t - 1), \qquad \text{for } q = 1, 2, 3, \ldots, p_d, \tag{40}$$

$$P_H(\Delta t) = (\Delta t - 1)P_L(\Delta t), \qquad \text{for } q = 0, 1, 2, 3, \ldots, p_d. \tag{41}$$

The integration terms associated with $\mathbf{B}(t)\mathbf{x}(t - \tau)$ are the $\mathbf{F}$ matrices which are concatenated in the corresponding matrices $\mathbf{F}_L = \{\mathbf{F}_{p_d+2}\ \mathbf{F}_{p_d+1}\ \cdots \mathbf{F}_{p_d+3-\text{size}(\mathbf{F}_L)}\}^T$ and $\mathbf{F}_H = \{\mathbf{F}_{p_d+2}\ \mathbf{F}_{p_d+1}\ \cdots \mathbf{F}_{p_d+3-\text{size}(\mathbf{F}_H)}\}^T$. These $\mathbf{F}$ matrices for the delayed state are

$$\mathbf{F}_{p_d+2-l} = [(p_d + 1 - l)\mathbf{F}_{p_d+1-l} - (\Delta t)^{p_d+1-l}\mathbf{I}]\mathbf{A}^{-1}, \qquad \text{for } p_d, p_d - 1, \ldots, 0. \tag{42}$$

Because p_c and p_d are considered to be independent and arbitrary, the presented method is referred to as rectangular unification which subsumes the square unification for which $p_c = p_d = p$. The unifications have been discussed in detail in [30]. Equations (17)–(42) define the monodromy matrices in a unified symbolic form requiring only the input of any combination of p_c and p_d (where $p_c = 0, 1, 2, \ldots, k + 1$ and $p_d = 0, 1, 2, \ldots, k$) to be programmed for computerized execution of chatter state interpolation, monodromy matrix construction, and stability identification/eigenvalue analysis. This departs from the existing literature, as seen in the earlier works [17, 18, 21, 23], of manual implementation of the first two tasks on case-by-case numerical specifications of p_c and p_d. Manual derivation of monodromy matrices through the first two tasks is difficult, and almost impossible at higher values of p_c and p_d, making the technological exploitation far from the mathematical understanding of the ordinary machinists. Therefore, this generalization offers the first platform, not only to eliminate all the associate manual analyses and hence fully computerize the process, but also to automatically investigate the parametric effect of interpolation order on the precision of thin-walled milling stability analysis. The approach as pioneered in [30] is not directly applicable to thin-walled workpiece and thus needs upgrading (as done here) by the incorporation of model order reduction for automatic computation of the reduced model matrices $\mathbf{A}$ and $\mathbf{B}(t)$ at every computational time step of stability analysis.

4 Numerical Results and Discussion

In this section, the full computerization of thin-walled milling stability analysis and its application to revealing the parametric effect of interpolation order on computational precision and time are demonstrated. The generalized monodromy matrix given in Eq. (17) is programmed in a MATLAB script which incorporates the parameters of the milling process and the workpiece adopted from [23]. The thin-walled workpiece is a 0.17 m × 0.03 m × 0.15 m steel material of density $\varrho = 7.8 \times 10^3\,\text{kgm}^{-3}$, Young's modulus $E = 210 \times 10^{-9}\,\text{Nm}^{-2}$, and Poisson's ratio $\upsilon_{pr} = 0.3$. These geometric and material properties were used to generate the modal matrices which were reduced with three eigenmodes. The two-tooth tool is subjected to up-milling at radial immersion $\rho = 0.25$. The feed per tooth is 0.1 mm, the tangential cutting coefficient is $C_t = 1.07 \times 10^8\,\text{Nm}^{-1-\gamma}$, the force law feed exponent is $\gamma = 0.75$, and the normal-to-tangential force ratio is $\chi = \frac{40}{107}$.

4.1 Rate of Convergence with Order

On a logarithmic scale to base ten, the spectral radii (SR) computation error expressed in percentages is given as

$$E_c = \log_{10}\left|100\frac{\mu_{\text{SR}}(k) - \mu_{\text{ESR}}}{\mu_{\text{ESR}}}\right|, \tag{43}$$

where μ_{SR} is the estimated spectral radius at k and μ_{ESR} is the exact spectral radius at a high value of k designated k_R. The high value of k_R is not feasible for practical application (tracking of stability boundaries) because of required high computational time but it is useful, by virtue of its much higher precision, as a benchmark for assessing the accuracy of the more practical lower values of k. The unidirectional workpiece chatter in the feed-normal direction, which is a very close approximation of the bidirectional chatter since the workpiece is almost rigid in the feed direction relative to the feed-normal direction, is considered first. The reduced unidirectional dynamics is defined with $\tilde{\mathbf{B}}_H(p_y) = \mathbf{V}^T(p_y)\mathbf{B}_H(:,2) \in \mathbb{R}^{d_R \times 1}$, $\tilde{\mathbf{C}}_H(p_y) = \mathbf{C}_H(2,:)\mathbf{V}(p_y) \in \mathbb{R}^{1 \times d_R}$ and $\mathbf{H}(t) = -wh_{yy}(t)$ while the rest of the quantities are identical with those of bidirectional model established in Sect. 2. While $\mathbf{B}_H(:,2)$ means the second column of $\mathbf{B}_H$, $\mathbf{C}_H(2,:)$ means the second row of $\mathbf{C}_H$. At the cutting process parameter coordinate $[\Omega, w] = [25000\,\text{rpm},\ 2\,\text{mm}]$, the surfaces for percent spectral radii computation

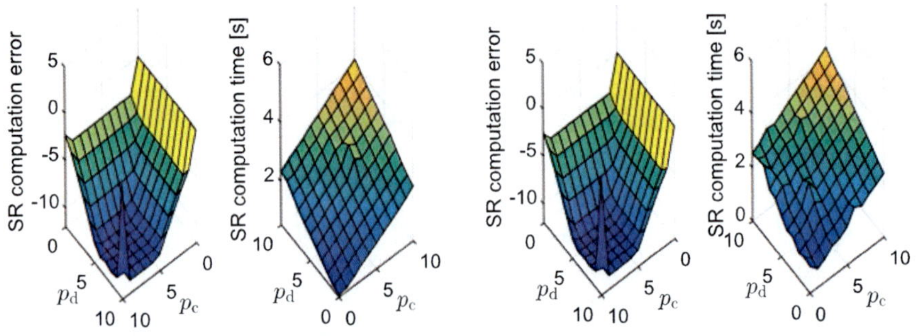

(a) The beginning of tool pass (b) The end of tool pass

Fig. 3 Spectral radii computation error and time for the unidirectional workpiece model

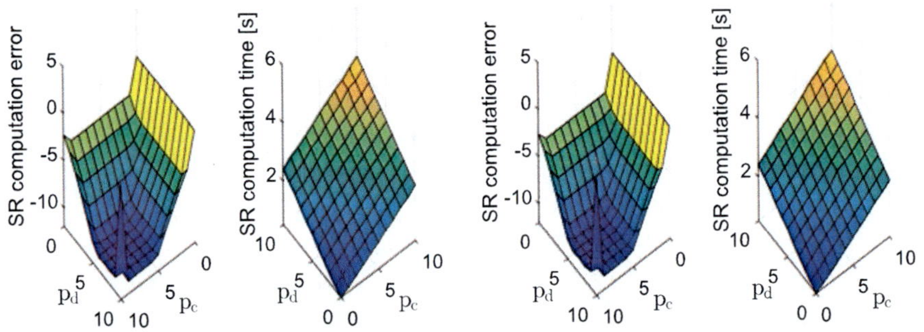

(a) Beginning of tool pass (b) End of tool pass

Fig. 4 Spectral radii computation error and time for the bidirectional workpiece model

error are presented for different interpolation orders of the cutting states $\mathbf{x}(t)$ and $\mathbf{x}(t-\tau)$ in Fig. 3a for the beginning of the tool pass and Fig. 3b for the end of the tool pass, respectively. Three eigenmodes were used for the modal truncation and $k = 40$ was used for the time-domain discretization of the system. The exact spectral radii μ_{ESR} were computed with $k_R = 200$ and the generalized FDM with $p_c = p_d = 3$. The figures suggest that error could rise dramatically when either p_c or p_d approaches either 0 or 10. Juxtaposed in the figures are the corresponding spectral radii computation time. The spectral radii computation time is seen to rise monotonically with the sum order $p_c + p_d$. The corresponding results for the bidirectional workpiece chatter are presented in Fig. 4a, b. The almost identical results for unidirectional and bidirectional models are clear confirmation that response of the workpiece in the feed direction is negligible.

4.2 Stability Diagrams

A 200 by 200 computational grid is created on the axes of spindle speed and depth of cut on which the spectral radius at each grid point is computed. Contour curves (which are the stability curves) are then placed on the plane to distinguish the spectral radii into those higher than unity in magnitude (the unstable ones) from those less than unity in magnitude (the stable ones). Since the best computational accuracy is expected in the range defined by $p_c = 1$ to 10 and $p_d = 1$ to 10, the square unification cases $p = 1, 2, \ldots, 10$ are considered first where square unification implies $p_c = p_d$. The full scale results for the unidirectional and bidirectional models of workpiece at the beginning of tool pass are shown in Fig. 5a, b. Each of the stability curves is identified with the discretization integer $k = 40$. The red reference stability curves are determined with the monodromy matrix $\Psi(3, 3)$ and the discretization integer $k = 100$. The identical results for the unidirectional and bidirectional models of the elastic flexible workpiece confirm the expectation that chatter in the feed direction is negligible because of the very high ratio of length to thickness typical of elastic flexible workpiece. Figure 6, which is an enlargement of the initial low speed domain of 1000–1800 rpm in Fig. 5a or Fig. 5b, shows that accuracy improved up to the third order and decreased beyond the third order. Figure 7 shows that all the methods within the stable range of interpolation order converge at the intermediate speeds but the methods beyond the ninth order fail to converge at the high speed range.

There is still a question of whether the sensitivity of stability lobes precision is affected the same way or differently by independent variations in p_c and p_d under rectangular unification where $p_c \neq p_d$. Figures 3 and 4 suggest a differing sensitivity when either p_c or p_d approaches zero. Error surface reflected higher magnitude when p_c approaches 0 than when p_d approaches 0. To investigate further, error surfaces are plotted for a low speed and an intermediate speed process parameter coordinates in Fig. 8 to clearly reveal a difference of sensitivities of stability lobe precision to variations in p_c and p_d. It should be recalled that the axes for spectral radii computation error in Fig. 8 are also given on logarithmic scale to base ten. Keeping p_c constant at 3 and varying p_d from 0 to 10, the stability boundary curves in Figs. 9 and 10 were generated. As presented in Figs. 11 and 12, the corresponding results were generated while keeping p_d constant at 3 and varying p_c from 0 to 10. Comparison of Figs. 9 and 11 confirms the suggestion of the error surfaces that stability lobes precision is more sensitive to variation in p_d than variation in p_c, especially at the low speed domain. Also, the comparison agrees with the error surfaces that $p_c = 0$ is numerically unstable while $p_d = 0$ is not only numerically stable but can be acceptable. The specific deductions from the comparison include the following:

- The precision of stability lobes is relatively resistant to change in p_c with the best accuracy expected for p_c in the range of 1 to 9.
- The precision of stability lobes is strongly sensitive to change in p_d, especially at the low speed range.

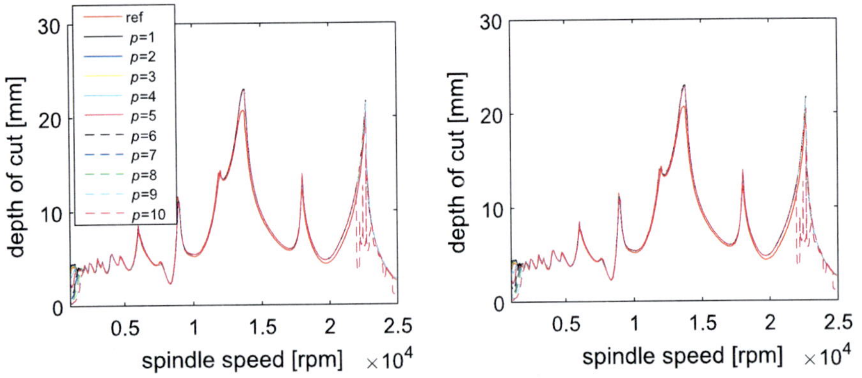

(a)Unidirectional model

(b)Bidirectional model

Fig. 5 Stability boundary curves of the workpiece

Fig. 6 An enlarged low speed portion of Fig. 5a or Fig. 5b showing the accuracy trend with interpolation order

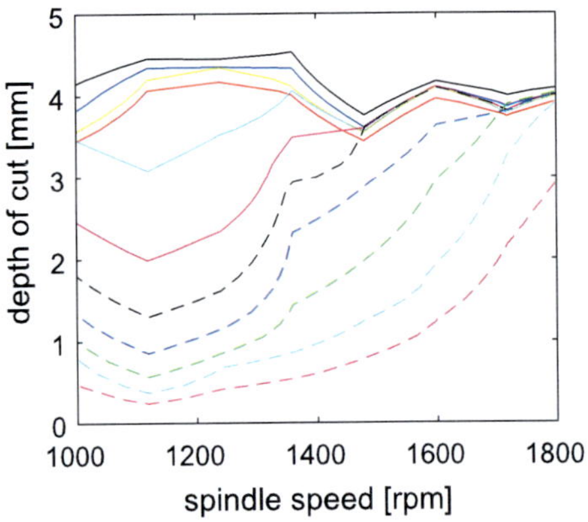

- At the numerically stable ranges of p_c and p_d, highest precision is attained with the variation of p_c.
- Though $p_d = 0$ is the least accurate in the low speed domain when p_c is fixed at 3, $p_d = 0$ becomes the most accurate in the high speed domain, see Fig. 10.

In order to compare the sensitivity of the stability lobes of the flexible workpiece to that of the flexible tool, the stability lobes of the 2DOF flexible tool system in [32] are plotted for the cases of variation in p_d from 0 to 10 with p_c fixed at 3 and variation in p_c from 0 to 10 with p_d fixed at 3. Comparison of Figs. 13 and 14 shows that, like the stability lobes of the flexible workpiece, the stability lobes precision of the flexible tool is more sensitive to variation in p_d than variation in p_c. This shows that interpolation order has similar effects irrespective of the flexible member in the interaction of tools and workpieces.

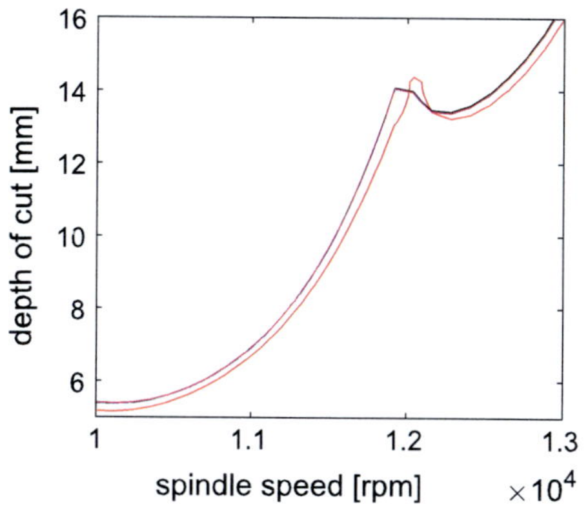

Fig. 7 An enlarged high speed portion showing convergence of different orders in a high speed range

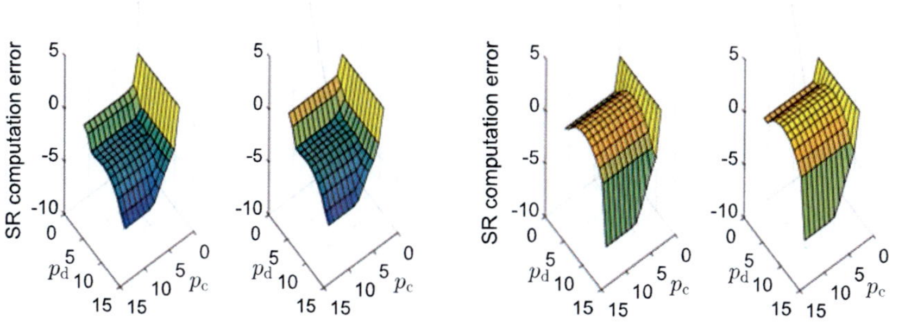

(a) [5000rpm, 2mm] and [5000rpm, 5mm] (b) [10000rpm, 2mm] and [10000rpm, 10mm]

Fig. 8 Spectral radii computation error at the indicated process parameter coordinates of the beginning of the tool pass

Fig. 9 Low speed stability lobes when the order p_c is fixed at 3 and the order p_d is varied

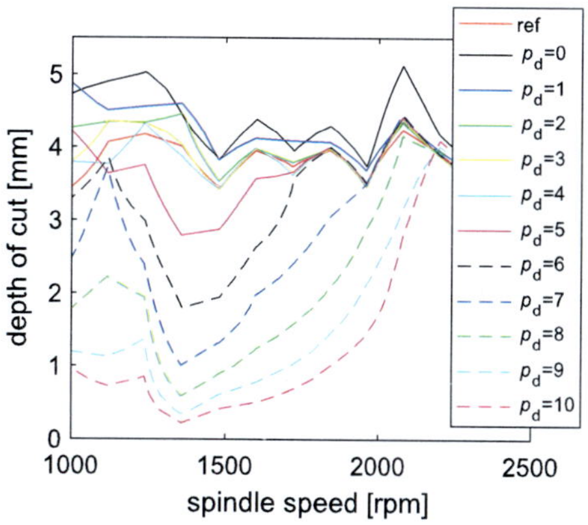

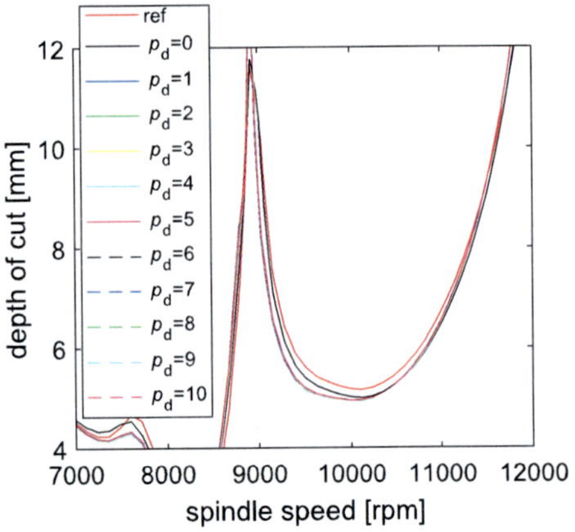

Fig. 10 High speed stability lobes when the order p_c is fixed at 3 and the order p_d is varied

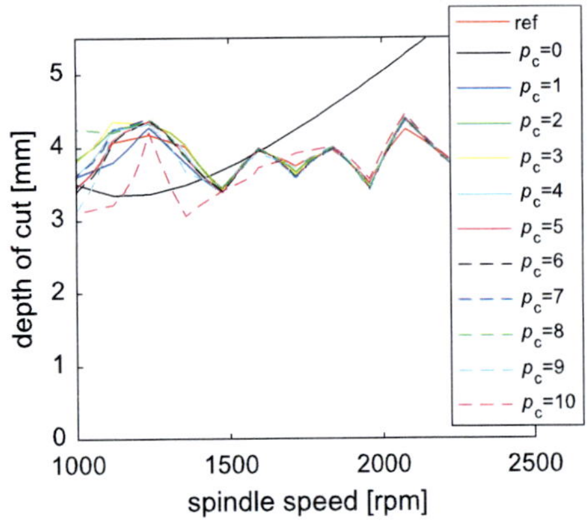

Fig. 11 Low speed stability lobes when the order p_d is fixed at 3 and the order p_c is varied

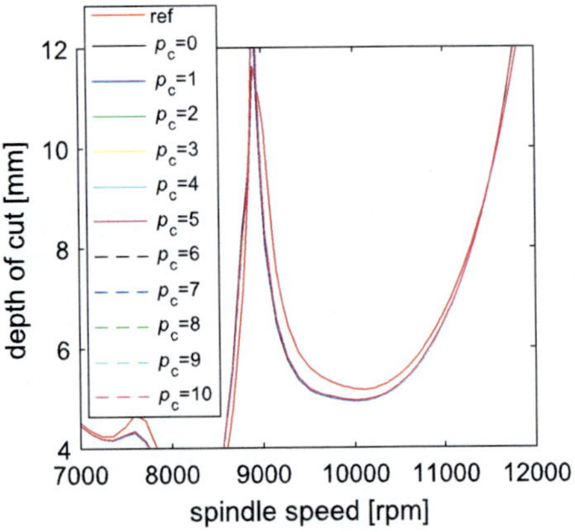

Fig. 12 High speed stability lobes when the order p_d is fixed at 3 and the order p_c is varied

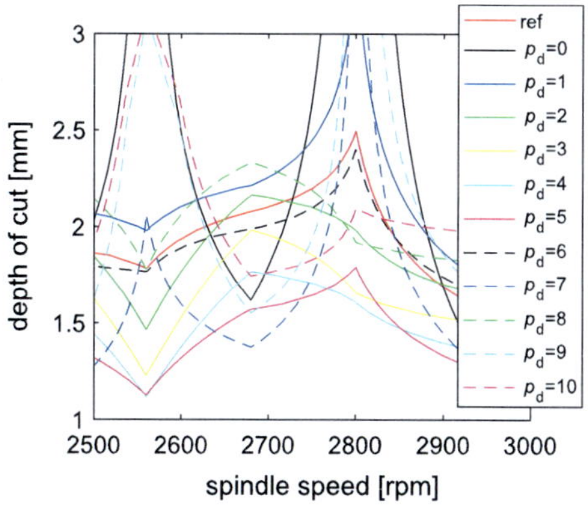

Fig. 13 Stability lobes of the flexible tool when the order p_c is fixed at 3 and the order p_d is varied

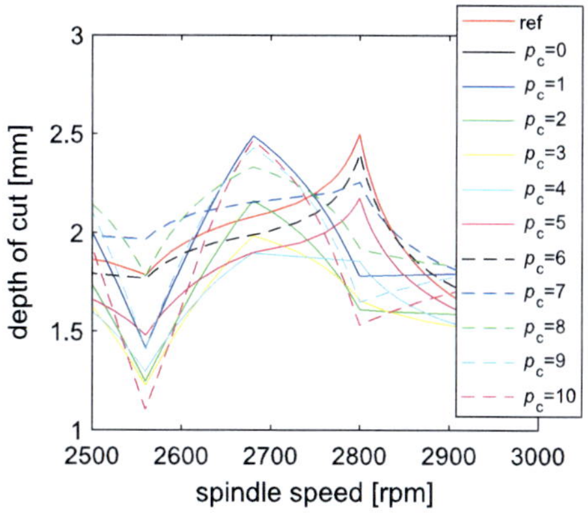

Fig. 14 Stability lobes of the flexible tool when the order p_d is fixed at 3 and the order p_c is varied

The stability curves of an elastic flexible workpiece depend on the spatial location of the tool and the extent of material removed; therefore, Figs. 15 and 17 are generated for the start and two-thirds of the first tool pass, respectively. The acronym CTRS in both Figs. 15 and 17 stands for Coordinate of Time Response Simulation. The stability curves are compared against a reference computed with $k = 200$ using the bidirectional model. Note that all the square unification curves agree on the presented scale because only a relatively high speed range is shown. Figure 16 shows a pair of the regenerative nodal responses of the workpiece at a point marked star in the region of disagreement between each of the curves and the reference curve. The pair of regenerative responses, which are composed of a nodal displacement and a nodal velocity, are simulated over a time interval of ten delays with the MATLAB integrator dde23. There are three of such pairs of nodal responses because three eigenmodes were used in model reduction of the workpiece FE model. The time-domain regenerative responses, which were simulated about the feed per tooth and the feed speed, agree with the reference curve since a rising trend which means

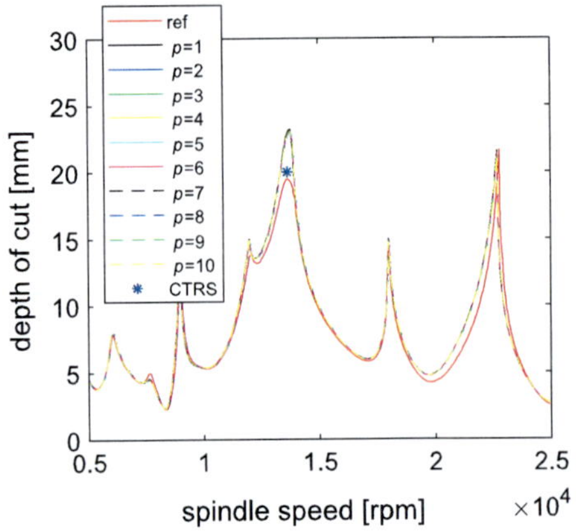

Fig. 15 Identified stability curves at the start of tool pass of the bidirectional model for the orders $p = p_c = p_d$

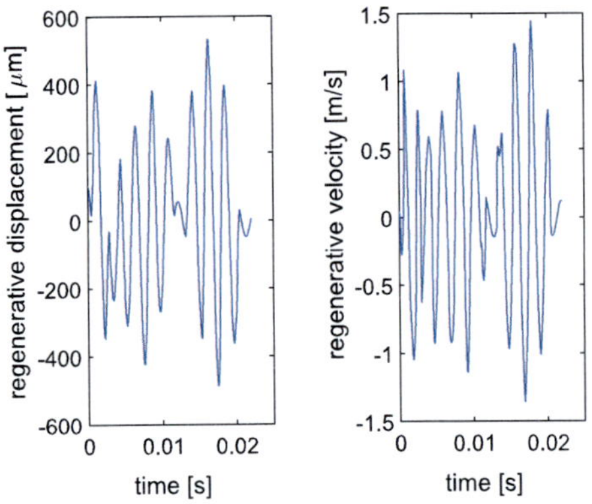

Fig. 16 Time-domain simulation of nodal responses at [13630 rpm, 20 mm] indicated by star in Fig. 15

instability is seen. The time-domain simulation validates the reference curve. As seen in Fig. 18, time-domain simulation also validates the reference stability diagram at the two-thirds of the tool pass.

Though the generalized computerization allows arbitrary choices of p_c and p_d for any given case study, such arbitrary choices show that there is a threshold interpolation order at which numerical instability becomes overbearing. Stability curves are generated for $p_c = p_d = p > 10$ and compared against the reference in Figs. 19 and 20. It is seen that $p_c = p_d = p = 9$ is the threshold interpolation order beyond which the FDM fails for the studied system. The failure starts from the high speed domain and extends deeper into the low speed domain for higher orders p, and the magnitude of the failure is higher for higher orders p at every spindle speed.

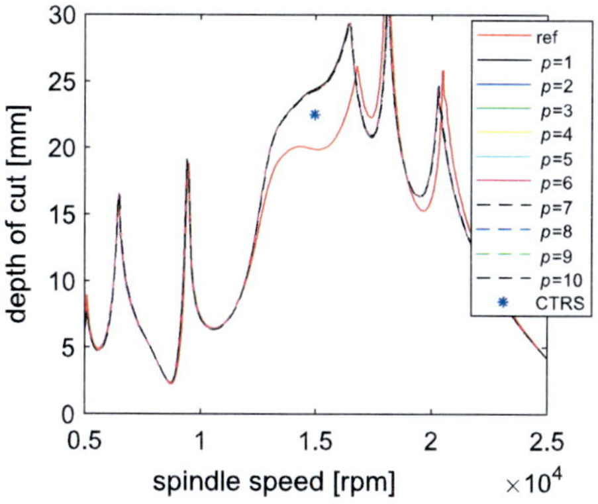

Fig. 17 The identified stability curves at two-thirds of tool pass of the bidirectional model for the orders $p = p_c = p_d$

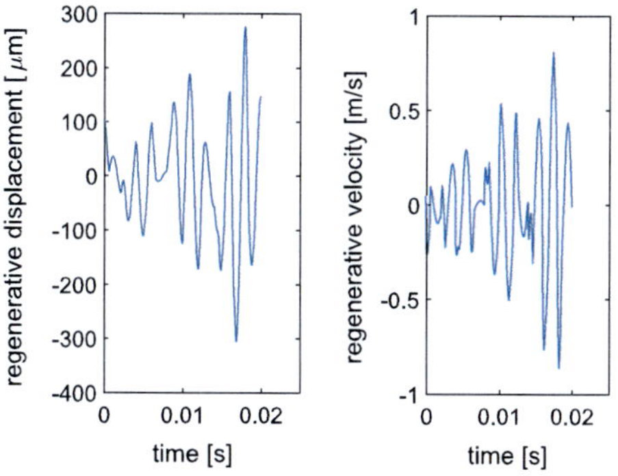

Fig. 18 The time-domain simulation of nodal responses at [15000 rpm, 22.5 mm] indicated by star in Fig. 17

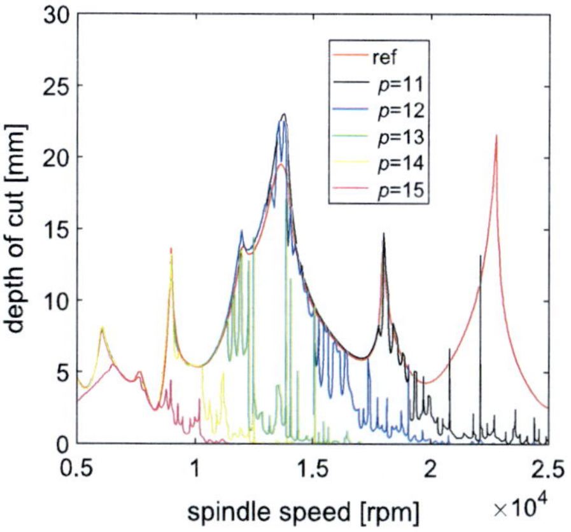

Fig. 19 Stability curves for $p = p_c = p_d$ compared against the reference at the start of tool pass

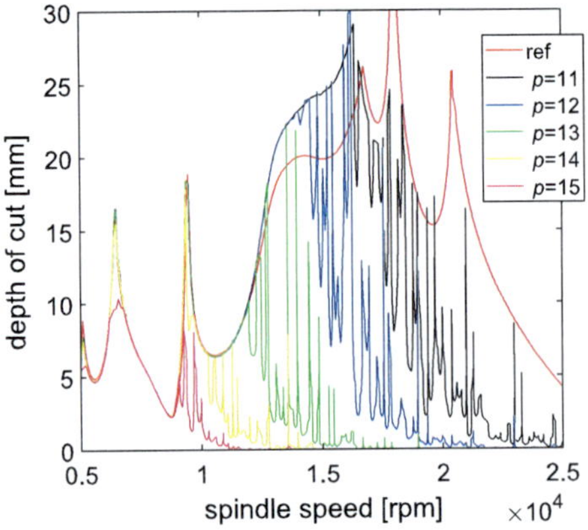

Fig. 20 Stability curves for $p = p_c = p_d$ compared against the reference at two-thirds of tool pass

5 Conclusions

A method which combines modal truncation and tensor-based general order FDM was developed to suppress all the case-by-case symbolic analyses associated with stability analysis of elastic thin-walled workpiece. Having avoided the barrier of symbolic complications it became possible to investigate the sensitivity of stability lobes precision within the feasible interpolation range. It was found that precision does not respond monotonically to rise or fall with the interpolation orders of the current and delayed chatter states justifying the developed method as a way of computerized search within the feasible range. Based on the studied system which showed almost identical results for unidirectional and bidirectional models, the following results were seen: (1) when current and delayed chatter states are interpolated with the same order tensor polynomial (that is, square unification), the best results are found around the third order; (2) stability lobes are mildly sensitive to the variation of the current state order but strongly sensitive to the variation of the delayed state order; and (3) no combination of current and delayed states orders outperforms the rest in all spindle speed range and thus a recommendation is made to keep the delayed state order at 3 while the current state order is varied to get the best results. These results raised the question of how the fact that different combinations of interpolation orders perform differently at different speed ranges can be exploited in maximizing stability lobes precision across a speed range of interest. A possible way is to consider hybridizing the proposed computerized algorithm with statistical/machine learning methods so that the error features of order combinations can be classified against speed segments and allow composite stability lobes with maximum possible precision for each speed segment to be identified. The automatic analysis would make usage of stability lobes for precise selection of productive chatter-free process parameters more accurate and user-friendly.

Acknowledgements The matrices for the elastic workpiece model were kindly provided by Dominik Hamann. This help is greatly appreciated. The described research was partially done while C.G. Ozoegwu visited the ITM at the University of Stuttgart in 2019. This stay was funded by the Priority Program SPP 1897 "Calm, Smooth, Smart" of the DFG (German Research Foundation). This support is highly appreciated.

References

1. Altinta, Y., Budak, E.: Analytical prediction of stability lobes in milling. CIRP Ann. Manuf. Technol. **44**(1), 357–362 (1995)
2. Insperger, T., Stépán, G.: Stability of the milling process. Period. Polytechn. Mech. Eng. **44**(1), 47–57 (2000)
3. Ding, Y., Zhu, L., Zhang, X., Ding, H.: Numerical integration method for prediction of milling stability. J. Manuf. Sci. Eng. **133**(3), 1–9 (2011)
4. Ozoegwu, C., Ofochebe, S., Omenyi, S.: A method of improving chatter-free conditions with combined-mode milling. J. Manuf. Process. **21**, 1–13 (2016)
5. Altintas, Y., Montgomery, D., Budak, E.: Dynamic peripheral milling of flexible structures. Precis. Eng. **13**(3), 137–145 (1991)
6. Bravo, U., Altuzarra, O., López De Lacalle, L.N., Sánchez, J.A., Campa, F.J.: Stability limits of milling considering the flexibility of the workpiece and the machine. Int. J. Mach. Tools Manuf. **45**(15), 1669–1680 (2005)
7. Herranz, S., Campa, F.J., De Lacalle, L.N., Rivero, A., Lamikiz, A., Ukar, E., Sánchez, J.A., Bravo, U.: The milling of airframe components with low rigidity: a general approach to avoid static and dynamic problems. Proc. Inst. Mech. Eng. Part B: J. Eng. Manuf. **219**(11), 789–801 (2005)
8. Thevenot, V., Arnaud, L., Dessein, G., Cazenave-Larroche, G.: Integration of dynamic behaviour variations in the stability lobes method: 3D lobes construction and application to thin-walled structure milling. Int. J. Adv. Manuf. Technol. **27**(7–8), 638–644 (2006)
9. Thevenot, V., Arnaud, L., Dessein, G., Cazenave-Larroche, G.: Influence of material removal on the dynamic behavior of thin-walled structures in peripheral milling. Mach. Sci. Technol. **10**(3), 275–287 (2006)
10. Adetoro, O.B., Sim, W.M., Wen, P.H.: An improved prediction of stability lobes using nonlinear thin wall dynamics. J. Mater. Process. Technol. **210**(6–7), 969–979 (2010)
11. Budak, E., Tunç, L.T., Alan, S., Özgüven, H.N.: Prediction of workpiece dynamics and its effects on chatter stability in milling. CIRP Ann. Manuf. Technol. **61**(1), 339–342 (2012)
12. Song, Q., Liu, Z., Wan, Y., Ju, G., Shi, J.: Application of Sherman-Morrison-Woodbury formulas in instantaneous dynamic of peripheral milling for thin-walled component. Int. J. Mech. Sci. **96–97**, 79–90 (2015)
13. Yang, Y., Zhang, W.H., Ma, Y.C., Wan, M.: Chatter prediction for the peripheral milling of thin-walled workpieces with curved surfaces. Int. J. Mach. Tools Manuf. **109**, 36–48 (2016)
14. Ahmadi, K.: Finite strip modeling of the varying dynamics of thin-walled pocket structures during machining. Int. J. Adv. Manuf. Technol. **89**(9–12), 2691–2699 (2017)
15. Song, Q., Shi, J., Liu, Z., Wan, Y.: A time-space discretization method in milling stability prediction of thin-walled component. Int. J. Adv. Manuf. Technol. **89**(9–12), 2675–2689 (2017)
16. Shi, J., Song, Q., Liu, Z., Ai, X.: A novel stability prediction approach for thin-walled component milling considering material removing process. Chin. J. Aeronaut. **30**(5), 1789–1798 (2017)
17. Song, Q., Ai, X., Tang, W.: Prediction of simultaneous dynamic stability limit of time-variable parameters system in thin-walled workpiece high-speed milling processes. Int. J. Adv. Manuf. Technol. **55**(9–12), 883–889 (2011)

18. Zhang, L., Gao, W., Zhang, D., Tian, Y.: Prediction of dynamic milling stability considering time variation of deflection and dynamic characteristics in thin-walled component milling process. Shock Vib. **2016**, 1–14 (2016)
19. Eksioglu, C., Kilic, Z.M., Altintas, Y.: Discrete-time prediction of chatter stability, cutting forces, and surface location errors in flexible milling systems. J. Manuf. Sci. Eng. **134**(6), 1–13 (2012)
20. Wan, M., Dang, X.B., Zhang, W.H., Yang, Y.: Optimization and improvement of stable processing condition by attaching additional masses for milling of thin-walled workpiece. Mech. Syst. Signal Process. **103**, 196–215 (2018)
21. Li, Z., Sun, Y., Guo, D.: Chatter prediction utilizing stability lobes with process damping in finish milling of titanium alloy thin-walled workpiece. Int. J. Adv. Manuf. Technol. **89**(9–12), 2663–2674 (2017)
22. Zhang, Z., Li, H., Liu, X., Zhang, W., Meng, G.: Chatter mitigation for the milling of thin-walled workpiece. Int. J. Mech. Sci. **138–139**, 262–271 (2018)
23. Hamann, D., Eberhard, P.: Stability analysis of milling processes with varying workpiece dynamics. Multibody Syst. Dyn. **42**(4), 383–396 (2018)
24. Ma, Y.-C., Yang, Y., Wan, M., Zhang, W.-H., Dang, X.-B.: An efficient decomposition-condensation method for chatter prediction in milling large-scale thin-walled structures. Mech. Syst. Signal Process. **121**, 58–76 (2019)
25. Dong, X., Zhang, W.: Stability analysis in milling of the thin walled part considering multiple variables of manufacturing systems. Int. J. Adv. Manuf. Technol. **89**(1–4), 515–527 (2017)
26. Ozoegwu, C.G.: Polynomial tensor-based stability identification of milling process: application to reduced thin-walled workpiece. In: Fehr, J., Haasdonk, B. (eds.) IUTAM Symposium on Model Order Reduction of Coupled Systems, Stuttgart, Germany, 22–25 May 2018, vol. 36, ch. 15, pp. 209–220. Springer International Publishing, Stuttgart, Germany (2020)
27. Tang, A., Liu, Z.: Three-dimensional stability lobe and maximum material removal rate in end milling of thin-walled plate. Int. J. Adv. Manuf. Technol. **43**(1–2), 33–39 (2009)
28. Qu, S., Zhao, J., Wang, T.: Three-dimensional stability prediction and chatter analysis in milling of thin-walled plate. Int. J. Adv. Manuf. Technol. **86**(5–8), 2291–2300 (2016)
29. Powałka, B., Jemielniak, K.: Stability analysis in milling of flexible parts based on operational modal analysis. CIRP J. Manuf. Sci. Technol. **9**, 125–135 (2015)
30. Ozoegwu, C., Eberhard, P.: Tensor-based automatic arbitrary order computation of the full-discretization method for milling stability analysis. In: Altenbach, H., Irschik, H., Matveenko, V. (eds.) Contributions to Advanced Dynamics and Continuum Mechanics, ch. 11, pp. 179 – 205. Springer International Publishing (2019)
31. Ozoegwu, C.G., Omenyi, S.N., Ofochebe, S.M.: Hyper-third order full-discretization methods in milling stability prediction. Int. J. Mach. Tools Manuf. **92**, 1–9 (2015)
32. Bayly, P.V., Mann, B.P., Peters, D.A., Schmitz, T.L., Stepan, G., Insperger, T.: Effects of radial immersion and cutting direction on chatter instability in end-milling. In: ASME International Mechanical Engineering Congress and Exposition, New Orleans, pp. 1–13 (2002)

Space Elevator—A Revolutionary Space Transportation System

Arun K. Misra and Stephen Cohen

1 Introduction

It is expected that space elevators will have a large impact on the method of deployment of Earth-orbiting satellites by offering a more efficient and elegant approach compared to the chemical rockets. A space elevator consists of one or more tethers (or ribbons) stretching from the surface of the Earth, somewhere on the equator, to a counterweight located beyond the geosynchronous altitude (Fig. 1). The tether will be in tension due to the gravitational and centrifugal loads. Climbers will be used to raise payloads from the surface of the Earth to space, similar to a regular elevator moving across floors of a building. The satellite placement, when accomplished in this manner, rather than by conventional rockets, is expected to be of the order of 50–100 times less expensive [1, 2]. For example, the current cost of placing a payload at the geostationary orbit is about $20,000/kg using chemical propulsion; this can be reduced to about $500/kg by utilizing a space elevator [2]. It is possible that several space elevators will operate simultaneously at any given time.

The idea of the space elevator was first conceived by Tsiolkovsky [3] in 1895, apparently inspired by the Eiffel Tower. Artsutanov [4], in 1960, discussed a similar concept in which a cable would be lowered from the geostationary altitude to the surface of the Earth utilizing a counterweight. In 1975, Pearson [5] conducted an in-depth technical study of the orbital tower proposed by Tsiolkovsky and Artsutanov. Since then, many technical investigations have been carried out by various researchers.

A. K. Misra (✉)
McGill University, Montreal, QC H3A 0C3, Canada
e-mail: arun.misra@mail.mcgill.ca

S. Cohen
Vanier College, Montreal, QC H4L 3X9, Canada
e-mail: cohens@vanier.college

© Springer Nature Singapore Pte Ltd. 2021

U. S. Dixit and S. K. Dwivedy (eds.), *Mechanical Sciences*,
https://doi.org/10.1007/978-981-15-5712-5_5

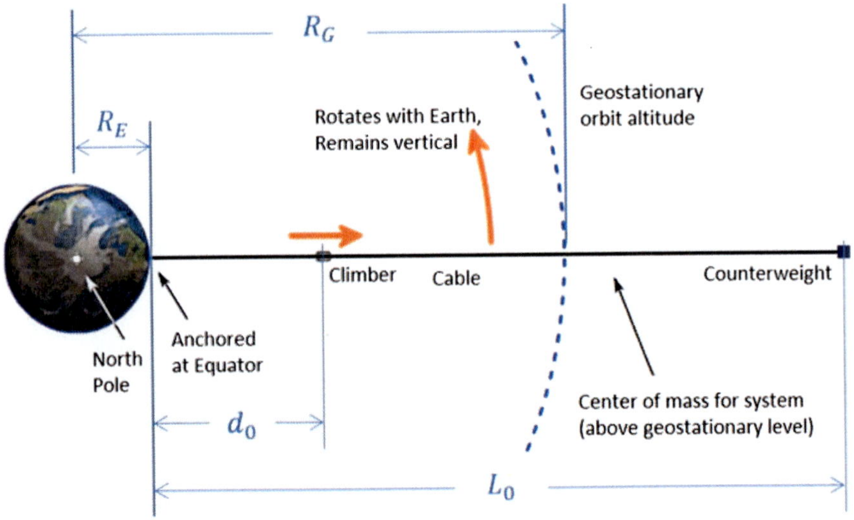

Fig. 1 Schematic diagram of the space elevator (modified from: https://en.wikipedia.org/wiki/Space_elevator)

Calculations show that the stress that will develop in a full space elevator is too high for conventional engineering materials, although advanced new materials, such as carbon nanotubes, are capable of handling such stress. Unfortunately, the current technology is not yet advanced enough to build such a large structure using carbon nanotubes. Hence, it is expected that a partial elevator will be constructed first using conventional materials, providing only partial benefits of a space elevator. A full space elevator will be built subsequently.

This article examines the mechanics of space elevators. To start with, a brief discussion of the satellite placement capability of a full space elevator is discussed. This is followed by the analysis of the statics and dynamics of a full space elevator.

2 Satellite Placement Using a Space Elevator

Figure 1 shows the main components of the space elevator. It consists of a ribbon, a counterweight, and one or more climbers, all in the equatorial plane of the Earth. The ribbon is in tension due to the gravity and centrifugal gradients, which act in opposite directions. The climber, propelled by an electric motor, will ascend the ribbon, transporting payloads to various altitudes.

Once a climber reaches a desired launch altitude, d_0, the satellite it contains may be released. If no additional velocity impulse is added to the satellite at its time of launch, it will have used no fuel to arrive in orbit. These orbits are called free Earth orbits.

Let us say that at the time of launch, the climber (and the satellite it contains) will have a radial distance of r_0 $(= d_0 + R_E)$, speed v_0, and flight path angle β_0. The values of v_0 and β_0 may be modified by an applied velocity impulse, if desired. The semi-major axis and eccentricity of the orbit that the satellite will fall into can be written as [6]

$$a = \frac{r_0}{2 - \frac{r_0 v_0^2}{\mu}} \tag{1}$$

and

$$e = \sqrt{\left(\frac{r_0 v_0^2}{\mu} - 1\right)^2 \cos^2 \beta_0 + \sin^2 \beta_0} \tag{2}$$

respectively, where $\mu = GM_E$ is the gravitational constant of the Earth.

Ideally, at the moment of satellite launch, the space elevator will be static (other than the nominal spin rate of the Earth), and in its nominal configuration consisting of a vertical ribbon. If no additional impulse is applied to the satellite, then $v_0 = \Omega(R_E + d_0)$ and $\beta_0 = 0$, where R_E and Ω are the radius and angular velocity of the Earth, respectively. The resulting semi-major axis and eccentricity pairings are plotted in Fig. 2 with all lengths nondimensionalized with respect to R_E. It is observed that the geosynchronous altitude is the only point on the ribbon that is in a natural circular orbit (shown by a bullet in Fig. 2). If a mass is released from any other altitude in the range given, it will be in an elliptical orbit.

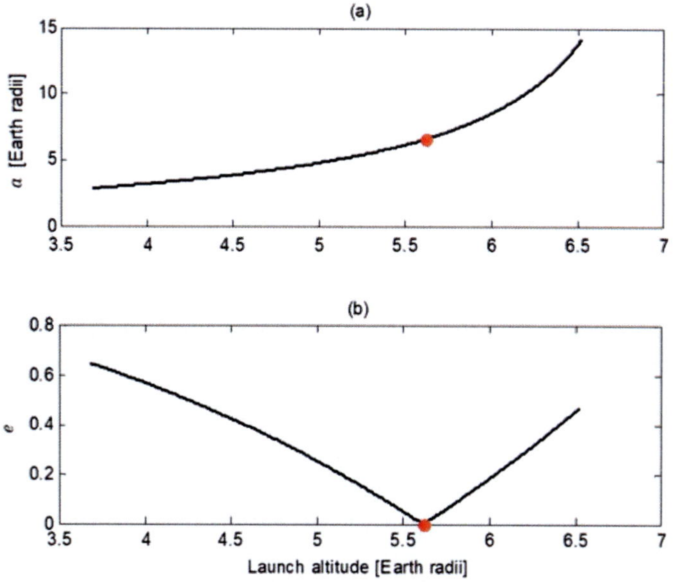

Fig. 2 Orbit parameters versus launch altitude: **a** semi-major axis, **b** eccentricity

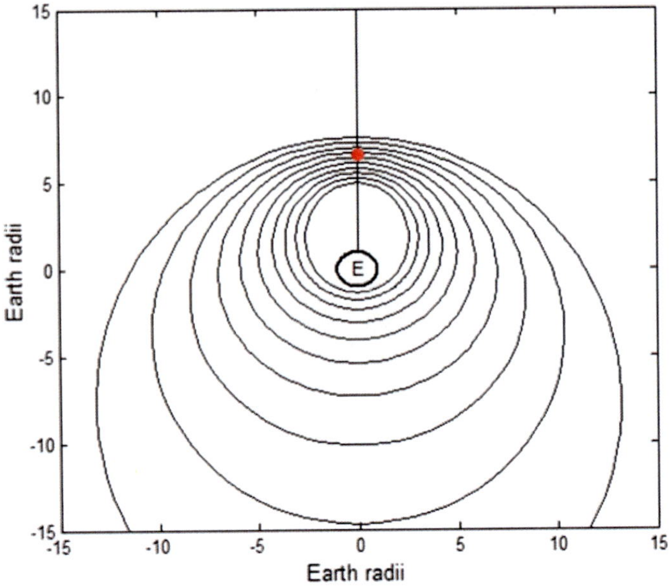

Fig. 3 Free Earth orbits using the space elevator

Though not shown in Fig. 2, $d_0 = 7.345\ R_E$, or about 46,850 km, is a critical launch altitude, which places payloads in a parabolic orbit. Therefore, for satellite placement in free Earth orbits, the launch altitude will be in the range given by $23,500 < d_0 < 46,850$ km. Also, any natural ($\Delta v = 0$) launch in the range given by $46,850$ km $< d_0 < L_0$, where L_0 is the total length of the ribbon, will send the payload into a hyperbolic orbit, as it will have a velocity greater than the escape velocity given by $v_{\text{esc}} = \sqrt{2\mu/r_0}$. These trajectories can be the starting point for planning interplanetary space missions using the space elevator.

A better picture of what has been called the free Earth orbits available to satellites using the space elevator is shown in Fig. 3. Since the flight path angle of the climber at the time of launch is zero, the point of launch can only be the apogee or the perigee of the orbit. Since the portion of the ribbon below R_G is traveling slower than it would in a natural circular orbit, launches below this radius commence at the apogee of the orbit. Conversely, for launches above R_G, the initial radius becomes the perigee of the orbit.

While Fig. 3 shows the spectrum of free orbits available to Earth satellites, there are a wide range of reasonably low cost orbits that may be reached by transferring from the free orbits with a small impulse, Δv. For example, if a particular circular orbit having $r_c \neq R_G$ (where R_G is the geosynchronous radius) is desired, where r_c is the radius of the circular orbit, a certain Δv will be required. The most efficient elliptical to circular orbit transfer occurs at the perigee of an elliptical orbit if $r_c < R_G$, and at its apogee if $r_c > R_G$. These transfers are equivalent to the second impulse of a

Hohmann transfer. For the case of $r_c < R_G$, the required impulse for circularization is given by

$$\Delta v = \sqrt{\frac{\mu}{a}}\left(\sqrt{\frac{1}{1-e}} - \sqrt{\frac{1+e}{1-e}}\right) \tag{3}$$

where a and e are the semi-major axis and eccentricity of the original elliptical orbit, respectively. For the case of $r_c > R_G$, the required impulse is given by

$$\Delta v = \sqrt{\frac{\mu}{a}}\left(\sqrt{\frac{1}{1+e}} - \sqrt{\frac{1-e}{1+e}}\right) \tag{4}$$

An analysis of the operational costs for satellite placement using a space elevator can be found in [2, 6] and is not presented here. However, design of a space elevator requires an investigation of the stresses generated in it as well as its dynamics analysis. These are presented in the following sections.

3 Static Analysis of the Ribbon

The ribbon connecting the counterweight to the Earth is subject to an axial stress due to the resultant of the gravitational force and centrifugal effect caused by the rotation of the Earth. This stress varies along the length of the ribbon and is larger for longer ribbons. An ideal ribbon design would have constant stress along the ribbon and would imply a variable cross section.

Consider an element dm at a distance r from the center of the Earth as shown in Fig. 4. The nominal strain in the ribbon is given by

$$\varepsilon_0 = \frac{\mathrm{d}u_0}{\mathrm{d}s} = \frac{\sigma_0}{E} \tag{5}$$

where u_0 is the static extension of the ribbon at a distance s from the surface of the Earth, σ_0 is the nominal stress, and E is the Young's Modulus of the ribbon material. The nominal strain ε_0 is given by σ_0/E. If the nominal stress (and hence strain) is desired to be maintained uniform along the ribbon, then

$$u_0(s) = \varepsilon_0 s \tag{6}$$

Thus, if the nominally stretched ribbon is to have a length L, then the original unstretched length must be $L_0 = L/(1 + \varepsilon_0)$.

The forces acting on the element are shown in Fig. 5. If $T(r)$ is the tension in the ribbon at a distance r from the center of the Earth and dF_g is the gravitational force acting on the element, then

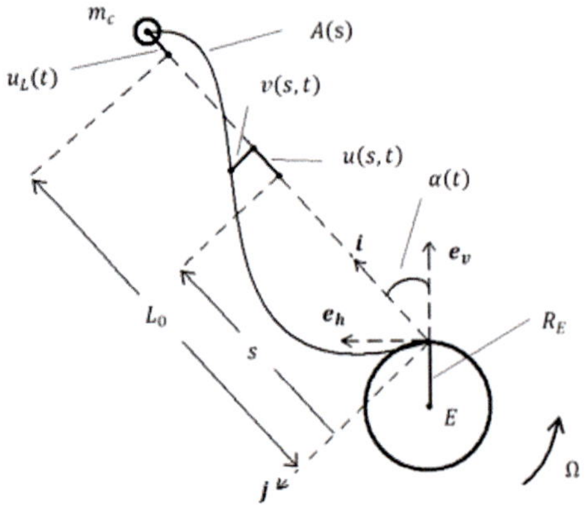

Fig. 4 A schematic diagram of the space elevator

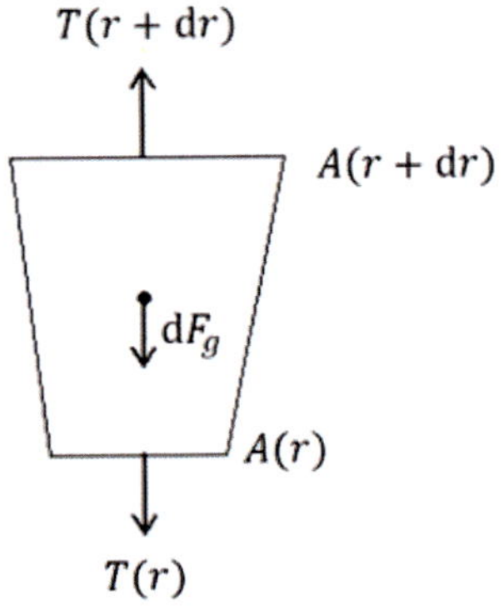

Fig. 5 Free-body diagram of a ribbon element

$$(\mathrm{d}m)\, a_r = \sum F_r = T(r + dr) - T(r) - \mathrm{d}F_g \tag{7}$$

Substituting values for tension, gravitational force, and acceleration,

$$\mathrm{d}m(-\Omega^2 r) = \sigma_0 A(r + \mathrm{d}r) - \sigma_0 A(r) - \mathrm{d}m(\mu/r^2) \tag{8}$$

where dm is the product of the bulk density, and the infinitesimal volume given by A ds and μ is the gravitational constant of the Earth. The spatial coordinate r and the ribbon coordinate s are related by

$$r = R_E + s + u_0(s) = R_E + s(1 + \varepsilon_0) \tag{9}$$

where R_E is the radius of the Earth. Making the preceding substitutions and then simplifying, Eq. (8) becomes

$$\gamma A(s)\, \mathrm{d}s\{\mu/[R_E + s(1 + \varepsilon_0)]^2 - \Omega^2[R_E + s(1 + \varepsilon_0)]\} = \sigma_0 \mathrm{d}A \tag{10}$$

It is now useful to introduce a quantity called the characteristic height of the ribbon, which is defined by $\bar{h} = \sigma_0/\gamma g_0$. Here, g_0 is the surface gravity of the Earth and is equal to μ/R_E^2. The characteristic height is a measure of the strength-to-density ratio of the ribbon material, scaled with respect to the surface gravity of the Earth to have a unit of length.

Substituting $\Omega^2 = \mu/R_G^3$ (where R_G is the geosynchronous orbit radius), $\mu = g_0 R_E^2$, and $\sigma_0 = \bar{h}\gamma g_0$ into Eq. (10), then simplifying, one arrives at

$$\frac{\mathrm{d}A}{A} = \frac{R_E^2}{\bar{h}}\left\{\frac{1}{[R_E + s(1+\varepsilon_0)]^2} - \frac{R_E + s(1+\varepsilon_0)}{R_G^3}\right\} \tag{11}$$

Integrating Eq. (11) results in

$$A(s) = c\exp\left(-\frac{R_E^2}{\bar{h}(1+\varepsilon_0)}\left\{\frac{1}{R_E + s(1+\varepsilon_0)} + \frac{[R_E + s(1+\varepsilon_0)]^2}{2R_G^3}\right\}\right) \tag{12}$$

where c is a constant of integration. The boundary condition for Eq. (12) is that the net force acting on the free end of the ribbon must be equal to the tension $(\sigma_0 A(s)|_{s=L_0})$ in it at that point. Because there is no net force acting at that point, this boundary condition could only be satisfied by having the area of cross section equal to zero there (this would ensure zero tension). However, from Eq. (12), it is clear that the area of cross section of the ribbon cannot be zero at any location for this case of constant stress. Thus, to satisfy the boundary condition at the tip of the ribbon, a mass m_c (the counterweight) must be attached there. The forces acting on the counterweight can be made equal to the tension at the tip by forcing

$$m_c\left[\Omega^2(R_E + L) - \mu/(R_E + L)^2\right] = \sigma_0 A(s)\big|_{s=L_0} \tag{13}$$

Through differentiation of Eq. (12), it may be shown that the maximum value of area of cross section occurs at the location $s = (R_G - R_E)/(1+\varepsilon_0)$, which corresponds to the radial position $r = R_G$. The area at this location may be set to the useful design parameter A_m, which is the maximum area of cross section of the ribbon and is a free design parameter. Then after some manipulation, the cross-sectional area profile may be expressed as

$$A(s) = A_m \exp\left[F(s)\right] \tag{14}$$

where

$$F(s) = \frac{R_E^2}{\bar{h}R_G(1+\varepsilon_0)}\left(\frac{3}{2} - \frac{R_G}{R_E + s(1+\varepsilon_0)} - \frac{[R_E + s(1+\varepsilon_0)]^2}{2R_G^2}\right) \tag{15}$$

or, equivalently,

$$A(r) = A_m \exp\left[\frac{R_E^2}{\bar{h} R_G (1+\varepsilon_0)}\left(\frac{3}{2} - \frac{R_G}{r} - \frac{r^2}{2R_G^2}\right)\right] \tag{16}$$

An almost identical solution as that given in Eq. (16) was obtained by Pearson [5]. The only difference is that the $(1 + \varepsilon_0)$ term did not appear in his solution. This is because the nominal strain was not considered in his taper function derivation. As the nominal strain is not negligible, this modification to the profile of the area of cross section is necessary. It is essential that there be a counterweight placed at the free end, and, from Eq. (13), it is given by

$$m_c = \frac{\sigma_0 A_m \exp[F(s)]|_{s=L_0}}{\Omega^2[R_E + L_0(1+\varepsilon)] - \frac{\mu}{[R_E + L_o(1+\varepsilon)]^2}} \tag{17}$$

The resulting taper ratio of the ribbon, which is defined here as the quotient of A_m and the area of cross section at the Earth's surface A_0, is given by

$$\frac{A_m}{A_0} = \exp\left[\frac{R_E}{(\bar{h}+\varepsilon_0)}\left(1 - \frac{R_E}{R_G}\right)^2\left(1 + \frac{R_E}{(2R_G)}\right)\right] \tag{18}$$

Hence, for a material with $\bar{h} = 2,762$ km and $E = 1,000$ GPa, the taper ratio is exactly 6. The taper function of a material having these values is plotted in Fig. 6. From [7], the density of carbon nanotubes is 1,300 kg/m^3, its Young's Modulus is

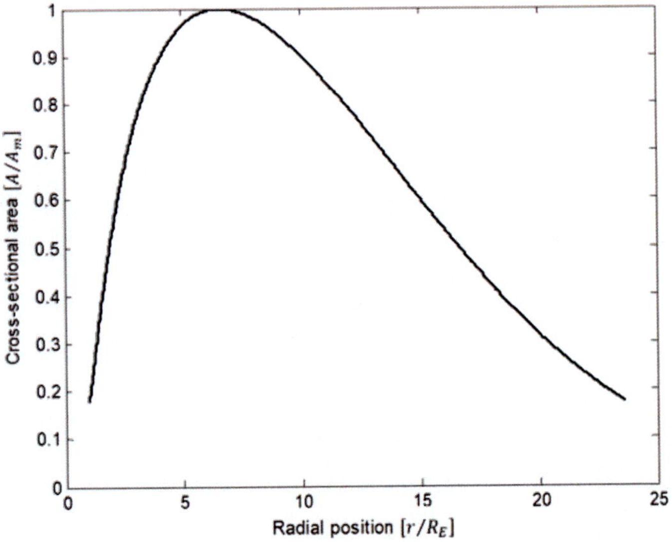

Fig. 6 Cross-sectional area profile of the ribbon for $\bar{h} = 2,762$ km

1000 GPa, and the theoretical value for its maximum tensile strength is 130 GPa. If the maximum strength of the ribbon material was only 70 GPa, its characteristic height would be 2,762 km using a safety factor of about two ($\sigma_0 = 35.2$ GPa), and the variation in area of cross section would be as shown in Fig. 6).

The nominal tension in the ribbon may now be expressed as

$$T_0(s) = \sigma_0 A(s) = \sigma_0 A_m \exp[F(s)] \tag{19}$$

The mass of the ribbon is given by

$$m_t = \gamma A_m \int_0^{L_0} \exp[F(s)]\mathrm{d}s \tag{20}$$

Through manipulation of Eq. (17), the required counterweight may be expressed as

$$m_c = \gamma A_m \bar{h} \frac{\exp\left[F(s)|_{s=L_0}\right]}{\left(\frac{R_E}{R_G}\right)^2 \left(\frac{\left[R_E + L_0(1+\varepsilon_0)\right]}{R_G} - \left\{\frac{R_G}{\left[R_E + L_0(1+\varepsilon_0)\right]}\right\}^2\right)} \tag{21}$$

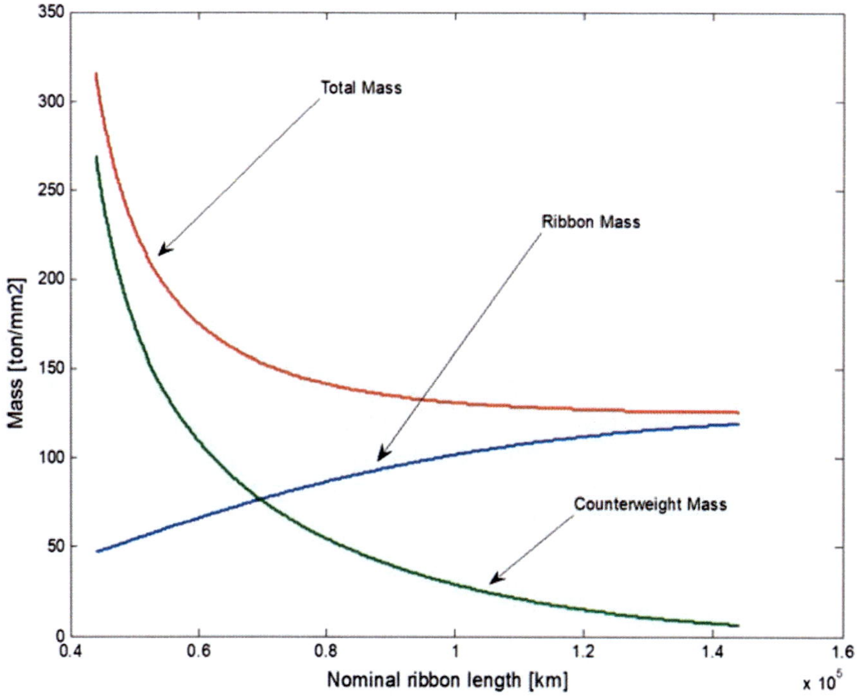

Fig. 7 Mass (per mm^2 of maximum cross-sectional area) of space elevator components versus nominal ribbon length

Clearly, the required counterweight is proportional to the maximum area of cross section of the ribbon, and is also dependent on other material and design parameters. In Fig. 7, the ribbon, counterweight, and total mass per maximum unit area of cross section of ribbon are plotted against nominal ribbon length, assuming a taper ratio of 6 (ribbon material parameters are as assumed above). A ribbon with L_0 = 100,000 km and A_m = 10 mm^2 would have a mass of about 1000 tons. The corresponding counterweight mass would be about 300 tons.

4 Elastic Oscillations of the Ribbon

Both longitudinal and transverse displacements of the ribbon, u and v, shown in Fig. 4, are expanded in series form as products of a set of generalized coordinates and spatial basis functions. The longitudinal extension is expressed as

$$u(s,t) = \varepsilon_0 s + \sum_{i=1}^{N} a_i(t)\psi_i(s) \tag{22}$$

where N generalized coordinates, $a_i(t)$, and appropriate basis functions, $\psi_i(s)$, are used to describe the extension of the ribbon from its nominal amount $u_0(s) = \varepsilon_0 s$. Since an energy method is used to derive the equations of motion, $\psi_i(s)$ need to be only admissible functions, not comparison functions [i.e., they need to satisfy only the geometric boundary condition, $\psi_i(0) = 0$].

The transverse displacement is represented by

$$v(s,t) = \sum_{i=1}^{M} b_i(t)\psi_i(s) \tag{23}$$

where M generalized coordinates $b_i(t)$ are used to describe this displacement. Again, $\psi_i(s)$ are appropriate admissible basis functions satisfying the geometric boundary conditions $\psi_i(0) = \psi_i(L_0) = 0$. Generalized coordinates $\alpha, a_i(1 = 1, 2, \ldots, N)$ and $b_i(1 = 1, 2, \ldots, M)$ fully define this $N + M + 1$ degree-of-freedom system. Here, α is the pitch rotation (or libration) of the ribbon in the equatorial plane.

The position vectors of a ribbon element and the counterweight with respect to the center of the Earth, respectively, are given by

$$\mathbf{r_R}(s) = (R_E\cos\alpha + s + u)\mathbf{i} + (v - R_E\sin\alpha)\mathbf{j} \tag{24}$$

and

$$\mathbf{r_C}(s) = (R_E\cos\alpha + L_0 + u_L)\mathbf{i} - R_E\sin\alpha\mathbf{j} \tag{25}$$

where

$$u_L = u(L_0, t) \tag{26}$$

The velocity vectors relative to the inertial frame are given by

$$\mathbf{v_R}(s) = [R_E\Omega \sin\alpha + \dot{u} - (\Omega + \dot{\alpha})v]\mathbf{i} + [R_E\Omega \cos\alpha + \dot{v} + (\Omega + \dot{\alpha})(s + u)]\mathbf{j} \tag{27}$$

and

$$\mathbf{v_C} = [R_E\Omega \sin\alpha + \dot{u}_L]\mathbf{i} + [R_E\Omega \cos\alpha + (\Omega + \dot{\alpha})(L_0 + u_L)]\mathbf{j} \tag{28}$$

The total kinetic energy of the system can then be written as

$$K = 1/2\gamma \int_0^{L_0} A(s)[\mathbf{v_R}(s) \cdot \mathbf{v_R}(s)]\mathrm{d}s + \frac{1}{2}m_c(\mathbf{v_C} \cdot \mathbf{v_C}) \tag{29}$$

The total potential energy of the system is given by

$$P = -\mu\gamma \int_0^{L_0} \frac{A(s)}{\sqrt{\mathbf{r_R}(s) \cdot \mathbf{r_R}(s)}} - \frac{\mu m_c}{\sqrt{\mathbf{r_C} \cdot \mathbf{r_C}}} + P_{st} \tag{30}$$

where the strain energy stored in the ribbon, P_{st}, is given by

$$P_{st} = \frac{1}{2}E\int_0^{L_0} A(s)\left[\left(\frac{\partial u}{\partial s}\right)^2 + \frac{\partial u}{\partial s}\left(\frac{\partial v}{\partial s}\right)^2 + \frac{1}{2}\left(\frac{\partial v}{\partial s}\right)^2 + \left(\frac{\partial u}{\partial s}\right)^2\left(\frac{\partial v}{\partial s}\right)^2\right]\mathrm{d}s \tag{31}$$

For a system without external excitations, Lagrange's equations can be written as

$$\frac{\mathrm{d}}{\mathrm{d}t}\left[\frac{\partial K}{\partial \dot{q}_i}\right] - \frac{\partial K}{\partial q_i} + \frac{\partial P}{\partial q_i} = 0 \tag{32}$$

In Eq. (32), q_i is a generalized coordinate. By substituting the energy expressions from Eqs. (29) and (30) into Eq. (32) and letting the generalized coordinates be the degrees of freedom of the space elevator, one obtains the $N + M + 1$ equations of motion of the system.

The details of the equations of motion obtained from Eq. (32) and their analysis can be found in Cohen and Misra [8]. Here, only a brief description of the analysis and the results are presented.

4.1 Longitudinal Oscillations

The frequencies of longitudinal oscillations of the ribbon can be obtained by solving the eigenvalue problem

$$\mathbf{M}^{\mathbf{A}}\mathbf{A}'' + \mathbf{K}^{\mathbf{A}}\mathbf{A} = \mathbf{0} \tag{33}$$

where the elements of the matrices $\mathbf{M}^{\mathbf{A}}$, $\mathbf{K}^{\mathbf{A}}$ are given by [8]

$$M_{ik}^{A} = M_p \int_0^1 \exp[F(\xi)]\psi_i\psi_k \mathrm{d}\xi + M_c\psi_{i1}\psi_{k1} \tag{34}$$

and

$$\begin{aligned} K_{ik}^{A} = {} & \bar{\Omega}^2 \int_0^1 \exp[F(\xi)]\left(\frac{\mathrm{d}\psi_i}{\mathrm{d}\xi}\right)\left(\frac{\mathrm{d}\psi_k}{\mathrm{d}\xi}\right)\mathrm{d}\xi \\ & - M_p \int_0^1 \exp[F(\xi)]\psi_i\psi_k \mathrm{d}\xi - M_c\psi_{i1}\psi_{k1} \\ & - 2\left(\frac{\beta}{\Lambda}\right)^3 \left\{ M_p \int_0^1 \frac{\exp[F(\xi)]\psi_i\psi_k}{[(1/\Lambda) + \xi(1+\varepsilon_0)]^3}\mathrm{d}\xi + M_c \frac{\psi_{i1}\psi_{k1}}{[(1/\Lambda) + (1+\varepsilon_0)]^3} \right\} \end{aligned} \tag{35}$$

where the indices i and k vary from 1 to N. The prime symbol appearing in Eq. (33) denotes differentiation with respect to nondimensional time τ. The nondimensional quantities appearing in Eqs. (33) to (35) are defined by

$$\tau = \Lambda t, \quad \xi = \frac{s}{L_0}, \quad M_p = \gamma A_m \frac{L_0}{m_{\text{tot}}}, \quad M_c = \frac{m_c}{m_{\text{tot}}} \tag{36}$$

$$\Lambda t = \frac{L_0}{R_E}, \quad \beta = \frac{R_G}{R_E}, \quad \bar{\Omega} = \frac{\left[\frac{EA_m}{m_{\text{tot}}L_0}\right]^{\frac{1}{2}}}{\Lambda} \tag{37}$$

and

$$F(\xi) = \frac{R_E^2}{\bar{h}R_G(1+\varepsilon_0)}\left\{\frac{3}{2} - \frac{R_G}{[R_E + \xi(1+\varepsilon_0)L_0]} - \frac{[R_E + \xi(1+\varepsilon_0)L_0]^2}{2R_G^2}\right\} \tag{38}$$

The generalized eigenvalue problem given by Eq. (33) is solved using MATLAB. Before solving, a set of basis functions $\psi_i(s)$ appearing in Eq. (22) must be chosen. They must all be zero at the base, i.e., at $s = 0$ or $\xi = 0$. Also, to allow the counter-

Table 1 Nondimensional frequencies of longitudinal modes (using sinusoidal basis functions)

Mode no.	Freq.	Mode no.	Freq.	Mode no.	Freq.	Mode no.	Freq.	Mode no.	Freq.
1	4.39	5	47.82	9	96.50	13	145.27	17	194.30
2	12.48	6	59.94	10	108.62	14	157.37	18	206.43
3	23.75	7	72.14	11	120.87	15	169.72	19	219.19
4	35.69	8	84.27	12	132.98	16	181.83	20	231.55

weight to move freely, the basis functions must be non-zero at $\xi = 1$. A reasonable choice is the polynomials, i.e., the nondimensional basis functions are given by

$$\psi_i(\xi) = \xi^i, \ i = 1, 2, \ldots \tag{39}$$

Although acceptable results for frequencies of longitudinal oscillations can be found for a small number of modes with this choice, numerical difficulties are encountered for a large number of modes considered in the calculations [8]. A more robust choice for the basis functions is

$$\psi_i(\xi) = \sin\left(i - \frac{1}{2}\right)\pi\xi, \ i = 1, 2, \ldots \tag{40}$$

Results were obtained for a space elevator having $L_0 =$ 100,000 km, $\gamma = 1300\,\text{kg/m}^3$, $E = 1000$ GPa, and a taper ratio of $A_m/A_0 = 6$. The corresponding values of the key nondimensional parameters are $M_p = 0.992$, $M_c = 0.228$, and $\bar{\Omega} = 3.8$. The first 20 longitudinal frequencies, nondimensionalized by dividing by the Earth's angular velocity, Ω, are given in Table 1. It may be noted that the lowest longitudinal frequency is 4.39 (i.e., 4.39 oscillations per day or a period of approximately 5.5 h).

4.2 *Transverse Oscillations*

The transverse displacement of the ribbon is now studied by considering only the nominal value of the longitudinal extension which has a second-order effect on transverse frequencies. However, the libration of the ribbon is included. Hence, if ribbon libration is included and if higher order terms and damping terms are neglected, the following eigenvalue problem is obtained:

$$\mathbf{M^B B}'' + \mathbf{K^B B} = \mathbf{0} \tag{41}$$

where

$$\mathbf{B} = [\alpha, B_1, B_2, \ldots, B_k]^T \tag{42}$$

and the elements of matrices $\mathbf{M^B}$ and $\mathbf{K^B}$ are given by

$$M_{11}B = (1+\varepsilon_0)^2\left(M_p \int_0^1 \exp[F(\xi)]\xi^2 \mathrm{d}\xi + M_c\right) \tag{43}$$

$$\begin{aligned} K_{11}^B = (1+\varepsilon_0)^2 \Bigg[\frac{1}{\Lambda}\left\{M_p \int_0^1 \exp[F(\xi)]\xi \mathrm{d}\xi + M_c\right\} \\ - \frac{\beta^3}{\Lambda^4}\left\{M_p \int_0^1 \frac{\exp[F(\xi)]\xi}{[(1/\Lambda)+\xi(1+\varepsilon_0)]^3}\mathrm{d}\xi + M_c \frac{1}{[(1/\Lambda)+(1+\varepsilon_0)]^3}\right\}\Bigg] \end{aligned} \tag{44}$$

$$M_{i1}^B = M_{1k}^B = M_p(1+\varepsilon_0)\int_0^1 \exp[F(\xi)]\xi\phi_i \mathrm{d}\xi \tag{45}$$

$$K_{i1}^B = K_{1k}^B = M_p\left\{\frac{1}{\Lambda}\int_0^1 \exp[F(\xi)]\phi_i \mathrm{d}\xi - \frac{\beta^3}{\Lambda^4}\int_0^1 \frac{\exp[F(\xi)]\phi_i}{[(1/\Lambda)+\xi(1+\varepsilon_0)]^3}\mathrm{d}\xi\right\} \tag{46}$$

$$M_{ik}^B = \int_0^1 \exp[F(\xi)]\psi_i\psi_k \mathrm{d}\xi \tag{47}$$

$$\begin{aligned} K_{ik}^B = \pi^2(\varepsilon_0 - \varepsilon_0^2)ik\bar{\Omega}^2 \int_0^1 \exp[F(\xi)]\frac{\mathrm{d}\phi_i}{\mathrm{d}\xi}\frac{\mathrm{d}\phi_k}{\mathrm{d}\xi}\mathrm{d}\xi \\ + M_p\left\{\left(\frac{\beta}{\Lambda}\right)^3 \int_0^1 \frac{\exp[F(\xi)]\phi_i}{[(1/\Lambda)+\xi(1+\varepsilon_0)]^3}\mathrm{d}\xi - \int_0^1 \exp[F(\xi)]\phi_i\phi_k \mathrm{d}\xi\right\} \end{aligned} \tag{48}$$

where indices i and k vary from 2 to $M+1$.

The basis functions are chosen as $\phi_i(\xi) = \sin(i\pi\xi)$, which vanish at the two ends of the ribbon. Again, the generalized eigenvalue problem defined by Eq. (46) may be solved using MATLAB. The system parameters used earlier for the analysis of the longitudinal oscillations are used again here. Table 2 contains the nondimensional natural frequencies for the first 20 modes of the transverse displacement of the ribbon (again, they are nondimensionalized with respect to Ω). The lowest frequency (zeroth

Table 2 Nondimensional frequencies of transverse modes

Mode no.	Freq.	Mode no.	Freq.	Mode no.	Freq.	Mode no.	Freq.	Mode no.	Freq.
0	0.16	4	8.92	8	17.74	12	26.57	16	35.41
1	2.37	5	11.13	9	19.95	13	28.78	17	37.62
2	4.53	6	13.33	10	22.15	14	30.99	18	39.84
3	6.72	7	15.53	11	24.36	15	33.19	19	42.05

mode) is that for the libration, which has a value of 0.16, and this corresponds to a period of about 6 days. The next mode of the system, the first for transverse vibration, has a nondimensional frequency of 0.27, i.e., it has a period of about 10 h. Again, the frequencies of the transverse modes (except the rigid-body mode) increase in a quasi-linear fashion.

The longitudinal-to-transverse frequency ratio of the first elastic mode is about two. This ratio increases for all subsequent modes. As was the case for the longitudinal motion, the modal frequencies and mode shapes of the transverse motion are independent of A_m.

5 Conclusions

This article presents an introduction to the concept of space elevators, which have the potential to revolutionize the way satellites will be placed in orbit in the future. The orbital parameters that can be achieved using a space elevator are briefly discussed. A static analysis is presented that relates the nominal length of the ribbon, material stress, variation of the cross-sectional area, along the length of the ribbon and the mass of the counterweight. These are essential considerations for the design of the space elevator.

An analysis is presented to determine the frequencies associated with the libration and elastic oscillations, both longitudinal and transverse. The libration (pendulum-type motion) has a period of about 6 days, while the first transverse mode has a period of about 10 h.

The effect of a climber on the dynamics of the space elevator is not discussed in this article. A climber not only changes the static deformation and frequencies of elastic oscillations to some extent, but also causes a Coriolis force on the elevator during its climbing, resulting in a librational motion. Since it is likely that multiple climbers will be used, proper phasing of the climbers can minimize this Coriolis effect.

Acknowledgements Some content in this chapter is based on the following paper: "Cohen, S.S. and Misra, A.K., 2007. Elastic oscillations of the space elevator ribbon. *Journal of guidance, control, and dynamics, 30(6)*, pp.1711–1717." Authors are grateful to **American Institute of Aeronautics & Astronautics (AIAA)** for permitting the inclusion of some content in this chapter.

References

1. Edwards, B.C., Westling, E.A.: The Space Elevator: A Revolutionary Earth-to-Space Transportation System (2003)
2. Swan, P.A., et al. (eds.): Space Elevators: An Assessment of the Technological Feasibility and the Way Forward. International Academy of Astronautics, Paris (2013)
3. Tsiolkovsy, K.E.: Speculations Between Earth and Sky. Akademiia Nauk SSSR (1895); Reprint Akademiia Nauk SSSR, Moscow (1959)
4. Artsutanov, Y.: To space by a locomotive. Komsomolskaya Pravda, July 1960
5. Pearson, J.: The orbital tower: a spacecraft launcher using the Earth's rotational energy. Acta Astronaut. **2**(9-10), 785–799 (1975)
6. Cohen, S.S.: Satellite placement using the space elevator. Climb **2**(1), 17 (2013)
7. Yakobson, B.I., Smalley, R.E.: Fullerene nanotubes: C1,000,000 and beyond. Am. Sci. **85**(4), 324 (1997)
8. Cohen, S.S., Misra, A.K.: Elastic oscillations of the space elevator ribbon. J. Guid. Control Dyn. **30**(6), 1711–1717 (2007)

Recent Advances in Free Surface Flows

Gautam Biswas and Kirti Chandra Sahu

1 Background

A free surface is an interface between liquid and gas [1]. A free-surface exhibits the surface tension force which is a consequence of the cohesive force acting at the interface. In many under-graduate books, free-surface flow is also termed as open-channel flow [2, 3]. In a moving air–water system, as the values of the density of water and air are approximately 998 kg m^{-3} and 1.2 kg m^{-3}, respectively, the inertia due to the air phase is much smaller than that of the water phase. Thus, one can think that water moves independently, or freely, with respect to the air phase. In other words, an air–water interface is free and the only coupling between the phases is due to the air pressure that exerts on the liquid surface. implest examples of free-surface flows are the waves created by throwing a small stone in a still-water surface and patterns formed at water surface due to air current. Although, we see these flows in our everyday life, understanding the associated physics is quite complicated [4]. If the free surface of a liquid is disturbed, waves are produced due to the interplay between gravity that acts to bring the disturbed surface back to its horizontal position and the momentum that causes the waves to overshoot. Due to this the free surface oscillates and the waves spread to the neighbouring regions. The surface tension force plays an important role to stabilize these waves/instabilities.

Free surface flows are encountered in a wide range of industrial applications and natural phenomena, such as jets, cavities, bubble columns, seepage of groundwater, ice melting, gravity waves, clouds and raindrops, to name a few (see for instance,

G. Biswas (✉)
Department of Mechanical Engineering, Indian Institute of Technology Kanpur, Kanpur 208016, UP, India
e-mail: gtm@iitk.ac.in

K. C. Sahu
Department of Chemical Engineering, Indian Institute of Technology Hyderabad, Sangareddy 502285, Telangana, India
e-mail: ksahu@che.iith.ac.in

© Springer Nature Singapore Pte Ltd. 2021

U. S. Dixit and S. K. Dwivedy (eds.), *Mechanical Sciences*,
https://doi.org/10.1007/978-981-15-5712-5_6

Refs. [1, 5]). Due to its complexity and rich underlying physics, this topic has been attracting the attention of many researchers since mid 1800. The interface between two immiscible liquids can also be treated as free surface.

The objectives of this chapter are to (i) review the numerical and experimental advances on free surface flows, (ii) highlight the mathematical models and common numerical techniques used to study interfacial flows and (iii) discuss few examples of complex phenomena involving free-surface flows. In the next section, we will discuss about the representation of a free surface mathematically and the associated boundary conditions at the interface.

2 Mathematical Representation

Consider an interface separating two fluids in a simple two-dimensional coordinate system (x, y) as shown in Fig. 1. The position of the interface between $fluid\ A$ and $fluid\ B$ is of the form $\vec{\eta}(x, y, t) = 0$, such that the height of the interface from x axis can be specified as $y = h(x, t)$. Thus,

$$\vec{\eta}(x, y, t) = \hat{h}(x, t) - y, \tag{1}$$

where $\hat{h}(x, t)$ is the perturbed interface.

It is easy to imagine that any fluid particle at the interface remains attached with the interface. This implies that the normal component of the velocity of the fluid particle at the interface is equal to the normal component of the interface velocity. Mathematically this condition can be expressed as

$$\frac{D\vec{\eta}}{Dt} = 0 \Rightarrow \frac{\partial \hat{h}}{\partial t} + u_i \frac{\partial \hat{h}}{\partial x} = v_i, \tag{2}$$

where u_i and v_i are the components of interface velocity in the x and y directions, respectively, and D is the substantial/material derivative. Equation (2) is commonly known as the kinematic condition for the interface/free surface.

The dynamics conditions for the interface state that the momentum must be conserved at the free surface. In other words, the normal forces on either side of the free surface should be equal and opposite in direction, and the tangential forces should

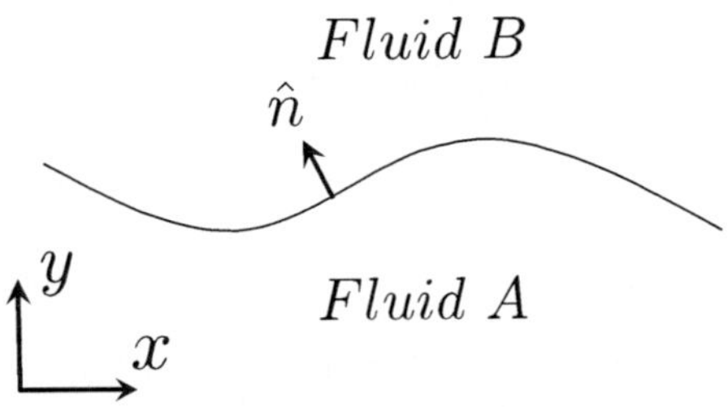

Fig. 1 Two-dimensional representation of an interface separating $fluid\ A$ and $fluid\ B$

be equal in magnitude and direction. Thus, both normal and tangential stresses must be balanced at the interface. The normal and tangential stress balance equations at the interface are given by

$$(\hat{n} \cdot \mathbf{T}_A) \cdot \hat{n} - \kappa\sigma = (\hat{n} \cdot \mathbf{T}_B) \cdot \hat{n}, \tag{3}$$

$$(\hat{n} \cdot \mathbf{T}_A) \cdot \hat{t} - (\hat{n} \cdot \mathbf{T}_B) \cdot \hat{n} = \nabla\sigma \cdot \hat{t}, \tag{4}$$

respectively. Equation (3) implies that the jump in normal stress across the interface must balance the curvature force per unit area. In Eq. (4), while the left-hand side represents the jump in tangential components of the hydrodynamic stress at the interface, the right-hand side represents the tangential stress associated with gradients in surface tension, σ, which may results from gradients in temperature or chemical composition at the interface. Here $\hat{n}$ and $\hat{t}$ are the unit normal and tangential vectors to the interface, respectively; $\kappa(= \nabla \cdot \hat{n})$ is the curvature of the interface. The stress tensors for $fluid\ A$ and $fluid\ B$ are given by

$$\mathbf{T}_A = -p_A + \mu_A \left[\nabla \mathbf{u}_A + (\nabla \mathbf{u}_A)^T\right], \tag{5}$$

$$\mathbf{T}_B = -p_B + \mu_B \left[\nabla \mathbf{u}_B + (\nabla \mathbf{u}_B)^T\right], \tag{6}$$

respectively. Here, p_A and p_B are the pressure, and $\mathbf{u}_A(u_A, v_A)$ and $\mathbf{u}_B(u_B, v_B)$ are velocity vectors associated with $fluid\ A$ and $fluid\ B$ at the interface; μ_A and μ_B are the dynamic viscosities of $fluid\ A$ and $fluid\ B$, respectively.

In addition, the velocity of $fluid\ A$ and $fluid\ B$ at the interface must be continuous, i.e.

$$u_A = u_B, \tag{7}$$

$$v_A = v_B. \tag{8}$$

Note that the extension of these interfacial conditions from two-dimensional to three-dimensional formulation is straightforward, and can be found in Ref. [6].

3 Solid-Surface Versus Free Surface

Stokes [7] derived an expression by balancing the weight, the buoyancy force, the drag force and the force of acceleration acting on a solid spherical object of radius R falling in a viscous fluid in the limit of vanishing Reynolds number (creeping flow regime). At steady state when the object reaches its terminal velocity, V_t, Stokes [7] proposed that the drag force, F_d is given by

$$F_d = 6\pi\mu R V_t. \tag{9}$$

The terminal velocity of a solid sphere falling in a viscous liquid is given by

$$V_t = \frac{2}{9}\frac{(\rho_{solid} - \rho_f)}{\mu_f} R^2 g, \tag{10}$$

where ρ_{solid} is the density of solid sphere, and ρ_f and μ_f are the density and dynamic viscosity of the surrounding fluid, respectively.

The fluid dynamics of a falling solid sphere is different from that of a liquid blob falling in air. At the solid-surface velocity field obeys the no-slip boundary condition. On the other hand, in case of a free-surface of a liquid sphere, the boundary conditions discussed in Sect. 2 will be applicable. This was first recognized by Hadamard [8] and Rybczynski [9], who proposed the following equation for the velocity for a blob of fluid (fluid 'i') falling in another fluid (fluid 's')

$$V_t = \frac{2}{3}\frac{(\rho_s - \rho_i)}{\mu_s}\frac{\mu_s + \mu_i}{2\mu_s + 3\mu_i} R^2 g, \tag{11}$$

where (ρ_i, μ_i) and (ρ_s, μ_s) are the density and viscosity of the dispersed phase (inner) and the continuous phase (surrounding medium), respectively. This equation is known as the Hadamard–Rybczynski equation. This can be considered as one of the first fundamental of contribution towards the understanding of the free-surface flows. The density and viscosity ratios can be defined as $\rho_r \equiv \rho_i/\rho_s$ and $\mu_r \equiv \mu_i/\mu_s$, respectively. Thus, for bubbles, $\rho_r \ll 1$ and for drops, $\rho_r \gg 1$. Recently, Tripathi et al. [10] investigated the differences between bubbles and drops using the concept of the Hadamard–Rybczynski flow. They defined bubble (drop) as a blob of fluid lighter (heavier) than the surrounding medium. They observed that the maximum vorticity prefers the lighter fluid; in bubbles and drops, the maximum vorticity lies inside and outside the dispersed phase, respectively, which in turn changes the dynamics completely. Thus there is no analogy between dynamics observed in the cases of bubbles and drops.

4 Governing Equations and Dimensionless Numbers

The equations governing the free-surface flows are the equations of mass and momentum conservation:

$$\nabla \cdot \mathbf{u} = 0, \tag{12}$$

$$\rho\left[\frac{\partial \mathbf{u}}{\partial \tau} + \mathbf{u}\cdot\nabla\mathbf{u}\right] = -\nabla p + \nabla\cdot\left[\mu(\nabla\mathbf{u} + \nabla\mathbf{u}^T)\right] + \delta(\mathbf{x} - \mathbf{x}_f)\sigma\kappa\mathbf{n} - \rho g\mathbf{j}. \tag{13}$$

Here, $\mathbf{u}$ denotes the velocity field and in three-dimensional representation, u, v and w represent the velocity components in the x, y and z directions, respectively; p is the pressure field; t denotes time; ρ and μ are the density and viscosity of the fluids; $\mathbf{j}$ denotes the unit vector along the vertical direction; $\delta(\mathbf{x} - \mathbf{x}_f)$ is the delta distribution function (denoted by δ hereafter) whose value is zero everywhere except at the interface, where $\mathbf{x} = \mathbf{x}_f$. The surface tension force is an interfacial force, so it appears in the normal stress balance boundary condition at the interface (see Eq. (3)). The surface tension force per unit volume can also be added as an interfacial body force term in the Navier–Stokes equations (see the term, $\delta(\mathbf{x}-\mathbf{x}_f)\sigma\kappa\mathbf{n}$ in Eq. (13)) using the continuum surface force (CSF) formulation proposed by Brackbill et al. [11].

The following scaling can be used to non-dimensionalize the above governing equations:

$$(x,y,z)=\mathcal{L}\left(\widetilde{x},\widetilde{y},\widetilde{z}\right),\ t=\widetilde{t}\mathcal{L}/V,\ \mathbf{u}=V\tilde{\mathbf{u}},\ p=\rho_s V^2\widetilde{p},\ \mu=\mu_s\widetilde{\mu},\ \rho=\rho_s\widetilde{\rho},\ \delta=\widetilde{\delta}/\mathcal{L}, \tag{14}$$

where $\mathcal{L}$ is the length scale, V is the velocity scale and μ_s and ρ_s are the reference viscosity and density, respectively. The tildes designate dimensionless quantities.

After dropping tildes from all nondimensional variables, the governing dimensionless equations are given by

$$\nabla\cdot\mathbf{u}=0, \tag{15}$$

$$\frac{\partial\mathbf{u}}{\partial t}+\mathbf{u}\cdot\nabla\mathbf{u}=-\nabla p+\frac{1}{\text{Re}}\nabla\cdot\left[\mu(\nabla\mathbf{u}+\nabla\mathbf{u}^T)\right]+\delta\frac{\nabla\cdot\mathbf{n}}{\text{ReCa}}\mathbf{n}-\frac{1}{\text{Fr}}\rho\mathbf{j}. \tag{16}$$

Here, $\text{Re}(\equiv\rho_s V\mathcal{L}/\mu_o)$ is the Reynolds number, $\text{Ca}(\equiv\mu_s V/\sigma)$ is the Capillary number and $\text{Fr}(\equiv V^2/g\mathcal{L})$ is the Froude number.

Other dimensionless numbers used in some of the free-surface flows, which can also be derived from the fundamental dimensionless numbers, can be obtained by setting $Fr=1$ (neutrally buoyant system). They are the Galilei number ($\text{Ga}(\equiv\rho_s g^{1/2}\mathcal{L}^{3/2}/\mu_s)$) and the Eötvös/Bond number, ($\text{Eo}(\equiv\rho_s g\mathcal{L}^2/\sigma)$). We will also discuss the other dimensionless numbers used in free-surface flows as and when required in this chapter.

5 Numerical Methods

It is very challenging to simulate free-surface flows due to its complex physics operational at the interface separating the phases of zero physical thickness. Two types of computational approaches, namely, interface tracking and interface capturing methods, have been commonly used. The interface tracking methods treat the free surface as a sharp interface. In this method, boundary-fitted grids are used and reconstructed at each time the free surface moves/deforms. The front tracking method [12] is one example of the interface tracking method, which was emerged as an effective approach to simulate free-surface flows in 1980s, and have been used to study coalescences and breakups in bubbly flows [13, 14]. On the other hand, the interface capturing methods have been used when the free surface undergoes large deformation and becomes too complex to track. In such situations, interface tracking becomes computationally very expensive. In the interface capturing methods, interface is not as sharp as that in the interface tracking methods, but the interface is determined from the fraction of each cell near the interface that is partially filled. The marker-

and-cell (MAC) scheme [15], the volume-of-fluid (VoF) method [16–19], level-set (LS) method and coupled level-set volume-of-fluid (CLSVOF) approach [20–22] are some examples of the interface capturing technique. In some cases, hybrid methods have been used to simulate interfacial flows. Here, we do not discuss all the methods, but in what follows, only highlight the coupled level-set and volume-of-fluid (CLSVOF) approach.

In the CLSVOF method, the interfacial dynamics is modelled using a level-set function (ϕ) and volume fraction (F) of $fluid\ A$ as follows:

$$\frac{\partial \phi}{\partial t} + \vec{V} \cdot \nabla \phi = 0, \tag{17}$$

$$\frac{\partial F}{\partial t} + \vec{V} \cdot \nabla F = 0, \tag{18}$$

where the volume fraction takes the values 0 for $fluid\ B$ (gas phase) and 1 for $fluid\ A$ (liquid phase). Similarly, the values of the level-set functions for fluid '1' and fluid '2' are ϵ and $-\epsilon$, respectively, and 0 at the interface separating the fluids. Here, ϵ is the numerical thickness of the interface; one can use $\epsilon = 0.5\Delta$, Δ being the grid size.

The functional dependences of the density (ρ) and the dynamics viscosity (μ) of the fluid with the level-set function ϕ *via* the Heaviside function $H(\phi)$ are given by

$$\rho(\phi) = \rho_A H(\phi) + \rho_B(1 - H(\phi)), \tag{19}$$

$$\mu(\phi) = \mu_A H(\phi) + \mu_B (1 - H(\phi)), \tag{20}$$

where

$$H(\phi) = \begin{cases} 1 & \text{if} \quad \phi > \epsilon, \\ \frac{1}{2} + \frac{\phi}{2\epsilon} + \frac{1}{2\pi}\left\{sin\left(\frac{\pi\phi}{\epsilon}\right)\right\} & \text{if} \quad |\phi| \le \epsilon, \\ 0 & \text{if} \quad \phi < -\epsilon. \end{cases} \tag{21}$$

6 Recent Advancement in Free-Surface Flows

There are many examples of free-surface flows. In what follows, we only discuss some specific examples which we frequently encounter in our day-to-day life and industrial applications.

6.1 Rayleigh–Taylor Instabilities and Gravity Currents

The Rayleigh–Taylor instability occurs at the interface separating two fluids with an unstable density stratification, i.e. when a heavier fluid overlays a lighter fluid. Due to the action of gravity, the heavier fluid penetrates into the region of the lighter fluid, while the lighter fluid moves up in order to maintain the conservation of mass, causing the interface to deform. The Rayleigh–Taylor instability has been a subject of considerable interest due to its relevance in many natural phenomena and practical applications [23, 24]. In nature, they are found in ocean tides, the accelerated interstellar clouds driven by newborn stars and black holes, supernova events, sinking of slabs of tectonic plates, in volcanic activities [25], to name a few.

A model problem to study the Rayleigh–Taylor instability is the transient mixing of two unstably-density-stratified fluids confined in a tilted enclosure, commonly known as the 'lock-exchange' problem as shown in Fig. 2. Such a problem has been studied experimentally by Séon et al. [28, 29], computationally by Sahu and coworkers [26] and Hallez and Magnaudet [30], in which a tilted tube was filled with two fluids of different densities and a plate separating them was suddenly removed. As a result of buoyancy the two fluids interpenetrated each other and mixed as shown in Fig. 3. The parameters characterizing such a mixing processes are the Atwood number ($\mathrm{At} \equiv (\rho_\mathrm{h} - \rho_\mathrm{l})/(\rho_\mathrm{h} + \rho_\mathrm{l})$), where ρ_h and ρ_l are the densities of the heavier and lighter fluids, respectively, the angle of tilt of the enclosure (θ) and the fluid viscosities. The Atwood number signifies the magnitude of the buoyancy force, while the tilt angle defines the two components of the gravity force. In situations where viscosity contrast accompany the density contrast, the flow pattern in the 'lock-exchange' configuration is due to the interplay between the Kelvin–Helmholtz and Rayleigh–Taylor instabilities. The flow dynamics of displacement flow of one

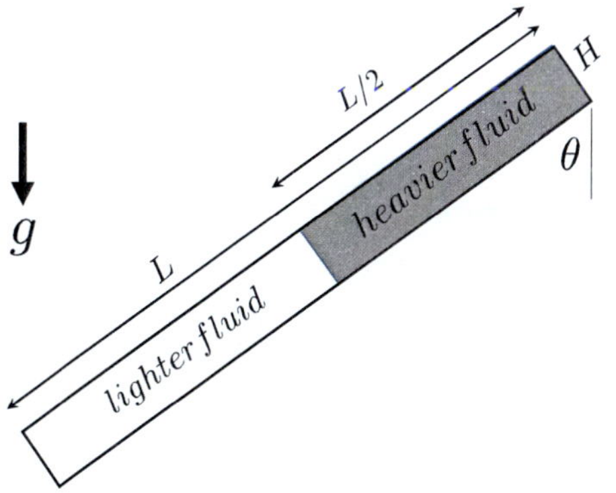

Fig. 2 The schematic diagram of the initial configuration of a 'lock-exchange' problem in an inclined confined channel of length L and height H. The lighter and the heavier fluids occupy $0 \leq x \leq L/2$ and $L/2 \leq x \leq L$, respectively. θ is the angle of inclination of channel to the vertical and g is the acceleration due to gravity, such that $g\cos\theta$ and $g\sin\theta$ act in the $-x$ and $-y$ directions, respectively

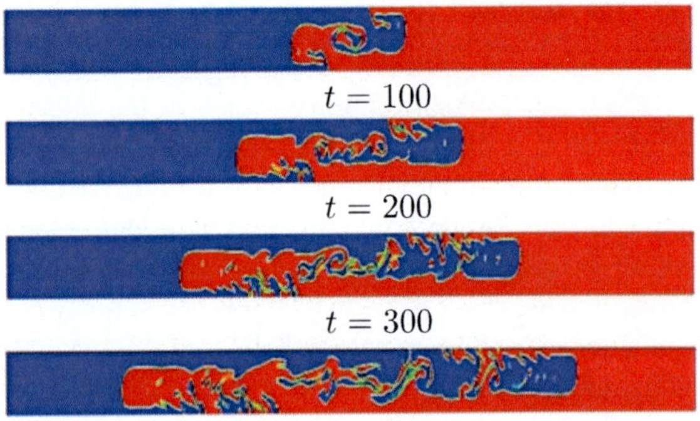

Fig. 3 Spatio-temporal evolution of the density contours for Re = 558.6, At = 0.01 and $\theta = 30°$. The red and the blue fluids represent the heavier and the lighter fluids, respectively. This figure is taken from Sahu and Vanka [26]

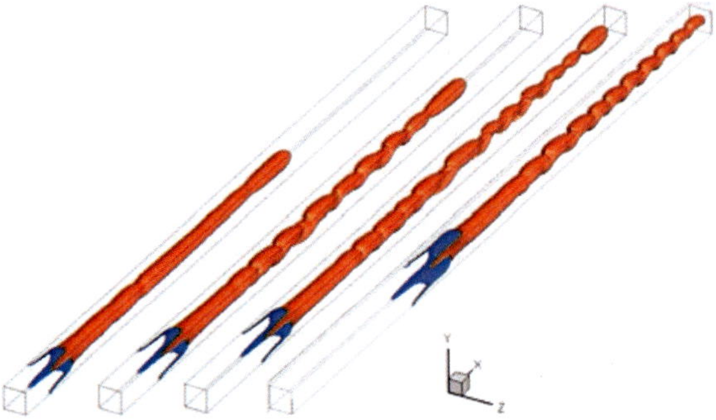

Fig. 4 Cork-screw mode in a displacement flow of one fluid by another immiscible fluid in a square duct. This figure is taken from Redapangu et al. [27]

fluid by another fluid and the resultant instabilities are discussed in Refs. [27, 31–33] via numerical simulations and linear stability analyses. Redapangu et al. [27] observed a cork-screw instability mode in case of displacement flow in a square duct (shown in Fig. 4). An extensive review of Kelvin–Helmholtz and Rayleigh–Taylor instabilities observed in shear flows can also be found in Ref. [4].

6.2 *Coalescence Dynamics of Droplets*

Another interesting free-surface phenomenon is the coalescence dynamics of a droplet falling in liquid pool [34–40]. A schematic diagram showing a spherical drop of fluid '1' (radius R) falling under the action of gravity, g on the free-surface of a pool of fluid '1' is plotted in Fig. 5. The surrounding medium is designated by fluid '2'. For some physical parameters based on droplet size and its impact velocity, three types of coalescence dynamics are observed, namely, partial coalescence, complete coalescence and splashing. In Refs. [34–36], different regimes from partial/complete coalescence to splashing have been observed based on the impact velocity of the primary droplet in the liquid pool. Figure 6 shows the variation of the diameter ratio, defined as $\zeta \equiv (D_s/D)$, versus D obtained by Chen et al. [41] exper-

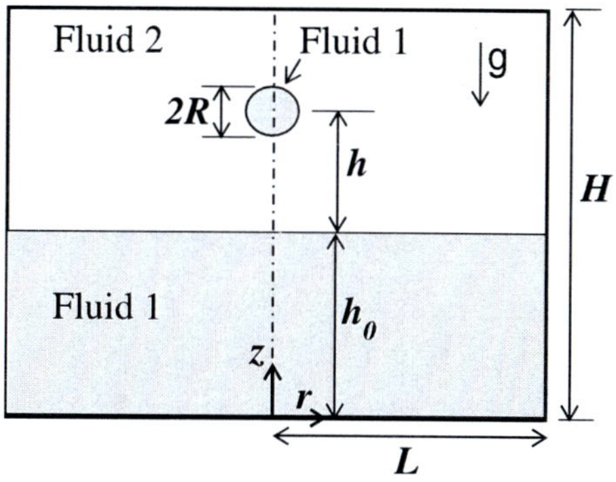

Fig. 5 Schematic diagram showing a droplet falling in a liquid pool

imentally and by Ray et al. [42] numerically. Here, D_s and D denote the diameter of the satellite droplet and primary droplet, respectively. They found different regimes dominated by viscous, inertio-capillary, gravity based on high, low and intermediate values of the Bond number, respectively. It can be seen that the diameter of the biggest satellite droplet is about 0.5 times of the primary droplet, which is obtained in the inertio-capillary regime. It can also be seen that for intermediate values of D only the partial coalescence dynamics is observed. In this case, the droplet floats on the free surface until the surrounding fluid trapped between the droplet and the free surface is drained out radially. Subsequently a neck is formed at the contact point of the droplet and the free surface expands rapidly due to high capillary pressure near the contact region. The resultant capillary waves move in the upward direction resulting in a liquid column at later time. The surface tension reduces the diameter of the neck and a satellite droplet is detached. This satellite droplet moves upward and decelerates due to gravity and comes back in the downward direction. This process gets repeated with a reduction of volume of the subsequent satellite droplets till entire volume is completely merged in the liquid pool. The phenomenon of partial coalescence is discussed in great detail in Refs. [42, 43]. For small and big primary droplet complete coalescence is observed. High impact velocities of the drops may lead to large bubble entrapment [36, 44]. Experimental evidence [45] of large bubble entrapment occurring outside the traditional small region on the impact velocity V-the diameter of the droplet (D) map, made the boundary of large bubble entrapment a topic of greater importance. A seminal contribution of the work [44] probes in to the redefined zone of large bubble entrapment and underlying physics (see Fig. 7).

Charles and Mason [46] argued that the partial coalescence of a droplet is due to inviscid instability [47]. However, later it was revealed that the dynamics of partial coalescence is primarily governed by gravity, viscosity and interfacial tension (see Refs. [34, 41–43]).

Recently in Ref. [48], the partial coalescence dynamics of a compound drop in a liquid pool has been investigated numerically. The dimensionless numbers used to describe the results were the Bond number, Bo $\left(\equiv \rho_1 g R_{\mathrm{eq}}^2/\sigma\right)$, the Ohnesorge numbers associated with fluid '1' and fluid '2', which are given by

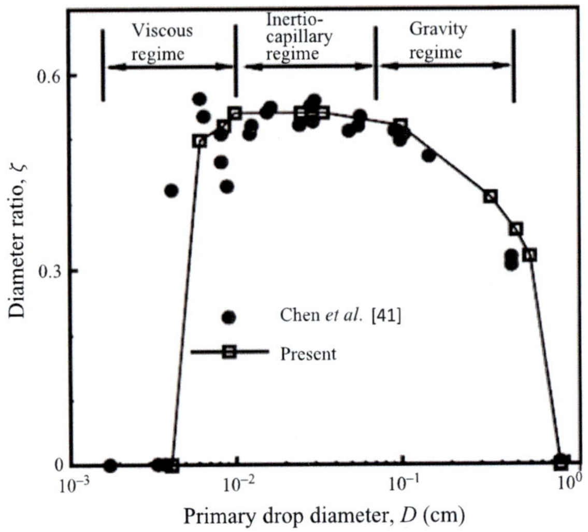

Fig. 6 Variation of the satellite droplet diameter/the primary droplet diameter (ζ) with D showing different regimes [41, 42]. This figure is taken from Ray et al. [42]

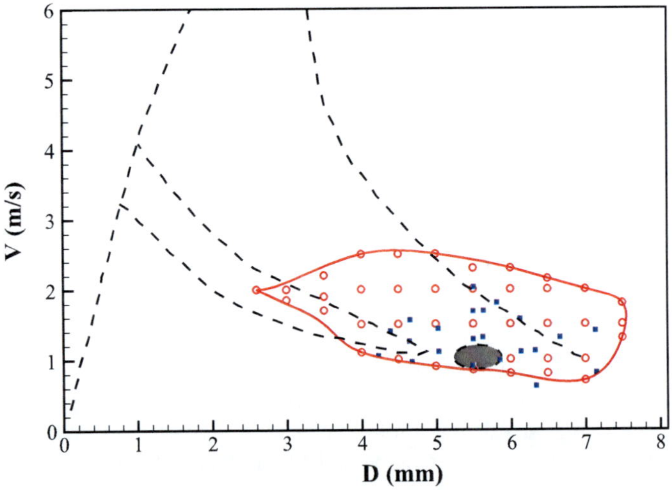

Fig. 7 The regime on the velocity of impact (V)- droplet diameter (D) map overlapped with the traditional V- D map (dashed lines). The red circles and the filled blue squares represent the points of the large bubble entrapment and the points where a large bubble entrapment was observed experimentally by Wang et al. [45]. This figure is taken from Deka et al. [44]

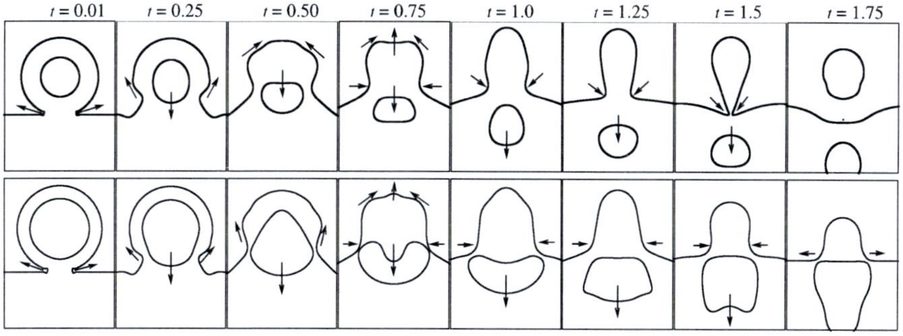

Fig. 8 The coalescence sequence of a compound ethanol droplet. The radius ratios in the top and bottom row are $R_r = 0.5$ and $R_r = 0.7$, respectively. Arrow marks indicate the direction of the momentum acting on the interface. The dimensionless parameters are $\mathrm{Oh}_1 = 1.09 \times 10^{-2}$, $\mathrm{Oh}_2 = 2.08 \times 10^{-4}$, At = 0.997 and Bo = 0.09. This figure is taken from Deka et al. [48]

$\mathrm{Oh}_1 \left(\equiv \mu_1/\sqrt{\rho_1 \sigma R_{eq}}\right)$ and $\mathrm{Oh}_2 \left(\equiv \mu_2/\sqrt{\rho_1 \sigma R_{eq}}\right)$, respectively, the Atwood number, At $(\equiv (\rho_1 - \rho_2)/(\rho_1 + \rho_2))$, and the radius ratio, R_r $(\equiv R_i/R_o)$. Here, R_{eq} is the equivalent spherical drop; (μ_1, ρ_1) and (μ_2, ρ_2) are the dynamic viscosity and density of the compound droplet, which is the same liquid as in the pool and surrounding fluid, respectively; R_i and R_o are the inner and outer radius of the compound droplet, respectively. They [48] found that the partial coalescence is surpassed for large radius ratio, $R_r > 0.6$. For $R_r > 0.6$ the inner bubble remains near the free surface and thus, prevents the necking of the liquid column. On the other hand, the partial coalescence dynamics of small R_r is similar to that of a 'normal' drop as shown in Fig. 8. It is found that the location of the inner bubble in the pool is found to play an important role in the pinch-off process. They also observed three different types of coalescence dynamics for different values of R_r, namely, (i) pinch-off of a satellite droplet without bursting of the bubble, (ii) pinch-off of a satellite droplet with the simultaneous bursting of the bubble and (iii) bubble bursting before the pinch-off.

6.3 *Path and Topology of Bubbles and Drops*

6.3.1 Single Bubble

The dynamics of an air bubble rising in a liquid has been an active area of research due to its relevance in many natural and industrial applications, such as carbon sequestration, bubble-column reactors, microfluidics, etc. (see for instance, Refs. [8, 19, 49, 50]). In dimensionless formulation, by conducting the Buckingham pi theorem, it can be shown that the behaviour of a rising bubble can be completely described by four dimensionless numbers: the Gallilei number $\left(\mathrm{Ga}(\equiv \rho_o g^{1/2} R^{3/2}/\mu_o)\right)$, the Eötvös number, $\left(\mathrm{Eo}(\equiv \rho_o g R^2/\sigma)\right)$, the density ratio $(\rho_r(\equiv \rho_i/\rho_o))$ and the viscosity ratio $(\mu_r(\equiv \mu_i/\mu_o))$. Here, R is the equivalent radius of the bubble, σ is the interfacial

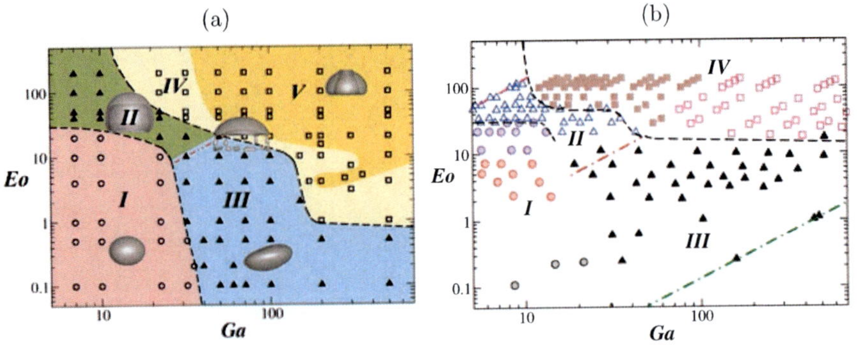

Fig. 9 Different regions of bubble shape and behaviour. **a** Computational and **b** experimental. The red dash-dotted line represents Mo $= 10^{-3}$, which separates region II and region III. The green and magenta lines in panel **b** represent Mo $= 2.52 \times 10^{-11}$ (pure water) and Mo $= 230.3$ (pure glycerol). The panels **a** and **b** are taken from Tripathi et al. [19] and Sharaf et al. [51], respectively

tension, while ρ_o, μ_o, and ρ_i, μ_i are densities and viscosities of the continuous and dispersed phases, respectively. An additional dimensionless parameter, the Morton number (Mo) can also be defined as $\mathrm{Eo}^3/\mathrm{Ga}^4 (\equiv g\mu_0^4/\rho_o\sigma^3)$, which is unique for a particular fluid as it depends on fluid properties alone.

Recently, Tripathi et al. [19] conducted three-dimensional numerical simulations by varying Ga and Eo for an air bubble rising in water such that $\rho_r = 10^{-3}$ and $\mu_r = 10^{-2}$. They identified five different regions of distinct bubble behaviours (namely, axisymmetric, skirted, zigzagging/spiralling, peripheral breakup and central breakup). They showed that an air bubble maintains its azimuthal symmetry (region I in Fig. 9a) for low Ga–low Eo, and is either spherical, oblate or dimpled. For low Ga, and high Eo (region II in Fig. 9a), skirted bubbles are observed, whereas for high Ga, and low Eo (region III in Fig. 9a), a bubble follows a spiral or zigzag path (wobbling motion). An air bubble with high Ga and high Eo breaks to form satellite bubbles (region IV in Fig. 9a) or undergoes topological changes to form a toroidal shape (region V in Fig. 9a; central breakup). Recently, Sharaf et al. [51] conducted experiments using different concentrations of aqueous solutions of glycerol and obtained a phase diagram presented in Fig. 9b. The experimentally obtained phase diagram qualitatively looks similar to that obtained from the numerical simulations (Fig. 9a). However, close inspection reveals that Sharaf et al. [51] did not get the central breakup region (region V) and also region III is slightly bigger than that in Fig. 9a. They have attributed these differences to the difficulty in creating a perfectly spherical bubble at high Ga and high Eo in their experiments. Also the actual density and viscosity ratios considered by Sharaf et al. [51] are also different from those taken by Tripathi et al. [19].

In this context, it is important to discuss the classical region map provided by Bhaga and Weber [49] (also see Ref. [52]), which has been used by several researchers, although Haberman and Morton [53] were probably the first to conduct experiment on rising bubble in viscous liquids. The authors of Refs. [49, 52] experimentally investigated the dynamics of an air bubble rising in aqueous sugar solutions of differing concentrations. They identified regimes of spherical, oblate, wobbling and skirted bubbles in the Reynolds and Eötvös numbers plane based on the bubble shapes and motion. This phase plot was prepared using three dimensionless parameters: the Reynolds number, the Eötvös number and the Morton number. The Reynolds number was defined based on terminal velocity of the bubble in Refs. [49, 52]. The Gallilei number used by Sharaf et al. [51] and Tripathi et al. [19] is similar to the Reynolds number, but uses $\sqrt{gR}$ instead, as the velocity scale. Consequently, the phase plots presented in Refs. [19, 51] also include unsteady bubbles, for which there is no terminal velocity. The use of Gallilei and Eötvös numbers gives another advantage as discussed by Landel et al. [54]. They showed that for the same volume of air (i.e. constant Gallilei and Eötvös numbers) spherical cap bubbles with a range of rise velocities (multiple Reynolds numbers) and volume of satellite bubbles can be produced. Therefore, use of the Reynolds and Eötvös numbers may provide a multivalued nature to the phase plot. The phase plot presented in Fig. 9 is a useful extension to the classical region map of Bhaga and Weber [49] and Clift et al. [52].

6.3.2 Two Bubbles

The interactions and trajectories of a pair of air bubbles rising side-by-side in a liquid was also studied by many researchers in the Stokes and the potential flow limits [55–57], and by performing simulations of the complete Navier–Stokes equations [57–60] and also experimentally [61–63]. Kok [56, 61] experimentally found that the two bubbles rising vertically in ultra-pure water tend to rotate to align themselves horizontally. Chen et al. [58] conducted two-dimensional simulations on two bubbles rising side-by-side and showed that the bubbles coalesce and the resultant bigger bubble exhibits shape oscillations [19, 51]. Duineveld [62] compared the rising dynamics of a single bubble and a pair of bubbles rising side-by-side and found that the amplitude of oscillations is higher in case of pair of bubbles as compared to that in case of single bubble. The coalescence and bouncing behaviours of two bubbles for different values of Reynolds and Weber numbers were investigated by Sanada et al. [63].

Recently, Tripathi et al. [64] conducted three-dimensional simulations and investigated the dynamics of a pair of air bubbles rising in water and investigated the influence of inertia on the dynamics. They found that interaction between the wakes of the bubbles, as shown in Fig. 10a, causes oscillatory motion/path instability as shown in Fig. 10b. Chakraborty et al. [65, 66] studied two bubbles rising in an inline configuration in stagnant liquid and studied deformation and coalescence dynamics using a CLSVOF approach (see Fig. 11).

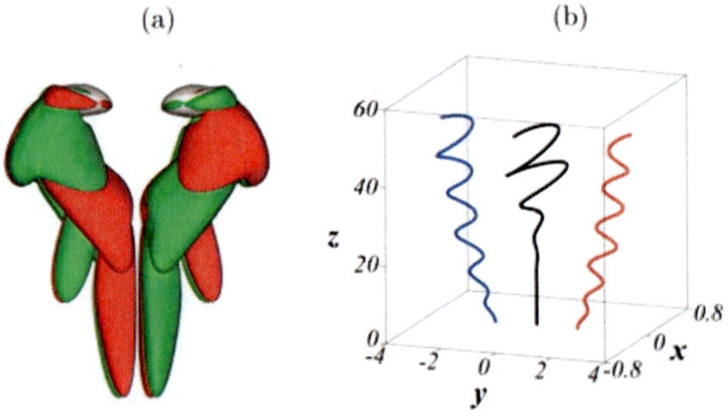

Fig. 10 **a** Isosurfaces for z vorticity as the bubbles rises up for Ga = 60. The positive and negative values are shown by red and green colours, respectively; $\omega_z = \pm 0.3$. **b** Trajectories of single bubble (shown in black) and bubble pair (shown in indigo and red) for Ga = 32. The rest of the parameter values are Eo = 4, $q = 3$, $\rho_r = 10^{-3}$ and $\mu_r = 10^{-2}$. The rest of the parameter values are Eo = 4, $q = 3$, $\rho_r = 10^{-3}$ and $\mu_r = 10^{-2}$. This plot is taken from Tripathi et al. [64]

6.3.3 Shape Oscillations in Drops

As noted by Tripathi et al. [10], in gas–liquid systems, bubbles and droplets behave differently due to different vorticity patterns observed in rising air bubbles and falling liquid droplets. It is also well known that, as a result of path instability, an initially spherical air bubble rising in a liquid (density ratio, $\rho_r \ll 1$) can either exhibits zigzagging or spiralling motion at high inertia and high surface tension [19, 68–70]. It undergoes an unsteady shape deformation resulting in vortex shedding behind the bubble during its wobbling motion. In contrast, a solid sphere or an initially spherical liquid drop ($\rho_r \gg 1$) falls in a straight path [71] in an air medium. In case of a falling leaf or flat/cylindrical solid objects (i.e. nonspherical), oscillatory motion is observed due to the associated aerodynamics/hydrodynamics (see, e.g Ref. [72]). A question that arises then is, can we observe path and/or shape instabilities by making the initial drop shape nonspherical in an air–liquid system. In order to answer this question, recently, Sahu and co-workers [73, 74] have investigated the dynamics of an initially nonspherical liquid droplet falling in air under the action of gravity. They observed symmetrical shape oscillations of the droplets which decay with time at low inertia. On the other hand, at high values of the Gallilei number the shape asymmetry in the vertical direction becomes prominent and the droplet undergoes breakup. The reason for this asymmetry has been attributed to the higher aerodynamic inertia. However, even for large inertia, unlike bubbles, no path deviations/oscillations were observed.

In liquid–liquid systems (with $\rho_r \leq 2.2$), Edge and Grant [75] experimentally found that a small liquid drop falls in a straight path, but a bigger drop falls in a zigzag path (i.e. wobbling motion). They observed a thread-like wake for small drops, whereas, vortex sheet are observed in the case of large drops undergoing wobbling motion and shape oscillations (oblate-prolate deformation). Later, Koh and Leal [76, 77] investigated the dynamics of an initially nonspherical liquid bubble rising in a quiescent liquid of slightly higher density by conducting numerical simulations

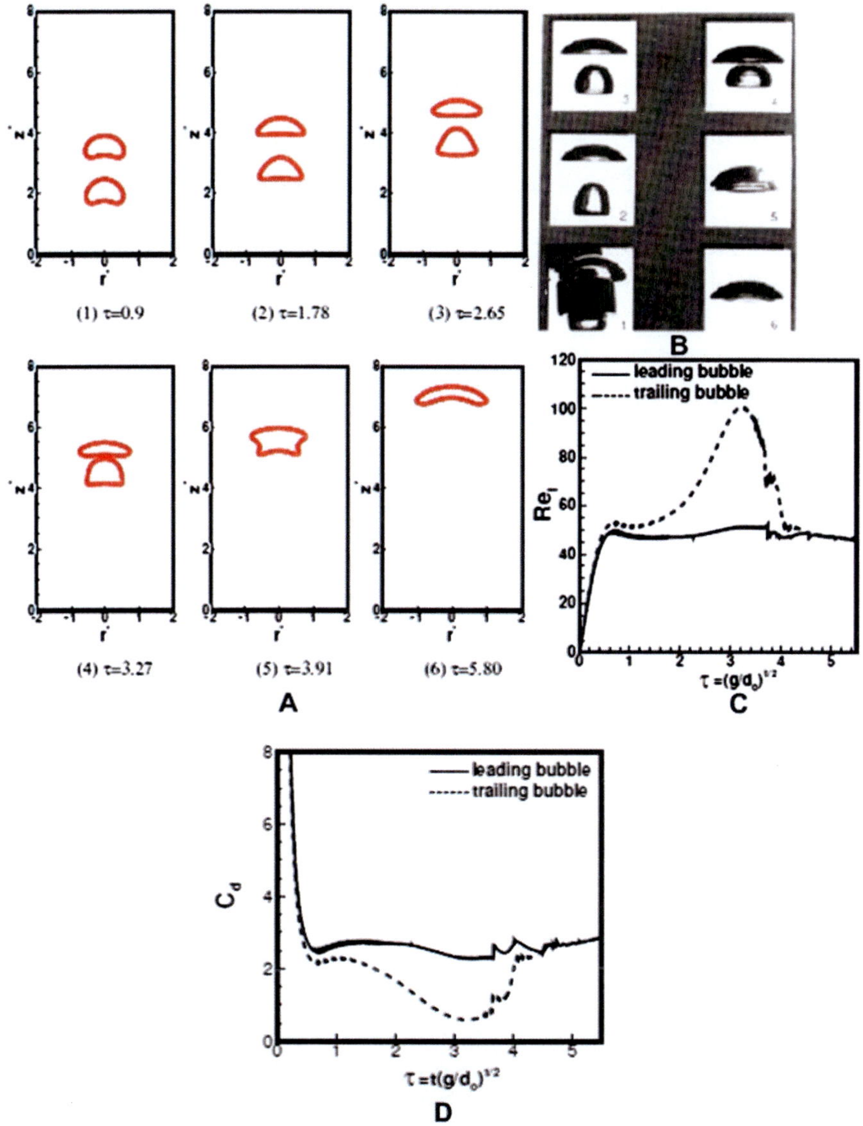

Fig. 11 The deformation and coalescence of the two inline gas bubbles rising in quiescent liquid. **a** Numerical method and **b** experiment [67]. **c** The variations of the instantaneous Reynolds numbers (Re_I) and **d** the instantaneous coefficient of drag of the two bubbles. The dimensionless parameters are used for the simulations according to data taken from experiment conditions $Eo = 16$ and $M = 2 \times 10^{-4}$. This plot is taken from Chakraborty et al. [65]

and experiments at low Reynolds numbers (Re < 0.01). They found that a bubble with a small degree of nonsphericity, eventually becomes spherical, but a bubble with a large degree of nonsphericity continues to deform with an increasing amplitude. Later Refs. [78, 79] focussed on the decay rate of the oscillations for a viscous drop in micro-gravity condition. Tsamopoulos and Brown [80] studied the effect of initial amplitude on frequency for nonlinear inviscid oscillations of droplets. The shape deformation of droplets in simple shear flows and in complex geometries with and without external forcing has also been investigated by several researchers (see for instance, Refs. [81, 82]).

6.4 Droplet Breakups

The phenomenon of liquid jet breakup is observed in many industrial applications, such as ink-jet printing, particle sorting, atomization, mixing, combustion, separation and spraying technologies [83]. Due to the Rayleigh–Plateau instability [84], a liquid jet issued from an orifice becomes unstable and breaks up leading to formation of satellite droplets due to the propagation of surface perturbations. The classical studies of Plateau [84], Savart [85] and Rayleigh [47] provides a fundamental understanding of the jet instabilities. Weber [86] investigated the influence of liquid viscosity and gas density on jet breakup phenomenon. Goedde and Yuen [87] investigated the capillary instability of water–glycerine interface. They showed that the nonlinear effects dominate the jet breakup process leading to formation of ligaments and satellite droplets. The nozzle characteristics can also affect the breakup length of the jet occurring due to the Rayleigh instability [88]. Unlike the aforementioned studies, which considered circular liquid jets, Kashyap et al. [89] and Farvardin and Dolatabadi [90] also investigated the dynamics of elliptical-orifice liquid jets. It is found that for different combinations of surface tension, inertia and aerodynamic interactions, four distinct breakup regimes can be observed. They are (i) the Rayleigh breakup regime (varicose perturbations), (ii) the first wind-induced breakup (sinuous perturbations), (iii) the second wind-induced breakup and (iv) the atomization (spray) regime [91–94]. The numerical simulations of liquid jets were conducted by Pan and Suga [95, 96] to investigate the characteristics of breakup. The breakup length of a turbulent liquid jet was studied by Lafrance [97].

In addition to the breakup of a liquid jet aligned with gravity (as discussed above), Borthakur et al. [98, 99] investigated the dynamics of curved slender jets due to their implications in the production of nanofibres from centrifugal spinning [100–102]. The variation of jet diameter along length for injections along the gravity (parallel injection) and orthogonal to gravity (perpendicular injection) at $t = 15$ and $t = 40$ is shown in Fig. 12a and b, respectively. It can be seen that the pinch-up occurs early in the case of perpendicular injection as compared to those in the parallel injection.

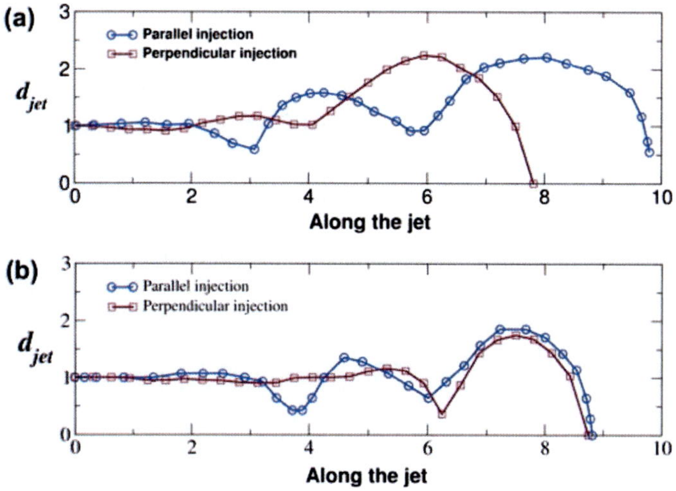

Fig. 12 The variation of jet diameter (d_{jet}) along length for the parallel and perpendicular injections at **a** $t = 15$ and **b** $t = 40$. These results are plotted upto the breakup locations. The rest of the dimensionless parameters are $\mathrm{Oh} = 4 \times 10^{-3}$, $\mathrm{Bo} = 0.06$ and $\mathrm{We} = 4$. This plot is taken from Borthakur et al. [98]

The various dimensionless used in their study to describe the results are the Weber, the Bond and the Reynolds numbers, defined as

$$\mathrm{We} = \frac{\rho_l v_{\mathrm{avg}}^2 D}{\sigma}, \quad \mathrm{Re} = \frac{\rho_l v_{\mathrm{avg}} D}{\mu_l}, \tag{22}$$

where ρ_l, μ_l, v_{avg} and D are the density, dynamic viscosity, average velocity of the liquid and diameter of the nozzle, respectively. They found that the jet breakup length increases with increase in the Weber and the Ohnesorge numbers. It is observed that increasing the Bond number significantly increases the curvature of the jet trajectory. The detached drops from the main jet exhibit rolling motion, as well as, shape oscillations while migrating along their trajectories. Uddin and Decent [103] also considered a jet injected perpendicular to gravity and performed a linear instability analysis to explore the nature of the most unstable wavenumber and the growth rate of an infinitesimally small perturbation at the liquid surface. Suñol and González-Cinca [104] carried out an experimentally studied liquid jets falling perpendicular to gravity. It was observed that the droplet size distribution was wider for the perpendicular injection as compared to the parallel injection.

6.5 Boiling and Phase Change

Multiphase flows undergoing phase change are ubiquitous and have several industrial applications, for instance, energy generation, manufacturing, combustion, etc. Phase change in multiphase flows can also take place due to chemical reaction, evapouration, melting, etc. In saturated film boilings, the heat transfer occurs through a thin vapour film via the formation of vapour bubbles, which transfer the heat from the vapour film to the bulk liquid. Experimental and computational studies in boiling have been dedicated to develop empirical correlations and to understand the detail of boiling phenomenon and transport processes. Welch and Wilson [105] were one of the pioneers to simulate phase change for wide range of density difference using the VOF formulation. This work was widely acknowledged and subsequently used by other researchers to implement the idea in other approaches and to study phase change and boiling (e.g. the CLSVOF approach of Refs. [106] and Electro hydro dynamics [107]). As the formation of bubbles, which follows a periodic pattern in space and time at the liquid–vapour interface, plays an important role in the heat transfer processes, several researchers have studied this phenomenon by conducting numerical simulations [22, 108–110]. They observed multi-mode bubble growth (as shown in Fig. 13) and showed that for higher super-heat ranges, the instability is guided by the Taylor–Helmholtz instability, whereas for the lower super-heat, the Rayleigh–Taylor instability dominates the phenomenon.

A plot showing the temporal evolution of bubble release cycle in water at 373 °C, 219 bar on an isothermal horizontal surface simulated using a volume of fluid (VOF) based numerical solver is shown in Fig. 14. It can be seen that as the bubble is released from the surface, the surface tension pulls the vapour film in the downward direction.

Fig. 13 Interface morphology for different wall super-heat temperatures. **a** 2 K, **b** 5 K, **c** 18 K and **d** 22 K. This plot is taken from Pandey et al. [110]

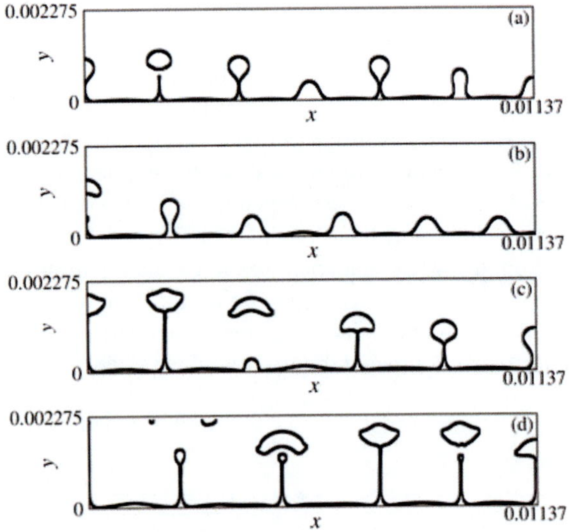

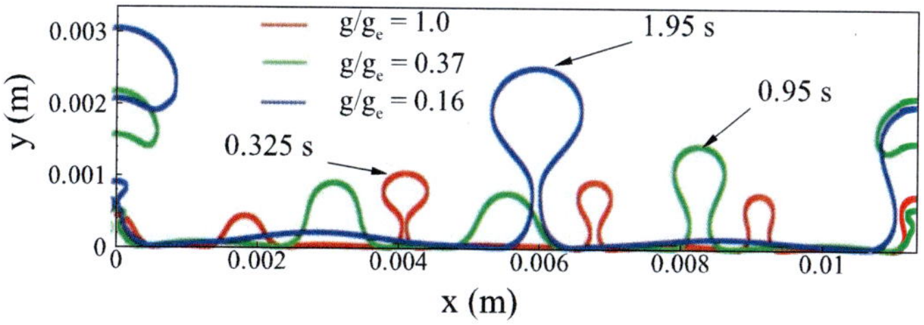

Fig. 14 Temporal evolution of bubble release cycle in water at 373 °C, 219 bar. This plot is taken from Pandey et al. [111]

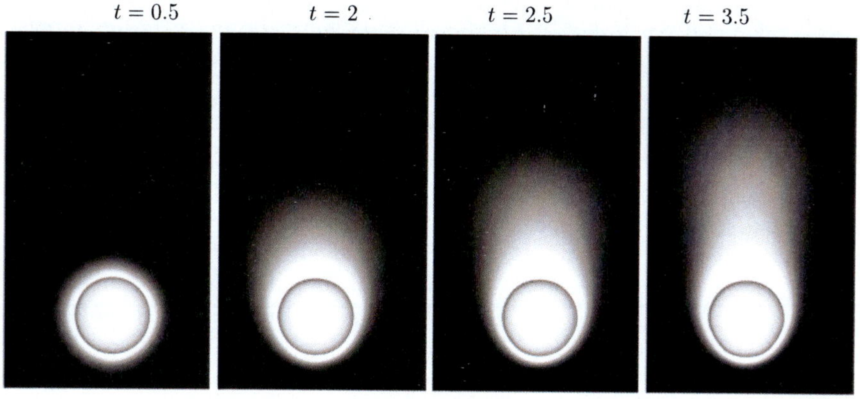

Fig. 15 The time evolution (left to right) of shape of the drop along with the vapour volume fraction. The time, t written in each panel is the dimensionless time. This plot is taken from Tripathi et al. [112]

Further the surface tension induces a flow to move the vapour packet towards the symmetric sidewalls. The vapour turns upward near the side walls to initiate an identical bubble release cycle. Pandey et al. [111] also found that reduced gravity decreases the temporal frequency of bubble release rate and there is a reduction in wave length pertaining to the bubble formation sites. They showed that applied electric field may be used to control the spatial and temporal frequencies. As such, the heat transfer rate deteriorates in reduced gravity conditions which can be recovered by the externally imposed electric field.

A falling liquid droplet in the context of raindrops is another situation where the phase change becomes important. Tripathi et al. [112] conducted three-dimensional numerical simulations to understand the evaporation dynamics in the case of a falling droplet. The time evolution of the droplet shapes, along with the contours of the vapour concentration generated due to evaporation, is shown in Fig. 15. It can be seen that due to the evaporation, a circular shaped envelope of liquid vapour is formed at the early time, which continues to increase with time. As the drop moves in the downward direction, a wake region is created at the upper part of the drop, which increases the evaporation rate in the wake region. A 'fore-aft' asymmetric in the vapour envelope is seen at later time as the vapour produced at the bottom part of the drop moves towards the wake region due to buoyancy. It can also be seen that as the drop moves in the downward direction, it has deformed slightly to an oblate shape.

7 Concluding Remarks

In the present chapter, some specific problems under the broad area of free-surface flows are analysed and discussed. Multiphase flows are characterized by the contrast in fluid properties and the surface tension acting at the interface separating the fluids. Due to the associated complex interfacial dynamics, free-surface flows are challenging to handle both computationally and experimentally. The review presented in this chapter is an attempt to highlight the non-intuitive phenomena and complex flow physics observed in some free-surface flows. In spite of the long history, this subject has been attracting the attention of many researchers due to its relevance in many industrial applications and heretofore unexplained natural phenomena. Few examples are ink-jet printing, particle sorting, atomization, mixing, combustion, separation and spraying technologies, carbon sequestration, bubble-column reactors, microfluidics, seepage of groundwater, ice melting, gravity waves, clouds and raindrops. Understanding of multiphase flows involving fluids with phase change, non-Newtonian rheology and external forcing has been the current trend due to its usefulness in the emerging areas of microfluidics and bioengineering in last two decades.

References

1. Yeung, R.W.: Numerical methods in free-surface flows. Annu. Rev. Fluid Mech. **14**, 395 (1982)
2. White, F.M., Corfield, I.: Viscous Fluid Flow, vol. 3. McGraw-Hill, New York, Boston (2006)
3. Som, S.K., Biswas, G., Chakraborty, S.: Introduction to Fluid Mechanics and Fluid Machines. McGraw-Hill, New Delhi (2017)
4. Govindarajan, R., Sahu, K.C.: Instabilities in viscosity-stratified flow. Annu. Rev. Fluid Mech. **46**, 331 (2014)
5. Biswas, G., Som, S.K., Gupta, A.S.: Instability of a moving cylindrical liquid sheet. J. Fluids Eng. **107**, 451 (1985)
6. Leal, L.G.: Advanced Transport Phenomena: fluid Mechanics and Convective Transport Processes, vol. 7. Cambridge University Press, Cambridge (2007)
7. Stokes, G.G.: On the effect of internal friction of fluids on the motion of pendulums. Trans. Camb. Philos. Soc. **9**, 8 (1851)
8. Hadamard, J.S.: Mouvement permanent lent d'une sphère liquid et visqueuse dans un liquide visqueux. CR Hebd. Seances Acad. Sci. Paris **152**, 1735 (1911)
9. Rybczynski, W.: Uber die fortschreitende Bewegung einer flussigen Kugel in einem zahen Medium. Bull. Acad. Sci. Crac. A **1**, 40 (1911)

10. Tripathi, M.K., Sahu, K.C., Govindarajan, R.: Why a falling drop does not in general behave like a rising bubble. Sci. Rep. **4**, 4771 (2014)
11. Brackbill, J.U., Kothe, D.B., Zemach, C.: Instability of a moving cylindrical liquid sheet. J. Comput. Phys. **100**, 335 (1992)
12. Unverdi, S.O., Tryggvason, G.: A front-tracking method for viscous, incompressible, multi-fluid flows. J. Comput. Phys. **100**, 25 (1992)
13. Mehrabani, M.T., Nobari, M.R.H., Tryggvason, G.: An efficient front-tracking method for simulation of multi-density bubbles. Int. J. Numer. Methods Fluids **84**, 445 (2017)
14. Mehrabani, M.T., Nobari, M.R.H., Tryggvason, G.: Accelerating Poisson solvers in front tracking method using parallel direct methods. Comput. Fluids **118**, 101 (2015)
15. Harlow, F.H., Welch, J.E.: Numerical calculation of time-dependent viscous incompressible flow of fluid with free surface. Phys. Fluids **8**, 2182 (1965)
16. Hirt, C.W., Nichols, B.D.: Volume of fluid (VOF) method for the dynamics of free boundaries. J. Comput. Phys. **39**, 201 (1981)
17. Popinet, S.: Gerris: a tree-based adaptive solver for the incompressible Euler equations in complex geometries. J. Comput. Phys. **190**, 572 (2003)
18. Popinet, S.: An accurate adaptive solver for surface-tension-driven interfacial flows. J. Comput. Phys. **228**, 5838 (2009)
19. Tripathi, M.K., Sahu, K.C., Govindarajan, R.: Dynamics of an initially spherical bubble rising in quiescent liquid. Nat. Commun. **6**, 6268 (2015)
20. Sussman, M., Puckett, E.G.: A coupled level set and volume-of-fluid method for computing 3D and axisymmetric incompressible two-phase flows. J. Comput. Phys. **162**, 301 (2000)
21. Tomar, G., Biswas, G., Sharma, A., Agrawal, A.: Numerical simulation of bubble growth in film boiling using a coupled level-set and volume-of-fluid method. Phys. Fluids **17**, 112103 (2005)
22. Hens, A., Biswas, G., De, S.: Analysis of interfacial instability and multimode bubble formation in saturated boiling using coupled level set and volume-of-fluid approach. Phys. Fluids **26**, 012105 (2014)
23. Benjamin, T.B.: Gravity currents and related phenomena. J. Fluid Mech. **31**, 209 (1968)
24. Joseph, D.D., Bai, R., Chen, K.P., Renardy, Y.Y.: Core-annular flows. Annu. Rev. Fluid Mech. **29**, 65 (1997)
25. Baird, M.H.I., Aravamudan, K., Rama Rao, N.V., Chadam, J., Pierce, A.P.: Unsteady axial mixing by natural convection in a vertical column. AIChE J. **38**, 1825 (1992)
26. Sahu, K.C., Vanka, S.P.: A multiphase lattice Boltzmann study of buoyancy-induced mixing in a tilted channel. Comput. Fluids **50**, 199 (2011)
27. Redapangu, P.R., Sahu, K.C., Vanka, S.P.: A lattice Boltzmann simulation of three-dimensional displacement flow of two immiscible liquids in a square duct. J. Fluids Eng. **135**, 121202 (2013)
28. Séon, T., Znaien, J., Perrin, B., Hinch, E.J., Salin, D., Hulin, J.P.: Front dynamics and macroscopic diffusion in buoyant mixing in a tilted tube. Phys. Fluids **19**(12), 125105 (2007)
29. Séon, T., Salin, D., Hulin, J.P., Hinch, E.J., Perrin, B.: Transient buoyancy-driven front dynamics in nearly horizontal tubes. Phys. Fluids **19**, 123603 (2007)
30. Hallez, Y., Magnaudet, J.: Effects of channel geometry on buoyancy-driven mixing. Phys. Fluids **20**, 053306 (2008)
31. Sahu, K.C., Ding, H., Valluri, P., Matar, O.K.: Linear stability analysis and numerical simulation of miscible two-layer channel flow. Phys. Fluids **21**, 042104 (2009)
32. Sahu, K.C., Ding, H., Valluri, P., Matar, O.K.: Pressure-driven miscible two-fluid channel flow with density gradients. Phys. Fluids **21**, 043603 (2009)
33. Redapangu, P.R., Vanka, S.P., Sahu, K.C.: Multiphase lattice Boltzmann simulations of buoyancy-induced flow of two immiscible fluids with different viscosities. Eur. J. Mech. B/Fluids **34**, 105 (2012)
34. Thoroddsen, S.T., Takehara, K.: The coalescence cascade of a drop. Phys. Fluids **12**, 1265 (2000)

35. Ray, B., Biswas, G., Sharma, A.: Regimes during liquid drop impact on a liquid pool. J. Fluid Mech. **768**, 492 (2015)
36. Deka, H., Ray, B., Biswas, G., Dalal, A.: Dynamics of tongue shaped cavity generated during the impact of high-speed microdrops. Phys. Fluids **30**, 042103 (2018)
37. Deka, H., Biswas, G., Chakraborty, S., Dalal, A.: Coalescence dynamics of unequal sized drops. Phys. Fluids **31**, 012105 (2019)
38. Thomson, J.J., Newall, H.F.: V. On the formation of vortex rings by drops falling into liquids, and some allied phenomena. Proc. R. Soc. Lond. **39**, 417 (1886)
39. Worthington, A.M.: A Study of Splashes. Longmans, Green, and Co., Harlow (1908)
40. Stone, H.A., Stroock, A.D., Ajdari, A.: Engineering flows in small devices: microfluidics toward a lab-on-a-chip. Annu. Rev. Fluid Mech. **36**, 381 (2004)
41. Chen, X., Mandre, S., Feng, J.J.: Partial coalescence between a drop and a liquid-liquid interface. Phys. Fluids **18**, 051705 (2006)
42. Ray, B., Biswas, G., Sharma, A.: Generation of secondary droplets in coalescence of a drop at a liquid-liquid interface. J. Fluid Mech. **655**, 72 (2010)
43. Blanchette, F., Bigioni, T.P.: Partial coalescence of drops at liquid interfaces. Nat. Phys. **2**, 254 (2006)
44. Deka, H., Ray, B., Biswas, G., Dalal, A., Tsai, P.-H., Wang, A.-B.: The regime of large bubble entrapment during a single drop impact on a liquid pool. Phys. Fluids **29**, 092101 (2017)
45. Wang, A.B., Kuan, C.C., Tsai, P.H.: Do we understand the bubble formation by a single drop impacting upon liquid surface? Phys. Fluids **25**, 101702 (2013)
46. Charles, G.E., Mason, S.G.: The mechanism of partial coalescence of liquid drops at liquid/liquid interfaces. J. Colloid Sci. **15**, 105 (1960)
47. Rayleigh, L.: On the instability of jets. Proc. Lond. Math. Soc. **1**, 4 (1878)
48. Deka, H., Biswas, G., Sahu, K.C., Kulkarni, Y., Dalal, A.: Coalescence dynamics of a compound drop on a deep liquid pool. J. Fluid Mech. **866** (2019)
49. Bhaga, D., Weber, M.E.: Bubbles in viscous liquids: shapes, wakes and velocities. J. Fluid Mech. **105**, 61 (1981)
50. Magnaudet, J., Mougin, G.: Wake instability of a fixed spheroidal bubble. J. Fluid Mech. **572**, 311 (2007)
51. Sharaf, D.M., Premlata, A.R., Tripathi, M.K., Karri, B., Sahu, K.C.: Shapes and paths of an air bubble rising in quiescent liquids. Phys. Fluids **29**, 122104 (2017)
52. Clift, R., Grace, J., Weber, M.: Bubbles, Drops and Particles. Dover Publications Inc., New York (1978)
53. Haberman, W.L., Morton, R.K.: Technical report, DTIC Document (unpublished)
54. Landel, J.R., Cossu, C., Caulfield, C.P.: Spherical cap bubbles with a toroidal bubbly wake. Phys. Fluids **20**, 122201 (2008)
55. Leal, L.G.: Laminar Flow and Convective Transport Processes. Butterworth and Heinemann, Stoneham, MA (1992)
56. Kok, J.B.W.: Dynamics of a pair of gas bubbles moving through liquid. Part I. Theory. Eur. J. Mech. B/Fluids **12**, 515 (1993)
57. Legendre, D., Magnaudet, J., Mougin, G.: Hydrodynamic interactions between two spherical bubbles rising side by side in a viscous liquid. J. Fluid Mech. **497**, 133 (2003)
58. Chen, R.H., Tian, W.X., Su, G.H., Qiu, S.Z., Ishiwatari, Y., Oka, Y.I.: Numerical investigation on coalescence of bubble pairs rising in a stagnant liquid. Chem. Eng. Sci. **66**, 5055 (2011)
59. Velez-Cordero, J.R., Sámano, D., Yue, P., Feng, J.J., Zenita, R. Hydrodynamic interaction between a pair of bubbles ascending in shear-thinning inelastic fluids. J. Non-Newtonian Fluid Mech. **166**, 118 (2011)
60. Islam, M.T., Ganesan, P., Cheng, J.: A pair of bubbles rising dynamics in a xanthan gum solution: a CFD study. RSC Adv. **5**, 7819 (2015)
61. Kok, J.B.W.: Dynamics of a pair of gas bubbles moving through liquid. Part II. Experiment. Eur. J. Mech. B/Fluids **12**, 541 (1993)
62. Duineveld, P.C.: Bouncing and coalescence of bubble pair rising at high Reynolds number in pure water or aqueous surfactant solutions. Appl. Sci. Res. **58**, 409 (1998)

63. Sanada, T., Sato, A., Shirota, M., Watanable, M.: Motion and coalescence of a pair of bubbles rising side by side. Chem. Eng. Sci. **64**, 2659 (2009)
64. Tripathi, M.K., Premlata, A.R., Sahu, K.C., Govindarajan, R.: Two initially spherical bubbles rising in quiescent liquid. Phys. Rev. Fluids **2**, 073601 (2017)
65. Chakraborty, I., Biswas, G., Ghoshdastidar, P.S.: A coupled level-set and volume-of-fluid method for the buoyant rise of gas bubbles in liquids. Int. J. Heat Mass Transf. **58**, 240 (2013)
66. Chakraborty, I., Biswas, G., Ghoshdastidar, P.S.: Bubble generation in quiescent and co-flowing liquids. Int. J. Heat Mass Transf. **54**, 4673 (2011)
67. Brereton, G., Korotney, D.: Coaxial and oblique coalescence of two rising bubbles. In: Dynamics of Bubbles and Vortices Near a Free Surface, vol. 119, p. 50 (1991)
68. Saffman, P.G.: On the rise of small air bubbles in water. J. Fluid Mech. **1**, 249 (1956)
69. Zenit, R., Magnaudet, J.: Path instability of rising spheroidal air bubbles: a shape-controlled process. Phys. Fluids **20**, 061702 (2008)
70. Cano-Lozano, J.C., Martínez-Bazán, C., Magnaudet, J., Tchoufag, J.: Paths and wakes of deformable nearly spheroidal rising bubbles close to the transition to path instability. Phys. Rev. Fluids **1**, 053604 (2016)
71. Vries, A.W.G.: Technical report, University of Twente, Netherlands. ISBN 90 365 15262 (unpublished)
72. Kolomenskiy, D., Schneider, K.: Numerical simulations of falling leaves using a pseudo-spectral method with volume penalization. Theor. Comput. Fluid Dyn. **24**, 169 (2010)
73. Balla, M., Tripathi, M.K., Sahu, K.C.: Shape oscillations of a nonspherical water droplet. Phys. Rev. E **99**, 023107 (2019)
74. Agrawal, M., Premlata, A.R., Tripathi, M.K., Karri, B., Sahu, K.C.: Nonspherical liquid droplet falling in air. Phys. Rev. E **95**, 033111 (2017)
75. Edge, R.M., Grant, C.D.: The terminal velocity and frequency of oscillation of drops in pure systems. Chem. Eng. Sci. **26**, 1001 (1971)
76. Koh, C.J., Leal, L.G.: The stability of drop shapes for translation at zero Reynolds number through a quiescent fluid. Phys. Fluids A **1**(8), 1309 (1989)
77. Koh, C.J., Leal, L.G.: An experimental investigation on the stability of viscous drops translating through a quiescent fluid. Phys. Fluids A **2**(12), 2103 (1990)
78. Miller, C., Scriven, L.: The oscillations of a fluid droplet immersed in another fluid. J. Fluid Mech. **32**, 417 (1968)
79. Prosperetti, A.: Free oscillations of drops and bubbles: the initial-value problem. J. Fluid Mech. **100**, 333 (1980)
80. Tsamopoulos, J.A., Brown, R.A.: Nonlinear oscillations of inviscid drops and bubbles. J. Fluid Mech. **127**, 519 (1983)
81. Nath, B., Biswas, G., Dalal, A., Sahu, K.C.: Cross-stream migration of drops suspended in Poiseuille flow in the presence of an electric field. Phys. Rev. E. **97**, 063106 (2018)
82. Konda, H., Tripathi, M.K., Sahu, K.C.: Bubble motion in a converging-diverging channel. J. Fluids Eng. **138**, 064501 (2016)
83. Eggers, J., Villermaux, E.: Physics of liquid jets. Rep. Prog. Phys. **71**, 36601 (2008)
84. Plateau, J.: Experimental and Theoretical Statics of Liquids Subject to Molecular Forces Only. Gauthier-Villas, Paris (1873)
85. Savart, F.: Mémoire sur la constitution des veines liquides lancées par des orifices circulaires en mince paroi. Ann. Chim. Phys **53**, 1833 (1833)
86. Weber, C.: Zum zerfall eines flüssigkeitsstrahles. Z. Angew. Math. Mech. **11**, 136 (1931)
87. Goedde, E.F., Yuen, M.C.: Experiments on liquid jet instability. J. Fluid Mech. **40**, 495 (1970)
88. Kitamura, Y., Takahashi, T.: Proceedings of ICLASS, vol. 78, pp. 1–7 (1978)
89. Kasyap, T.V., Sivakumar, D., Raghunandan, B.N.: Breakup of liquid jets emanating from elliptical orifices at low flow conditions. At. Spray **18**(7) (2008)
90. Farvardin, E., Dolatabadi, A.: Numerical simulation of the breakup of elliptical liquid jet in still air. J. Fluids Eng. **135**, 071302 (2013)
91. Lin, S.P., Reitz, R.D.: Drop and spray formation from a liquid jet. Annu. Rev. Fluid Mech. **30**, 85 (1998)

92. Strasser, W., Battaglia, F.: Identification of pulsation mechanism in a transonic three-stream airblast injector. J. Fluids Eng. **138**, 111303 (2016)
93. Strasser, W., Battaglia, F.: The influence of retraction on three-stream injector pulsatile atomization for air-water systems. J. Fluids Eng. **138**, 111302 (2016)
94. Kulkarni, V., Sivakumar, D., Oommen, C., Tharakan, T.J.: Liquid sheet breakup in gas-centered swirl coaxial atomizers. J. Fluids Eng. **132**, 011303 (2010)
95. Pan, Y., Suga, K.: A numerical study on the breakup process of laminar liquid jets into a gas. Phys. Fluids **18**, 052101 (2006)
96. Pan, Y., Suga, K.: Capturing the pinch-off of liquid jets by the level set method. J Fluids Eng. **125**, 922 (2003)
97. Lafrance, P.: The breakup length of turbulent liquid jets. J. Fluids Eng. **99**, 414 (1977)
98. Borthakur, M.P., Biswas, G., Bandyopadhyay, D., Sahu, K.C.: Dynamics of an arched liquid jet under the influence of gravity. Eur. J. Mech. B/Fluids **74**, 1 (2019)
99. Borthakur, M.P., Biswas, G., Bandyopadhyay, D.: Formation of liquid drops at an orifice and dynamics of pinch-off in liquid jets. Phys. Rev. E. **96**, 013115 (2017)
100. Zhmayev, Y., Divvela, M.J., Ruo, A., Huang, T., Joo, Y.L.: The jetting behavior of viscoelastic Boger fluids during centrifugal spinning. Phys. Fluids **27**, 123101 (2015)
101. Noroozi, S., Alamdari, H., Arne, W., Larson, R.G., Taghavi, S.M.: Regularized string model for nanofibre formation in centrifugal spinning methods. J. Fluid Mech. **822**, 202 (2017)
102. Riahi, D.N.: Modeling and computation of nonlinear rotating polymeric jets during force-spinning process. Int. J. Non Linear Mech. **92**, 1 (2017)
103. Uddin, J., Decent, S.P.: Instability of non-Newtonian liquid jets curved by gravity. Prog. Ind. Math. ECMI **2008**, 597 (2010)
104. Suñol, F., González-Cinca, R.: Liquid jet breakup and subsequent droplet dynamics under normal gravity and in microgravity conditions. Phys. Fluids **27**, 77102 (2015)
105. Welch, S.W.J., Wilson, J.: A volume of fluid based method for fluid flows with phase change. J. Comput. Phys. **160**, 662 (2000)
106. Gerlach, D., Tomar, G., Biswas, G., Durst, F.: Comparison of volume-of-fluid methods for surface tension-dominant two-phase flows. Int. J. Heat Mass Transf. **49**, 740 (2006)
107. Welch, S.W.J., Biswas, G.: Direct simulation of film boiling including electrohydrodynamic forces. Phys. Fluids **19**, 012106 (2007)
108. Tomar, G., Biswas, G., Sharma, A., Welch, S.W.J.: Multimode analysis of bubble growth in saturated film boiling. Phys. Fluids **20**, 092101 (2008)
109. Agarwal, D.K., Welch, S.W.J., Biswas, G., Durst, F.: Planar simulation of bubble growth in film boiling in near-critical water using a variant of the VOF method. J. Heat Transf. **126**, 329 (2004)
110. Pandey, V., Biswas, G., Dalal, A.: Effect of superheat and electric field on saturated film boiling. Phys. Fluids **28**, 052102 (2016)
111. Pandey, V., Biswas, G., Dalal, A.: Saturated film boiling at various gravity levels under the influence of electrohydrodynamic forces. Phys. Fluids **29**, 032104 (2017)
112. Tripathi, M.K., Sahu, K.C.: Evaporating falling drop. Procedia IUTAM **15**, 201 (2015)

Fluid Dynamics in Deformable Microchannels

M. Kiran Raj and Suman Chakraborty

1 Deformable Microchannels—Motivation of Hemodynamic Investigation

The human body consists of many systems in which the circulation of fluids comes to picture. Both liquid and gaseous exchanges form the vital component of life. Cardiovascular system consists of loops of vessels through which blood flows from heart to cells and back. With lungs as the driving organ, respiratory system deals with the exchange of gases that ensures the supply oxygen to the cells. Table 1 provides a glimpse of the size of the vessels in the human body.

The vessel pressure is extremely important as it indicates the health as measured by a physician using a sphygmomanometer. The high- and low-pressure conditions are termed as hypertension and hypotension, respectively. Both can lead to serious consequences if not detected and dealt within the early stages. Most of these vessels are made up of convective tissues, consisting of fibrous elastin. Due to the soft and elastic nature, they can easily deform under the action of fluid flow. From the clinical perspective, both qualitative and quantitative information about the deformation gives clues about the severity of the condition. The physicians are trained to examine the sound signatures from the pulse characteristics in the circulatory system which are basically the result of the pressure waves traveling through vessels. More deformable walls can act like stabilizers and absorb the energy to dampen the pressure wave. However, as the age progresses, the flexibility of the tissues deteriorates and it can impede the normal functioning of the circulatory system. These arguments point out

M. Kiran Raj
Department of Mechanical and Mechatronics Engineering, University of Waterloo, Waterloo N2L 3G1, Ontario, Canada

S. Chakraborty (✉)
Department of Mechanical Engineering, Indian Institute of Technology, Kharagpur, Kharagpur 721302, India
e-mail: suman@mech.iitkgp.ernet.in

© Springer Nature Singapore Pte Ltd. 2021

U. S. Dixit and S. K. Dwivedy (eds.), *Mechanical Sciences*,
https://doi.org/10.1007/978-981-15-5712-5_7

Table 1 Vessels in human body and their typical sizes

Vessel	Size (mm)
Aorta	25
Artery	4
Vein	5
Vena Cava	30
Arteriole	0.03
Capillary	0.008
Venule	0.02

the necessity of a clear understanding of flow through such vessels. It is imperative for accurate clinical diagnosis and to determine the further course of many treatments. This signifies the importance of biofluid dynamics within the domain of biomedical science.

2 Origin and History of Biofluid Mechanics

The origin of biofluid dynamics dates to the time when people started to take interest in the circulatory system and its relation to the diseases. Many pioneers contributed to the field of biofluid mechanics, which include both scientists and engineers. Major contributions of some stalwarts are elucidated in the following subsections

2.1 William Harvey (1578–1657)

William Harvey was an English physician who was instrumental in the development of anatomy and physiology as we see it today. He is credited with the complete description of the circulation and properties of blood that is pumped by the heart. Before Harvey, ancient Greek physician Galen proposed that the arterial and venous systems were different and they came in contact through unseen pores. In 1628, Harvey published his seminal work titled "On the Motion of the Heart and Blood in Animals" in which he proved that arteries and veins are functionally connected to the lung and peripheral tissues (Fig. 1). He also explained the nature of blood circulation within the body using experimental evidence and deductive logic. Though faced with resistance from the fellow scientists of the time, his theories gained a gradual acceptance over years.

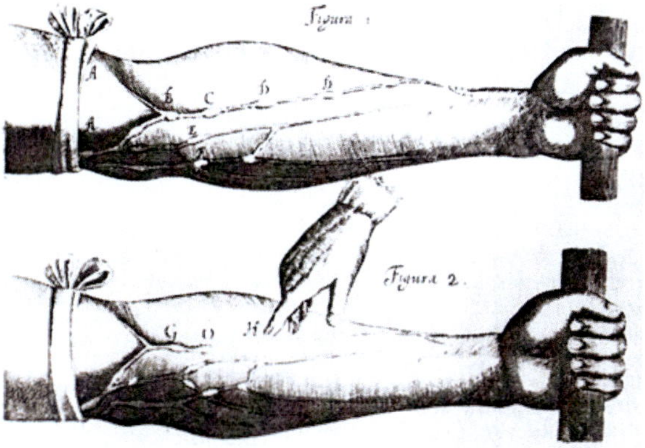

Fig. 1 Willaim Harvey's experiments to demonstrate the circulation of blood from his book published in 1628

2.2 *Jean Léonard Marie Poiseuille (1797–1869)*

Poiseuille was a French physician by training, but more popular in the field of fluid dynamics due to the well-known Hagen–Poiseuille equation named after him (Gotthilf Hagen (1797–1884) a German civil engineer, independently discovered the same relation). With a curiosity toward the nature of blood flow in the narrow vessels of the human body, he designed his own experiments to characterize the basic quantities in circular tubes with remarkable precision. He characterized the pressure and flowrate for laminar flows and published his result in 1846, thereby introducing the simple formula

$$\Delta P = \frac{8\mu L Q}{\pi R^4}. \tag{1}$$

Here, ΔP is the pressure loss, L is the length of pipe, μ is the dynamic viscosity, Q is the volumetric flow rate, and R is the radius of the tube. He is also credited for the invention of U-tube mercury manometer, which he utilized to measure the blood pressure in horses and dogs (Fig. 2). The modified version of this came to be known as hemodynamometer which quickly became popular among physiologists and is considered as a key milestone in the history of biofluid mechanics with significant clinical implications.

2.3 *John R. Womersley (1907–1958)*

Womersley was a British mathematician and one of the earliest computer scientists. He served the British government on a number of practical problems with his keen applied mathematical skills, especially during the Second World War. Later, he joined one of the pioneering biofluid dynamics groups led by physiologist McDonald

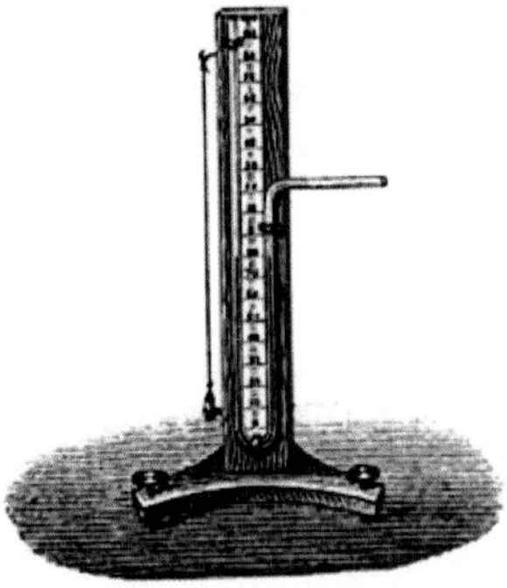

Fig. 2 Mercury manometer designed by Poiseuille

at St Bartholomew's Hospital in London which led to the foundational theoretical work on pulsatile flow in blood vessels. His model predicted the velocity profiles at different phase angles and was published in 1955. The non-dimensional parameter which characterizes the nature of unsteady flow was subsequently renamed as the Womersley number (Wo). It is the time-varying counterpart of Reynolds number in a steady flow and is given by

$$\mathrm{Wo} = R\sqrt{\frac{\rho\omega}{\mu}}. \tag{2}$$

Here, ω is the frequency of pulsation and ρ is the density of the fluid. Wo denotes the ratio of oscillatory inertia force to the shear force. At a lower Wo, the oscillations are slow and the flow becomes momentarily fully developed during the oscillation cycle. At higher Wo, the flow is slower at the central portion and the fluid flows as in a plug flow as shown in Fig. 3. The analytical form of the velocity profile in cylindrical coordinates (r, θ, z, t) is given by the following equation:

$$u(r,t) = \sum_{n=-N}^{N} \frac{iP_n}{\rho n\omega}\left[1 - \frac{J_0\left(\mathrm{Wo}\frac{r}{R}n^{\frac{1}{2}}i^{\frac{3}{2}}\right)}{J_0\left(\mathrm{Wo}\,n^{\frac{1}{2}}i^{\frac{3}{2}}\right)}\right]\mathrm{e}^{in\omega t} \tag{3}$$

where J_0 represents the Bessel function of zeroth order and P_n is the coefficient of periodic pressure function.

2.4 *Otto Frank (1865–1944)*

Otto Frank was a German physiologist credited for providing the mathematical foundation for one of the most important characteristics of arterial mechanics—the Windkessel effect. It is loosely translated as an air chamber and has the idea of an elastic reservoir which can store the fluid temporarily, analogous to a spring or flywheel in terms of storing energy in a cyclic operation. It is responsible for damping the

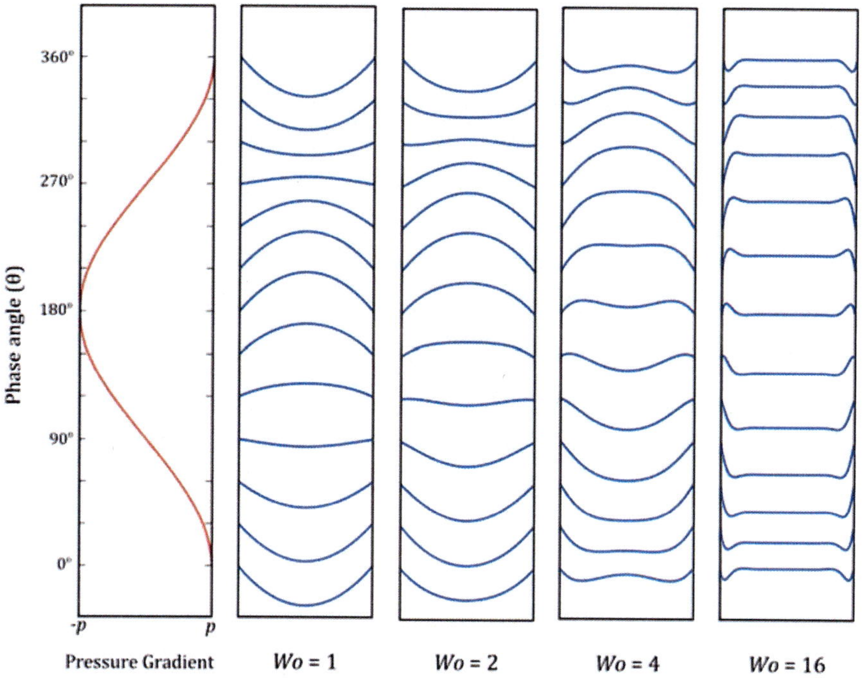

Fig. 3 Velocity profile in unsteady flow in a tube at different phase angles and Womersley numbers

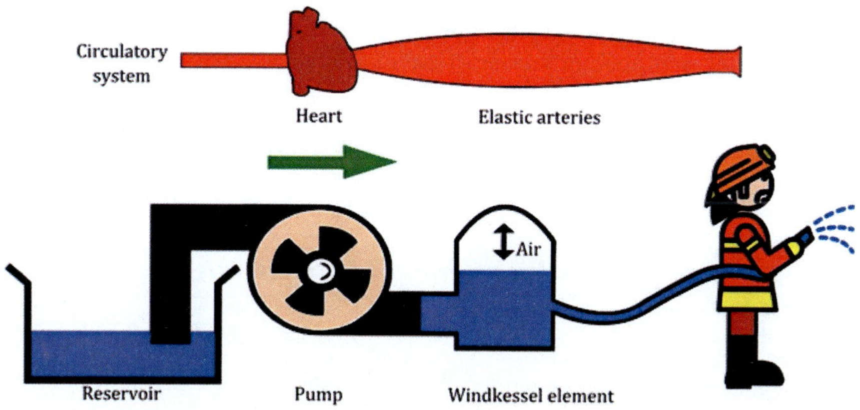

Fig. 4 The pumping of blood from the heart to arteries analogous to the Windkessel effect

fluctuations in blood pressure during the cardiac cycle and helps to maintain organ perfusion in the human body. The effect explains the interaction between the stroke volume and the compliance of arteries. Figure 4 shows the Windkessel mechanism and the analogous physiological system in the body.

3 State of the Art—Biomicrofluidics

Modern technology has embraced miniaturization in the last century as indicated by the proposition from Richard Feynman: "There's plenty of room at the bottom" during his lecture at Caltech in 1959. Like any other field of science and engineering, fluid mechanics too pushed its limits to explore the potential at the micro- and nanoscales and resulted in the inception of microfluidics and nanofluidics. With the advancements in characterization and probing techniques at small scales, it is now possible to manipulate the materials at nano level or even less. Following sections deal with the important developments in the area microengineering at small scales, pertinent to the study of biofluid mechanics.

3.1 Microfluidics and Lab-on-Chip Technology

In the past few decades, microfluidics has emerged as an extensive and multidisciplinary area of research. The rise in technological advancements, especially in the field of electronics and instrumentation overlapped interests with many other fields including the biomedical sciences. This resulted in the inception of biomedical engineering to cater the needs of physicians using engineering techniques including diagnosis, vital monitoring, support systems, and therapeutic devices. The contribution of biofluid dynamics to biomedical engineering is noteworthy as many critical appliances like blood pumps and dialyzing units work on the principles of fluid dynamics. Microfluidics came handy in the development of diagnostics devices and methods with huge advantages during sensing, sample requirement, time of detection, and the affordability to the masses. Figure 5 indicates the salient aspects of microfluidics that are directly related to biomedical engineering, manifesting its cross-disciplinary nature.

3.2 Microfabrication—Emergence of Lithographic Techniques

Microchannel fabrication is one of the most important aspectsof microfluidics. The capability of fabricating very small channels limited the studies until a few decades back. Though conventional machining methods like milling have been employed to create channels of a few hundred micron size, soft lithography has emerged as the most popular choice to fabricate channels up to nanometer dimensions. It is called 'soft' because of the elastomeric materials used in the fabrication process. The soft lithography follows a molding technique similar to that found in a foundry

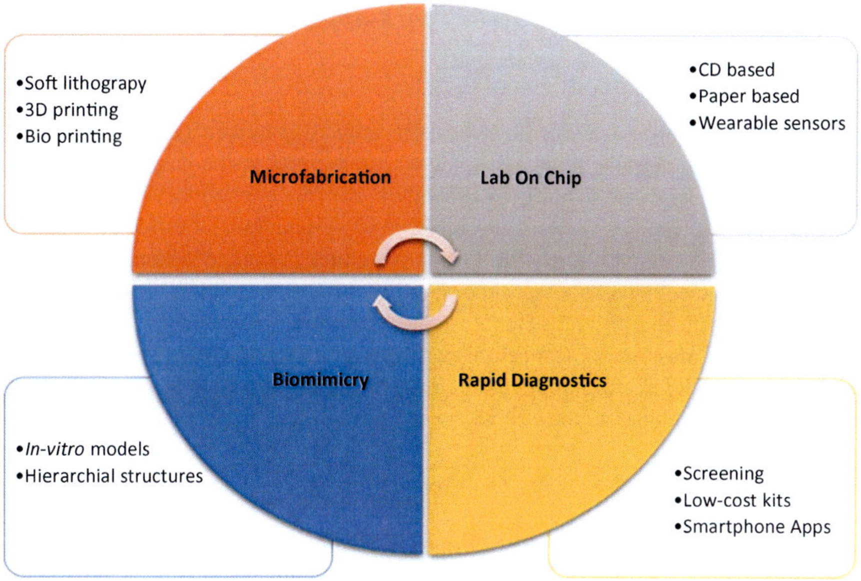

Fig. 5 Aspects of microfluidics for biomedical applications

workshop. The most commonly used material for casting is PDMS (Polydimethylsiloxane) which is a silicone elastomer, which is popular among the scientists due to its biocompatibility, transparency, low cost, and ease of usage. The preceding step for making the mold is called photolithography and is an offshoot from lithographic fabrication techniques of p–n junction in the semiconductor industry. There are now variants of soft lithography for more sophisticated requirements like Nanolithography, Microcontact printing, and Multilayer soft lithography. Due to the use of elastomeric materials, microchannels can easily be tuned for its deformability. A detailed characterization of the material, including the mechanical chemical and surface properties can be used in theoretical modeling of the flows and as parameters in numerical simulations.

Generally, the geometry of the channels fabricated using soft lithography is rectangular, but many other shapes and multidimensional structures can also be fabricated using stereolithographic techniques. As the biological conduits are cylindrical in shape, cylindrical molds are employed to fabricate channels of the circular cross section. However, they can pose problems during imaging due to refractive index mismatches between the channel material and the flowing fluid. A common issue as the channels get smaller is the collapse of the vessel when the walls are too soft. This failure of proper replication using the lithography technique compromises the structural integrity and thus the function of the model to be fabricated.

3.3 The Electronic Analogy—Programmable Microfluidic Networks

One of the fascinating aspects of flow-through channels is its direct analogy with the electronic circuit components which is extremely helpful in the design and prototyping of microfluidic circuits. Table 2 shows some of the key analogical components. When having multiple operations and material to be handled, an efficient network of microchannels are required with an aim to reduce the input power in order to increase the efficiency and the resistances involved in the path. Optimizing the analogous circuit will help the fabricators to choose among the tradeoffs in the viable production at a mass scale. Capacitor as an analogue to a deformable membrane has led to many interesting applications including flexible sensors. The following derivation will demonstrate the effectiveness of this simple technique to model flow in a deformable vessel.

For a deformable channel flow, the mass flow rate is not constant with time within the channel. Here, the pressure can cause the wall to either expand or contract. This will change the diameter of the channel as shown in the figure and thus the Hagen–Poiseuille equation for a constant diameter is not valid here. Consider an incompressible, unidirectional flow; the volume change can be expressed as

$$\frac{\mathrm{d}V}{\mathrm{d}t} = Q_{\mathrm{i}} - Q_{\mathrm{o}}. \tag{4}$$

Table 2 Electronic components and their hydraulic analogy in a fluidic network

Circuit Element		Electronic Analogy	Hydraulic analogy
Node	●	Junction	Junction
Path	—	Wire	Rigid pipe
Resistance		Resistor	Constriction
Compliance	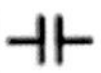	Capacitor	Diaphragm
Inertance		Inductor	Paddlewheel
Valve		Diode	Check valve
Pressure source	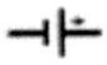	Voltage source	Pump
Flow source		Current source	Pump

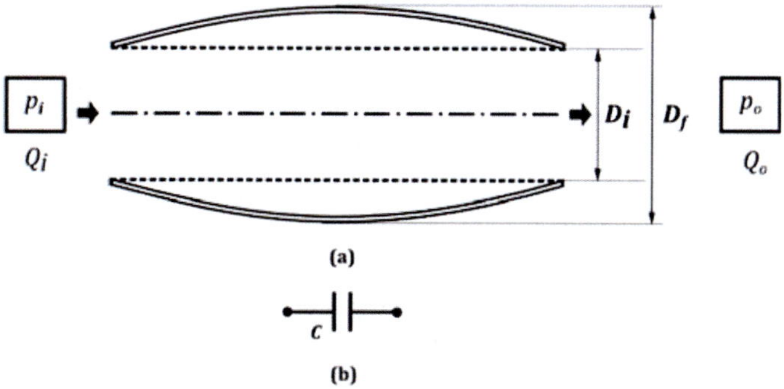

Fig. 6 **a** Schematic representation of deformable channel which deformed underflow with a pressure gradient p_o–p_i from an initial diameter $D_{\rm i}$ to final diameter $D_{\rm f}$. **b** Analogous capacitor element

Here, $Q_{\rm i}$ is the inlet and $Q_{\rm o}$ is the outlet flowrate. For a rigid tube, $\frac{{\rm d}V}{{\rm d}t} = 0$. It is evident that the deformed tube can store more fluid and this ability is termed as the compliance (C) with a unit of m^3/Pa. It is the effective change in volume per change in pressure expressed as

$$C = \frac{{\rm d}V}{{\rm d}p} \tag{5}$$

Now, the effective rate of the volumetric change for stored fluid is

$$Q_{\rm s} = Q_{\rm i} - Q_{\rm o} = \frac{{\rm d}V}{{\rm d}t} = \frac{{\rm d}V}{{\rm d}p}\frac{{\rm d}p}{{\rm d}t} = C\frac{{\rm d}p}{{\rm d}t} \tag{6}$$

Analogous to the capacitance shown in Fig. 6, we can write the flowrate–pressure relationship for a one-dimensional case as

$$Q = C\frac{{\rm d}(\Delta P)}{{\rm d}t} \tag{7}$$

where $\Delta P = p_0 - p_i$.

In a general sense, both compressibility of the liquid and the wall deformability contributes to the compliance. However, in the context of physiologically relevant fluids, compressibility plays no perceivable role unless there is a presence of gases in the form of bubbles.

The aforementioned formulation presents a convenient first-hand tool though a practical approach includes a complete impedance analysis to estimate the quantities of interest. For a network of such channels, tools of circuit analysis like Kirchhoff's law and Thevenin's theorem can be used. For the most relevant case of time-varying flows, the telegrapher's equation can be reduced to a one-dimensional diffusion to model the diffusive spreading of the pressure waves in a deformable channel with

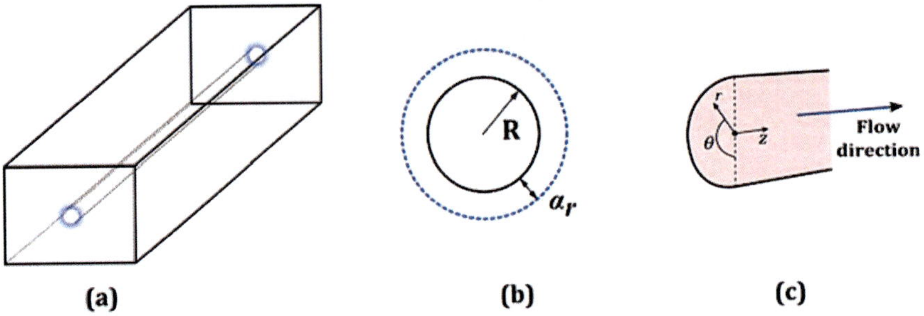

Fig. 7 **a** Rectangular domain housing a cylindrical channel. **b** Cross section showing radial deformation (α_r). **c** Cylindrical coordinates (r, θ, z) and the direction of flow

the aid of Fourier analysis. Much like a resonance in an RLC circuit, resonating microfluidic channels can be employed as a switching mechanism in a lab-on-a-chip platform, thus making it easy to automate and control remotely.

4 Some Theoretical Aspects and Governing Equations of Deformable Channels

From a theoretical to model the physiological fluid flow in vessels, this is formidable since most of the problems that both fluid dynamicists and engineers deal with models consist of rigid tubes with negligible wall deformation. Here, the velocity of the flow affects the wall deformation and vice versa. This coupled phenomenon is termed as fluid–structure interaction (FSI) and solving such problems requires special attention. Theoretical studies start with a non-deformable case and gradually add the complexities associated with the wall deformation. While many of these can be difficult to solve analytically due to the nature of equations, many simple formulations and scaling arguments can predict the outcomes of experiments, thereby understanding the elementary context of the problem. It should be noted that there are far more complex models accounting for the pulsatile flow in deformable channels. Readers are directed to one of the foundational texts on this subject by Fung [1] for more detailed topics and an online course (www.coursera.org/learn/fluid-solid-interaction) for a visual exploration of this otherwise complex phenomenon. The following two derivations discuss the two simple geometries that include the deformability of the wall. Both of them are experimentally verified in their respective geometries.

4.1 Cylindrical Channel [2]

Figure 7 shows the schematic of the microchannel of radius R in a rectangular block of dimensions L, W, and H. Note that our configuration is different from a thin-walled tube configuration for which there is a well-established relationship for the deformation. The mechanical equilibrium equations are given by

$$\nabla \cdot \sigma = 0, \tag{8}$$

where σ is the stress tensor. As the channel is axisymmetric and slender, the radial component of the mechanical equilibrium equations reduces to

$$\frac{1}{r}\frac{\partial}{\partial r}(r\sigma_{rr}) - \frac{\sigma_{\theta\theta}}{r} = 0, \tag{9}$$

where σ_{rr} and $\sigma_{\theta\theta}$ are the normal components of the stress tensor in the radial and the angular directions, respectively. Note that the term involving σ_{rz} scales out because we assume zero axial displacement and because of the smallness of the radial displacement compared to the characteristic axial length scale. These reductions are equivalent to the lubrication approximation for fluid flow through narrow confinements. For the solid material, we use linear, elastic, isotropic behavior given by

$$\sigma = \lambda(\nabla \cdot \alpha)\mathrm{I} + 2G\left\{\frac{1}{2}(\nabla\alpha) + (\nabla\alpha)^T\right\} \tag{10}$$

where λ and G are the Lamé parameters, I is the identity tensor, and α is the displacement vector in the cylindrical coordinates given by $\alpha = \begin{bmatrix}\alpha_r & \alpha_\theta & \alpha_z\end{bmatrix}^T$. Using the same considerations that were used to reduce the mechanical equilibrium equations, we obtain

$$\sigma_{rr} = \lambda\frac{1}{r}\frac{\partial}{\partial r}(r\alpha_r) + 2G\frac{\partial\alpha_r}{\partial r} \tag{11}$$

$$\sigma_{\theta\theta} = \lambda\frac{1}{r}\frac{\partial}{\partial r}(r\alpha_r) + 2G\frac{\alpha_r}{r}. \tag{12}$$

We use Eqs. 11 and 12 in Eq. 9 to obtain the equation that governs the displacement along the radial direction as

$$\frac{1}{r}\frac{\partial}{\partial r}\left[r\left\{\lambda\frac{1}{r}\frac{\partial}{\partial r}(r\alpha_r) + 2G\frac{\partial\alpha_r}{\partial r}\right\}\right] - \frac{1}{r}\left[\lambda\frac{1}{r}\frac{\partial}{\partial r}(r\alpha_r) + 2G\frac{\alpha_r}{r}\right] = 0. \tag{13}$$

For the boundary conditions, we first require that the radial displacement should not be infinite even for extremely large r, thus

$$\text{as } r \to \infty, \alpha_r \text{ is finite.} \tag{14}$$

Secondly, we require that at the wall (where the solid material interfaces with the liquid), the normal stress in the radial direction is dictated by the fluid pressure; thus

$$\text{at } r = R, \;\; \sigma_{rr} = -p(z) \tag{15}$$

where, importantly, the fluid pressure (p) varies along the axial (z) direction. Solving Eq. 10 subject to the boundary conditions (14) and (15), we obtain

$$\alpha_r(z) = \frac{p(z)R^2}{4G}\frac{1}{r}. \tag{16}$$

If the pressure gradient is assumed to be constant along z, the deformation is also linear in z.

Finally, we arrive at the expression for the wall deformation (at $r = R$) along the length of the channel with p_1 and p_2 as the inlet and the outlet pressure of the channel.

$$\alpha_r(z) = \frac{\left\{p_1 + \left(\frac{p_2 - p_1}{L}\right)\right\}zR}{4G}. \tag{17}$$

The velocity field will remain same as the parabolic profile according to the Hagen–Poiseuille equation with the new deformed diameter $D + 2\alpha$. The magnitude of the average velocity increase along the length of the channel as the deformation dies out.

4.2 Rectangular Channel [3]

Figure 8 shows the schematic of the microchannel of width w and undeformed height h_0. Assuming steady, incompressible, laminar, and Newtonian flow, the Navier–Stokes equation for the system can be written as

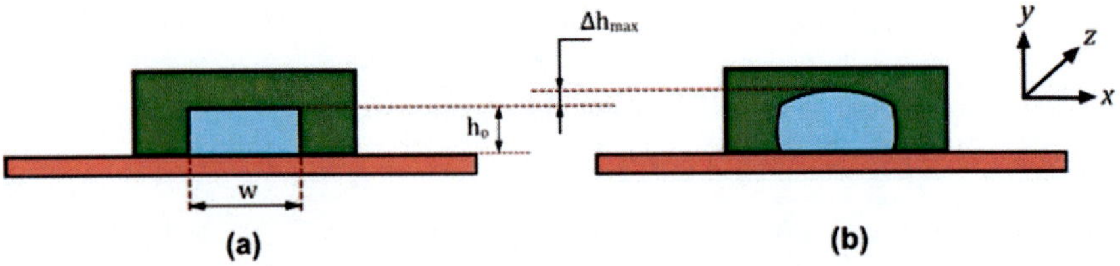

Fig. 8 Rectangular channel deformation. **a** Before deformation with a width w and height h_0. **b** In the deformed state with a maximum deformation Δh_{max}

$$\rho\left(\frac{\partial u}{\partial t} + u \cdot \nabla u\right) = -\nabla p + \mu \nabla^2 u \tag{18}$$

Assuming the flow to be fully developed, a much simple solution to the equation for the case of rectangular channel geometry is given by

$$\Delta p = Q\frac{12\mu L}{h_0^3 w}\left[1 - 0.630\frac{h_0}{w}\right]^{-1} \tag{19}$$

It is evident that the pressure drop in a channel with $w \gg h_0$ scales inversely with the third power of the channel height h_0. This drives the strong coupling between the flow solution and the structural deformation of the channel walls. The imposed flow will cause the three walls to deform with a positive change in the cross-sectional area, which in turn modifies the local velocity and pressure distribution. This again drives the deformation of the channel and essentially becoming an FSI paradigm. The deformation of the top wall is proportional to the channel width and not to the channel height. Thus, we get

$$\Delta h_{\max} = \frac{3}{2}\frac{\beta w}{E}p \tag{20}$$

Assuming a parabolic deflection of the top wall, the effective channel height is obtained as

$$h(z) = h_0\left(1 + \frac{2}{3}\frac{\Delta h_{\max}}{h_0}\right) \tag{21}$$

Using the relation from Eq. 20, the deformation at any axial location z can be computed by

$$h(z) = h_0\left(1 + \beta\frac{p(z)w}{Eh_0}\right) \tag{22}$$

With this simple formulation of the wall deformation, the Δp–Q relationship and the velocity profile along follows as

$$Q = \frac{h_0^4 E}{48\alpha\mu(L - z)}\left(\left(1 + \beta\frac{p(z)w}{Eh_0}\right)^4 - 1\right) \tag{23}$$

$$u(z) = \frac{Q}{wh_0}\left(\frac{48\beta\mu(L - z)Q}{h_0^4 E} + 1\right)^{-\left(\frac{1}{4}\right)} \tag{24}$$

In the above derivations, β is called the dimensionless deformation parameter which can only be calculated through fluid–structure computations but is approximately constant for a given channel geometry.

4.3 Special Case of Non-Newtonian Fluids

To realize the full potential of mimicking the blood vessels, it is important to consider the fluid and its constituents. As it is well established in the scientific literature, the whole blood consists of a myriad of components like blood cells and platelets and thus cannot be simply approximated as a Newtonian fluid. These components can alter the very nature of the flow itself. In blood, this is mostly dictated by RBCs, which forms the major portion of the volume fraction (also known as the Hematocrit). As an elementary analysis, blood is considered as a shear-thinning liquid where the apparent viscosity is expressed as a function of the applied shear rate as $\mu = m(\dot{\gamma})^{n-1}$ where $m = 3$–4 mPas and $n = 0.5$–0.8. Note that, in a cylindrical channel, it modifies the Hagen–Poiseuille equation as

$$Q = \frac{\pi R^3}{\frac{1}{n} + 3}\left(\frac{\Delta P}{2mL} R\right)^{\frac{1}{n}} \tag{25}$$

For the velocity profile, the relation takes the following form:

$$u(r) = \left(\frac{\Delta P_{\text{exp}}}{2mL} R\right)^{\frac{1}{n}} \frac{R}{\frac{1}{n} + 1}\left(1 - \left(\frac{r}{R}\right)^{\frac{1}{n}+1}\right) \tag{26}$$

and essentially makes the profile a blunt one compared to that of a Newtonian flow, thereby affecting both the maximum and average velocity. For $n = 1$ and $m = \mu$, the original Hagen–Poiseuille equation is recovered.

5 Experimental Techniques for Flow Investigation in Deformable Channels

One of the notable earliest experimental models to study the biofluid mechanics within deformable vessels is the Starling resistor. It was invented by English physiologist Ernest Starling and consists of a fluid-filled elastic tube mounted inside a chamber filled with air which has provision to expand and collapse. The static pressure inside the chamber (p_e) is used to control the degree of expansion and collapse of the tube, thereby acting like a variable resistor with a potential difference $p_2 - p_1$ (Fig. 9). This resistance can be used to simulate the total peripheral resistance (TPR) in vascular flow. Apart from a model in the study of a range of physiological phenomena (e.g., collapse of the pharynx during breathing difficulties or obstructive sleep apnea), it is also widely employed as a source of rich physical phenomena by itself due to highly non-linear characteristics while in operation. Two such non-linear characteristic behaviors are the "waterfall effect" in which, subsequent to collapse,

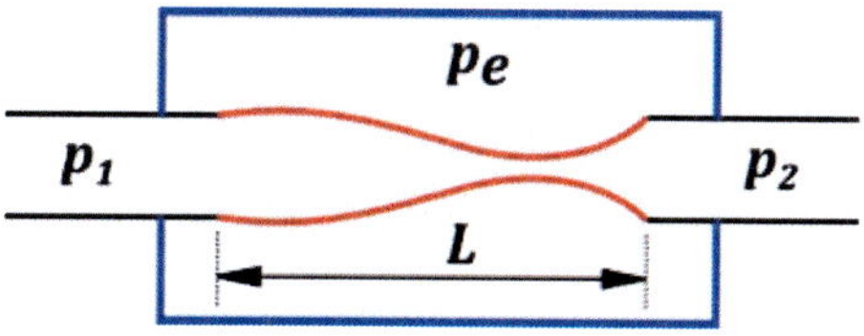

Fig. 9 Starling resistor with a chamber pressure of p_e

the flow through the tube becomes independent of the downstream pressure and the self-excited oscillations, both having pronounced implications in the domain of FSI.

5.1 Modern Measurement and Analysis Techniques

With the advancement of technological innovations in instrumentation and imaging, we have evolved far from pitot tubes for measuring pressure and dyes for visualization. Today, Particle Image Velocimetry (PIV) techniques are used for whole-field visualization, while pressure drop experiments using piezoelectric sensors and edge detection techniques based on image processing are employed to track the wall deformation. Figure 10 shows a typical setup to study a time-varying flow in deformable channels. To achieve steady and time-varying flow, a positive displacement pump like a syringe pump and a solenoid pinch-off valve is used. The syringe pump gives a steady fluid flow by mechanically pushing the plunger of a fluid-filled syringe at a constant rate. The tube that connects the microchannel inlet and the syringe pump has a small section of flexible tubing which is inserted into the solenoid pinch-off valve. Due to the plunging action, the tubing is pinched momentarily, producing a pulsed flow in the circuit. The plunger in the valve is actuated using a specific peak-to-peak

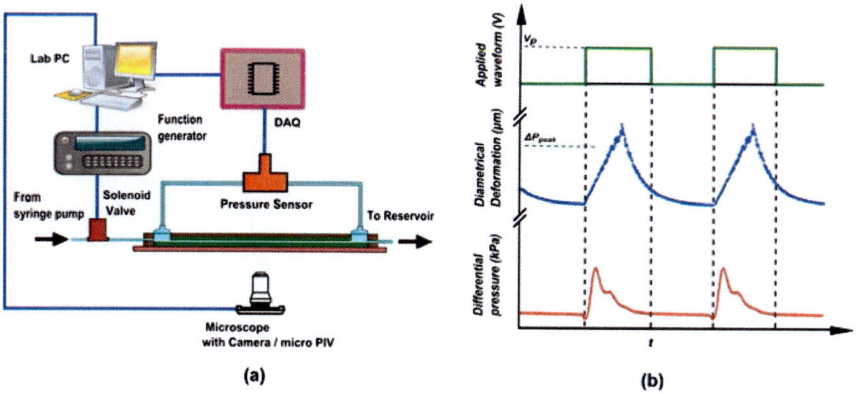

Fig. 10 **a** Basic experimental setup for the experiment to investigate time-varying flow in deformable channels. **b** Waveforms showing the applied voltage (V_p) in the function generator, deformation and pressure profile over two cycles

voltage (limited by the inertia of the solenoid valve) waveform of a specific duty cycle with the desired frequency sourced from a function generator.

The differential pressure drop across the microchannel length can be measured using a differential pressure sensor connected to a data acquisition system (DAQ) which is interfaced with a computer. The DAQ continuously acquires the raw voltage from the pressure sensor and is converted to physical units (Pa) based on the instrument calibration chart. Usually, a high sampling rate is employed in the DAQ which is matched with the response time of the pressure sensor.

Estimation of the wall deformation is the most sensitive measurement in such a study. A three-dimensional scanning technique like the confocal microscope is ideal for the deformation measurement in usual rectangular glass-PDMS microchannels [3]. Further, indirect methods utilizing the fluorescent microscopy and dyed fluid are also used for estimating the deformation of the top wall [4]. However, in a cylindrical channel, due to the axisymmetric nature of the wall, deformation can be observed directly under the microscope with a simple phase-contrast microscopy. Standard edge detection algorithms are applied to the captured images that will give an accurate location of the wall position. A high-speed camera is used to capture the real time will deformation using high magnification zoom lenses to focus near the wall.

For flow visualization, PIV apparatus is utilized. Fluorescent seeding particles like polystyrene beads of 1–10 μm nominal diameters are used as tracers which are neutrally buoyant and faithfully follow the streamlines in the flow. The synchronizer of the PIV is synced with the pressure sensor with Transistor–Transistor Logic (TTL) signal from a computer for simultaneous measurement of pressure drop and the velocity field. An iterative multi-pass cross-correlation algorithm using a Fast Fourier Transform (FFT) is employed to evaluate the final velocity field. For steady flow, ensembles averaging of a large number of pairs are used and for the pulsatile flow, phase averaging over multiple cycles is analyzed. Representative images for micro-PIV and deformation analysis are given in Fig. 11. All the measurements for pressure drop, deformation, and micro-PIV are recorded after reaching a steady state.

6 Numerical Techniques to Solve FSI Problems

The origin and development of FSI lie in the ocean engineering and aerospace engineering where large-scale structures are subjected to flow of water and air. It is also crucial in understanding the design considerations of many structures like bridges and skyscrapers. Failing to foresee the oscillatory interactions with high-speed wind can be catastrophic for these structures which are subjected to fatigue (e.g., the tragic failure of Tacoma Bridge in 1940). The equations that define this phenomenon are too complex in general to solve analytically and experiments and numerical solutions are the only way out. It can be perceived as a merger between Computational Fluid Dynamics (CFD) techniques and computational structural dynamics.

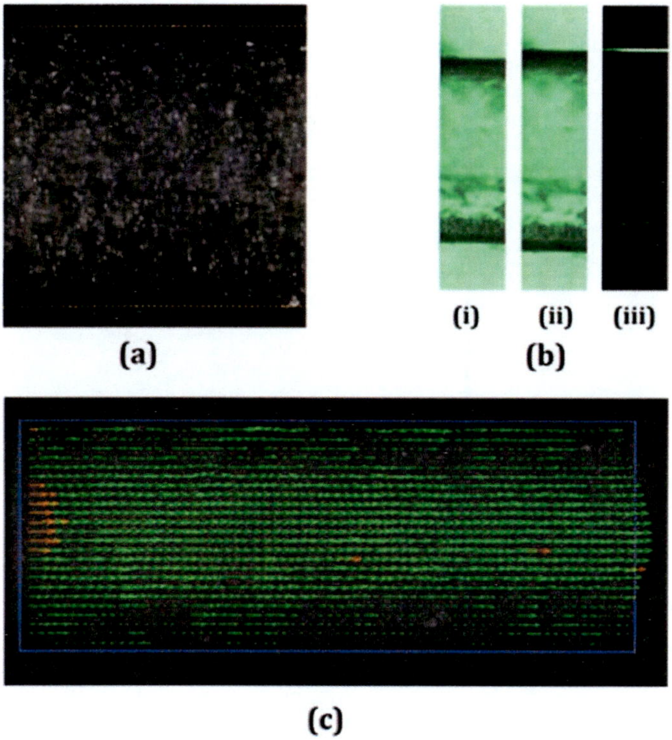

Fig. 11 Typical PIV and deformation analysis. **a** Fluorescently labeled particles flowing through the channel. **b** (i) Initial undeformed state (ii) Deformed state (iii) Edge detection applied to detect the wall position. **c** PIV analysis showing the velocity vector field

The governing equations are represented using a combination of Eulerian and Lagrangian approaches. The computational domain consists of χ_{f} as the fluid domain and χ_{s} as the solid domain Γ representing the fluid–solid interface as shown in Fig. 12. The superscripts s and f represent solid and fluid domains, respectively. Using D'Alembert's principle,

$$\rho\dot{v}_i - \sigma_{ij,j} + b_i = 0 \tag{27}$$

Here, v is the velocity field, σ is the stress tensor, and b is the body force, usually the gravity. For brevity, the Einstein summation convention is followed. Now, the fluid domain is represented by superscript f and the structural part is represented by s. For the fluid domain:

$$\rho^f\dot{v}_i^f - \sigma_{ij,j}^f + b_i^f = 0 \,\text{in}\, \chi_{\mathrm{f}}. \tag{28}$$

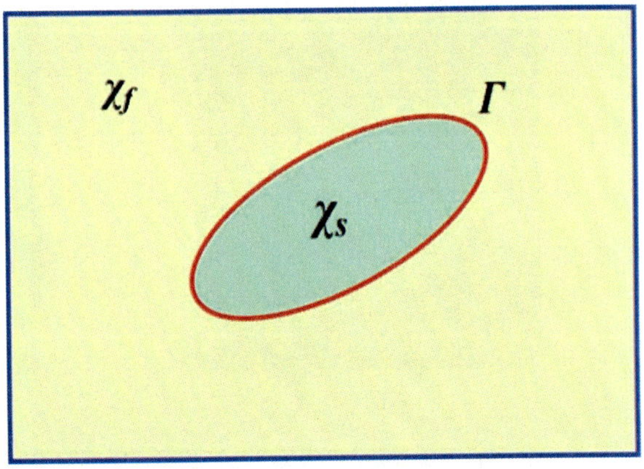

Fig. 12 Domain for FSI for a solid body within a fluid domain with an interface Γ

It is represented by the Eulerian formulation as in conventional fluid mechanics and in the first term representing the inertia, the velocity field is given by

$$\dot{v}_i^f = \frac{dv_i^f}{dt} = \frac{\partial v_i^f}{\partial t} + v_j^f v_{i,j}^f \tag{29}$$

For a Newtonian fluid with incompressibility condition, the fluid stress is given by

$$\sigma_{ij}^f = -p\delta_{ij} + \tau_{ij} \tag{30}$$

where the stress tensor is given by

$$\tau_{ij} = 2\mu\left(e_{ij} - \frac{\delta_{ij}e_{kk}}{3}\right), \text{ where } e_{ij} = \left(v_{j,i}^f + v_{i,j}^f\right) \tag{31}$$

Here, p is the pressure acts as the necessary enforcing condition to maintain the incompressibility of the fluid, i.e., $v_{i,i}^f = 0$ and δ is the Kronecker delta function.

Now for the structural part, the governing equation is the same

$$\rho^s \dot{v}_i - \sigma_{ij,j}^s + b_i^s = 0 \text{ in } \chi_s \tag{32}$$

Here, the velocity $\dot{v}_i^s$ is the total (or material) derivative of the displacement field u_i^s or $v_i^s = \dot{u}_i^s$. As conventional with the solid mechanics, Eq. 32 is represented in a Lagrangian framework in which the first term represents inertia and the second the internal stresses. For a simple linearly elastic material, as discussed in Sect. 4.1, the stresses follow the Hooke's law and thus a function of the strains (ϵ) and the Lamé parameters λ and G are given by

$$\sigma_{i,j}^s = \lambda\delta_{ij}\epsilon_{ll} + 2G\epsilon_{ij} \tag{33}$$

$$\epsilon_{ij} = \frac{1}{2}\left(u_{i,j} + u_{j,i}\right) \tag{34}$$

$$G = \frac{E}{2(1+\nu)} \tag{35}$$

$$\lambda = \frac{E\nu}{(1+\nu)(1-2\nu)} \tag{36}$$

Here, E is the Young's modulus and ν is the Poisson ratio. The usual no-slip boundary condition is imposed as the following Dirichlet condition:

$$v_i^s = v_i^f, \text{ on } \Gamma \tag{37}$$

The other condition is based on the fact that the displacement condition is same for both fields, i.e.,

$$x_i^s = x_i^f \text{ on } \Gamma \tag{38}$$

Differentiating the above equation yields the Neumann boundary condition given by

$$\sigma_{ij}^s n_i = \sigma_{ij}^f n_i, \text{ on } \Gamma \tag{39}$$

Depending on the smoothness of the spatial and temporal domains, Eq. 38 is sometimes preferred over Eq. 37 as the Dirichlet condition by some FSI methods.

In terms of the types of solvers used for solving FSI problems, there are two main approaches: monolithic and partitioned. The former tries to solve the governing flow equation and the displacement of the structure simultaneously with a single solver, while the latter solves them separately with two distinct solvers. Further, in terms of the mesh treatment in FSI problems, there are body conforming mesh methods and non-conforming mesh methods depending on the accommodation of interface (Γ) within the domain as shown in Fig. 13 for the perimeter of a circular body. The operation of these methods in turn depends on how the boundary conditions (Eqs. 37–39) are enforced. The conforming mesh methods require mesh updates in each time step since it tracks the motion of the interface and explicitly enforce Eqs. 38 and 39 on Γ. This is convenient for the partitioned approach. The most popular non-conforming mesh method is the Immersed Boundary Method (IBM) which enforces the Dirichlet condition Eq. 37. They are computationally inexpensive and gaining a wider audience in recent years. There are more advanced methods that combine both IBM and LBM (Lattice Boltzmann Method) that exploit the parallel computing platforms to find solutions in faster ways.

Though the mathematical implementations of FSI formulations are extremely involved, there are a number of CFD packages that makes the life easier for practicing scientist and engineers. Nowadays it is very common in the industry to generate data

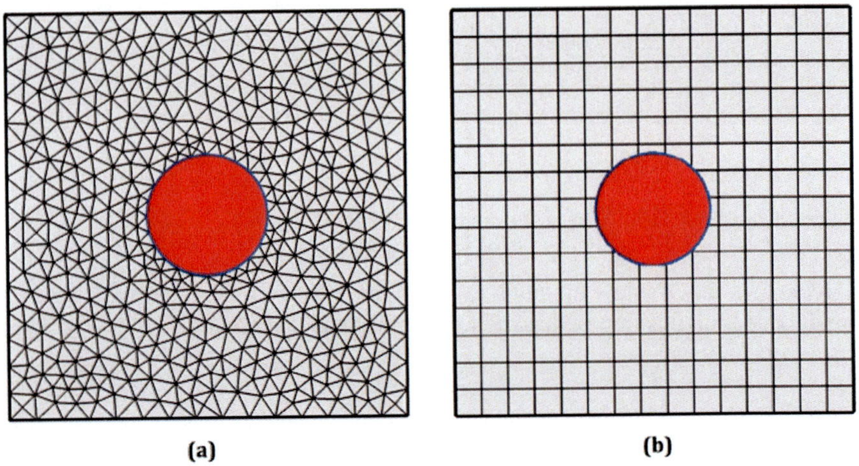

Fig. 13 Types of meshes used in FSI studies. **a** Body conforming. **b** Non conforming

from numerical simulations as the first step in the design of a new product. One of the most fascinating aspects of numerical simulations from a practical point of view is that it can give clues about the impending failures about a prototype model before it is set to the assembly line for actual production, without any sophisticated testing procedures. The emergence of clusters and supercomputers and its availability to a wider scientific community has led to many breakthrough results. A plethora of commercial or open-source packages is available for this which serves the researchers and industry professionals alike. However, many scientists consider this as a 'black-box' approach that impairs the physical insight into the problem. Hence, it is always advisable for beginners to first follow the theoretical derivations and to implement it numerically in a standard programming language like C or FORTRAN. For advanced numerical computations, there are numerous commercial packages like ANSYS and COMSOL though open-source packages like openFOAM attract a big audience in both academia and industry.

7 Future Directions

The prospects of microfluidics for engineering applications are very bright. It embraces the power to mitigate many real-life issues, especially in the realm of biomedical and pharmaceutical sciences. Apart from a purely translational perspective, it is a powerful tool to investigate fundamental problems in fluid mechanics as well. The following sections summarize some of the potential future directions in which fluid dynamics at the small scale will play a critical role.

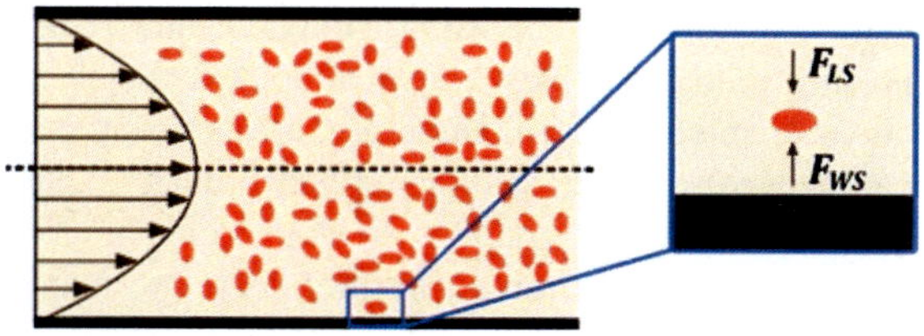

Fig. 14 Forces pertinent to inertial microfluidics for a flow involving RBCs

7.1 RBC Dynamics in Deformable Microchannels

For the ease of analysis, blood is considered as a homogenous fluid, without considering the effects of individual components like RBCs to the flow dynamics. The forces acting on particles in flow have been studied separately under the domain of microhydrodynamics for many industrial applications like emulsification. Considering the individual components in a biomicrofluidic flow leads us to an entirely different picture, now mostly dealt in the paradigm of inertial microfluidics. There are two competing forces acting on a particle like RBC in a confined flow, namely the wall lift force F_{LW} pushing it toward the center, and the shear-induced lift force F_{LS} that pushes it towards the wall as depicted in Fig. 14. Eventually, an equilibrium position is reached at a location of 0.6 times the radius of the channel once the focusing length is achieved, known as the Segré–Silberberg annuli. The scenario becomes more interesting if we consider the deformation of the wall. Now, lift forces can be altered by the presence of a deformable wall thereby affecting the particle motion along the transverse direction. This is attributed to the fact that streamlines around the particles are modified which may further be influenced by an increase in the particle concentration and when the rigidity of the particles itself is varied. This technique is useful in sorting, focusing, and selective trapping of cells and microparticles for Lab-on-Chip applications.

7.2 Elastocapillary Flows

Though not so manifested at the macro scales, surface effects, especially the surface tension becomes important as the length scale becomes smaller. Capillary forces generated by surface tension are associated with droplets, bubbles, and wetting and imbibition phenomena. They become dominant over the gravitational forces when the length scale reduces below the capillary length given by

$$L_{\mathrm{c}} = \sqrt{\frac{\gamma}{\rho g}} \tag{40}$$

where γ is the surface tension. However, under special conditions, they can also deform a solid object like a thin membrane and extremely soft gels. Interesting applications on micro- and millistructures are demonstrated to control the geometries, for example, in the case of capillary origami which skilfully employs the bending, buckling, and coiling deformations on simple membranes to generate unique structures without any external forces. Further, there are many micro- and nano-engineered biomimetic systems based on interesting phenomena in nature like hummingbird's tongue.

7.3 Wearable Sensors

Deformable microchannels constitute an integral constituent in the wearable sensors. They are fabricated by filling deformable microchannels with conductive liquids. With the soft lithographic techniques, lightweight and highly biocompatible backbones for sensors are developed which are seamlessly integrated with the organs. Upon acted by strain or any mechanical stimuli, they respond by transferring the output to the adjoining electronic circuitry which then decides the action to be taken based on the task at hand. They find extensive application in tactile sensing, fitness tracking, disease monitoring, and prosthesis which are increasingly becoming 'smart' with the integration of in-built computers. Needless to say, these are highly interdisciplinary in nature and experts from a myriad of backgrounds from clinical biology to electronics are involved in the development to work in a streamlined manner to achieve the end product to hit a competitive market like healthcare.

7.4 Artificial Organs and AI

The organ-on-chip concept is envisaged to do the functions of organs on tiny chips for diagnostic as well as therapeutic benefits which can be implanted in a body and monitored live. Advanced prototyping techniques like 3D bioprinting can aid in the fabrication of biocompatible implants and organ parts. The latest developments are in the bioinformatics and AI (Artificial Intelligence)-based healthcare monitoring utilizing big data analytics for patient information database to strategize the treatments specifically for each individual. Genetic mapping can tailor the medicines and targeted drug delivery will ensure its accurate delivery with minimum side effects. With the right information about the end-user, we can now generate the most suitable form of the organ for the subject. Especially in the case of patient data like Computed Tomography (CT) and Magnetic Resonance Imaging (MRI) scans, AI proved to a game-changer employing the tools from machine learning, big data analytics, fuzzy logic, evolutionary algorithms, and neural networks. The entire flow system in a body can be mapped accurately, thereby identifying the problems like blockage and stenosis. The AI engines coupled with the cloud computing platform running the previous cases in a global database can make judicious decisions, thereby providing better diagnostics than a human intervention with a low error margin.

8 Conclusions

Microfluidics offers a great advantage due to its inherent capabilities of addressing many fundamental issues that affect the biofluid mechanics in physiological conduits. In this chapter, an attempt is made to summarize some of these aspects starting with the history of biofluid mechanics to the state of the art of microfluidics that is finding its way to the customers, quicker than ever, or any other technology. With the advancement in microfabrication technology, one can fabricate micro- and nanochannels easily, whereas, with the power of computers, one can solve many physical problems using numerical simulation. A careful discussion is carried out for some of the elegant and simple mathematical models that form the basics of FSI phenomena. In the present world where the boundaries become thinner between nations as well as the different domains of science and technology, we can hope these tools will aid in the empowerment of humanity to have a healthy and productive life.

References

1. Fung, Y.: Biomechanics: Mechanical Properties of Living Tissues. Springer Science & Business Media (2013)
2. Raj, M.K., et al.: Flow-induced deformation in a microchannel with a non-Newtonian fluid. Biomicrofluidics **12**(3), 034116 (2018)
3. Gervais, T., et al.: Flow-induced deformation of shallow microfluidic channels. Lab Chip **6**(4), 500–507 (2006)
4. Raj, M.K., et al.: Hydrodynamics in deformable microchannels. Microfluid. Nanofluidics **21**(4), 70 (2017)

Further Reading

5. McDonald, D.A.: Blood Flow in Arteries. Williams & Wilkins, Baltimore (1974)
6. Chakraborty, S.: Mechanics Over Micro and Nano Scales. Springer Science & Business Media (2011)
7. Bruus, H.: Theoretical Microfluidics, vol. 18. Oxford University Press, Oxford (2008)

Current Status of the Development of Blood-Based Point-of-Care Microdevices

Vijai Laxmi, Siddhartha Tripathi, and Amit Agrawal

1 Introduction

Blood may be considered as a storehouse of health information of the human body and can be easily accessible from an individual. It is one of the most common fluids with applications in point-of-care testing. Testing of blood components is a common procedure employed for the diagnosis of diseases. Due to the presence of a large number of analytes or indicators, blood serves as an index to various diseases and testifies its importance in disease diagnostics and therapeutics. Blood is composed of cells and the liquid portion known as plasma; cells are 45–50% while plasma is 50–55% (v/v). The blood cells constitute red blood cell (RBCs) or erythrocytes, white blood cells (WBCs) or leukocytes, and platelets or thrombocytes [1, 2]. The constituents of blood are shown in Fig. 1. Each component of blood carries important analytes and information linked with human health condition. For example, blood plasma is required for a wide variety of clinical tests, such as in blood sugar and cholesterol tests while red blood cells, white blood cells, and platelets are required for the detection of malaria, human immunodeficiency virus (HIV), and cardiovascular diseases.

Conventional diagnostic techniques require specialized equipment and setups. Hospitals and clinics are also needed for clinical diagnostics of the specimen. In a common scenario, a patient consults a doctor during illness phase and the doctor recommends a diagnostic test (commonly blood based). Later, the patient visits a hospital or a clinic where blood is extracted from the patient employing venipuncture and the sample is sent to a centralized laboratory for further assessment. Finally, the results are conveyed to the doctor for further analysis and medication. Overall, this

V. Laxmi · A. Agrawal (✉)
Department of Mechanical Engineering, IIT Bombay, Powai, Mumbai 400076, India
e-mail: amit.agrawal@iitb.ac.in

S. Tripathi
Birla Institute of Technology and Science Pilani, Goa Campus, Goa 403726, India

© Springer Nature Singapore Pte Ltd. 2021
U. S. Dixit and S. K. Dwivedy (eds.), *Mechanical Sciences*,
https://doi.org/10.1007/978-981-15-5712-5_8

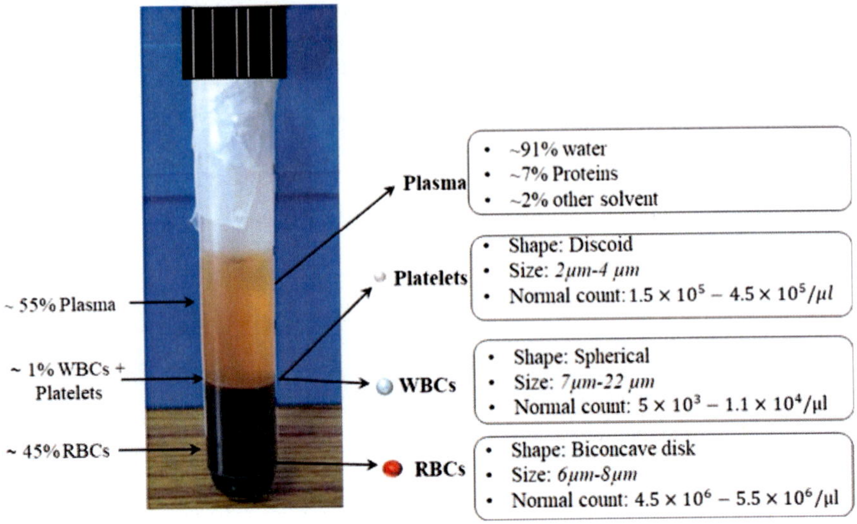

Fig. 1 Centrifuged blood sample representing plasma and blood cells

process is not only time consuming but it also delays the timely treatment. Also, in several emergency cases immediate and rapid decision-making is requisite to save the life of the patient. Therefore, the medical community in conjunction with engineers started looking for techniques to reduce the turn-around time and facilitate rapid detection of biomarkers or analytes at the site of the patient. This requirement gave birth to the concept of point-of-care testing (POCT) technology.

POCT enables diagnostics near the site of the patient. Using devices based on the point-of-care technology diagnostics and assessment is possible at patient's location in a short span of time. A schematic of conventional method of testing and POCT-based testing method is shown in Fig. 2. The major advantage with its usage is very early detection of diseases, monitoring of disease progression, portability, ease of operation, and interpretation of results. Further, the POCT devices are valuable in places where access to healthcare facilities is limited. The most common example of a point-of-care device is the handheld glucometer. A point-of-care device constitutes sample preparation and processing, mixing or separation, sensing, and detection of analytes and communication of data [3] integrated into a single entity, as shown in Fig. 2B.

Point-of-care devices can be classified into two categories: handheld devices and benchtop devices. Although both handheld devices and benchtop devices qualify as POCT devices, handheld devices have better maneuverability as they can be employed in doctor's cabin, emergency rooms, and at home. World health organization (WHO) recommends the ASSURED criteria for the ideal POC test: affordable, sensitive, specific, user-friendly, rapid and robust, equipment-free, and deliverable to users. Ideally, a point-of-care device should be minimally invasive which requires small amount of unprocessed blood sample directly from patients and produces

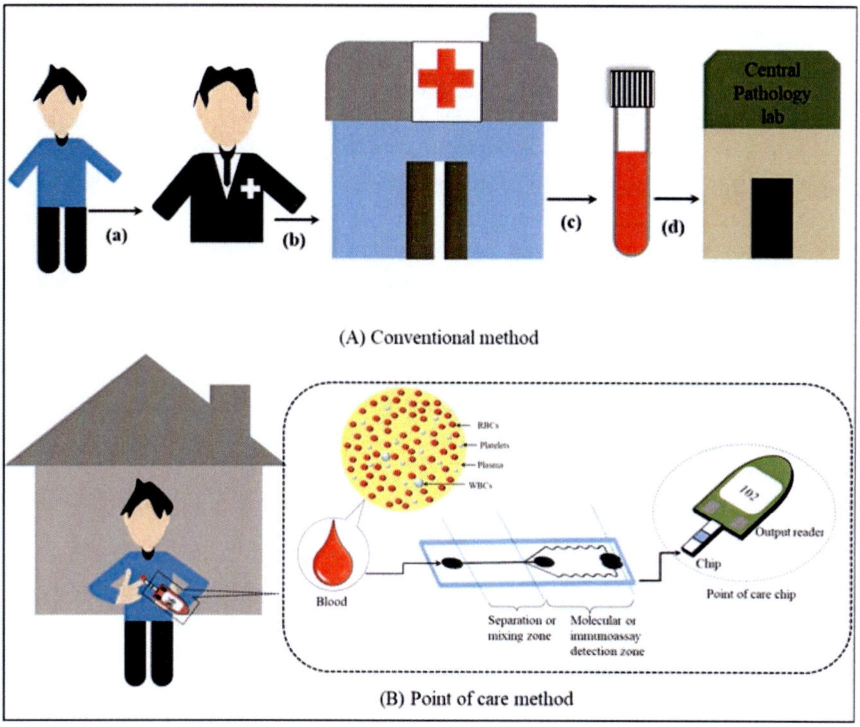

Fig. 2 Schematic of diagnostic tests: **A** conventional method of testing involving several steps (a) patient consults a doctor, (b) doctor refers to a hospital for required test, (c) blood is collected for the test, (d) sample is sent to centralized labs or pathology for testing. **B** Schematic of point-of-care (POC) testing demonstrating blood test at home by finger prick on POCT device based on microfluidics

instant test results on the site. In addition, the POC device should have good accuracy, good selectivity, low cost, and applicability in all environmental conditions. Development of a POCT device is a challenging task; the major factors responsible for the un-expected slow growth of POCT are lack of training, poor standardization in obtaining samples, and insufficient quality assessment.

Point-of-care technology relies on the development of lab-on-chip technology or micro total analysis system (μ-TAS). The lab-on-chip technology integrates several functionalities of a laboratory into a single miniaturized chip enabling automation of several tasks [3, 4]. Microfluidics is an emerging technology for the lab-on-chip devices. A simple definition of microfluidics would be the manipulation of fluids at length scales of the order of microns to submillimeters. At these scales, the behavior exhibited by fluids tends to greatly differ from the behavior exhibited at macroscale. Microfluidics is a vast field that provides tremendous approach to elucidate various complex traditional processes, especially in the areas of biology and clinical diagnostics [5]. The excitement surrounding microfluidics rests on the fact that it is highly capable of providing advantages in terms of a significant reduction in the sample processing time, sample volumes, consumption of costly reagents, parallel processing, high throughputs, low power consumption, increased sensitivity as well as precision in controlling fluid assays among others.

Development of blood-based point-of-care microdevices for disease diagnostics is an area of tremendous growth and possibilities. In this study, we will focus on blood-based point-of-care microdevices for the diagnosis of diseases. Although several review articles are available on point-of-care technology [3, 4], this particular study is dedicated to the current status and development in the field of blood-based microfluidic devices. Here, we will discuss functioning of lab-on-chip microdevices, which are dedicated to point-of-care technologies for blood-based diagnostics available in the literature. First, we briefly discuss the importance of a particular blood component, and this is followed by a discussion on prominent microfluidic devices which have been reported in the literature including the common tests being carried out for the disease diagnostic purposes. We also report and discuss lab-on-chip microdevices for studying circulating tumor cells (CTCs). Commercialized devices existing in the market are also reported in this study. Finally, we highlight the current status, challenges, and the future of point-of-care testing technology.

2 Plasma

Human blood plasma is the liquid medium in which cells are suspended. It makes about 55% of the total blood volume. It contains approximately 91% water, 7% proteins, 1% inorganic ions, and other organic substances. Common analytes found in plasma are proteins, inorganic and organic compounds, metabolites, bacteria, fungi, microorganisms, viruses, and circulating nucleic acids [6–8]. Due to the presence of large number of analytes or indicators, blood plasma serves as an index to various diseases and testifies its importance in disease diagnostics and therapeutics. Blood plasma is separated from other constituents on a routine basis. The conventional method of plasma separation is via centrifugation using a benchtop centrifuge. Though commonly employed, it involves manual and time-consuming steps. Several testing procedures demand immediate separation of plasma from the blood sample. Prolonged exposure of plasma with blood cells may deteriorate the quality of analytes present in plasma and can adversely affect the test results which will lead to inaccurate diagnosis. In order to detect the analytes effectively high-quality cell-free plasma is desired. Use of plasma is preferred over whole blood in several diagnostic tests due to clogging, cell lysis, and cell interference issues associated with whole blood testing [9, 10]. Plasma holds incredible clinical potential. It is required for a wide variety of clinical tests, such as glucose test for detecting diabetes mellitus, cholesterol, brain natriuretic peptide (BNP), troponin T (CTnI) for detecting heart diseases, C-reactive protein (CRP) for inflammation, Prostate-specific antigen (PSA) for Prostate cancer, cytokines for cancers, alanine aminotransferase/aspartate aminotransferase (ALT/ASP) for liver disorders, and several other analytes for diagnosis of various diseases.

In the recent past, microfluidic technology has rapidly progressed toward the development of lab-on-chip technologies and has contributed to point-of-care microdevices for blood plasma separation. Microfluidic methods for blood plasma separation can be broadly classified as active and passive methods. Active methods of plasma separation depend on external forces, these forces could be electrical [11–13], magnetic [14], acoustic [15, 16], optical, or gravity [17, 18]. The passive separation methods do not depend on external forces and rely on flow properties and geometry of the device. The passive methods can be further classified as filtration [19–23] sedimentation [24–27], deterministic lateral displacement [28, 29], and hydrodynamic methods [7, 30–32]. The major advantages offered by the passive hydrodynamic flow separation methods are simplicity of design, ease of fabrication, continuous operation, use of relatively high flow rates, and ease of integration with a biosensor [6, 10]. In this section, we discuss microfluidic devices which have been developed exclusively for point-of-care use. We also report devices, which can be used for plasma separation and analyte detection along with already available point-of-care devices in the market.

Dimov et al. [27] innovated a highly efficient and automated microdevice named self-powered integrated microfluidic blood analysis system (SIMBAS) and demonstrated its capabilities by detecting biotin with high sensitivity within 10 min using a drop of blood. This device incorporates most of the desired point-of-care features. The device design is shown in Fig. 3, the device integrates various components together, i.e., the volumetric metering, plasma separation, and the immunoassay. The device is stored in vacuum conditions, whole blood (5 μL) is loaded in the inlet port and blood flows through the channel due to the pressure differential created by the occlusion of

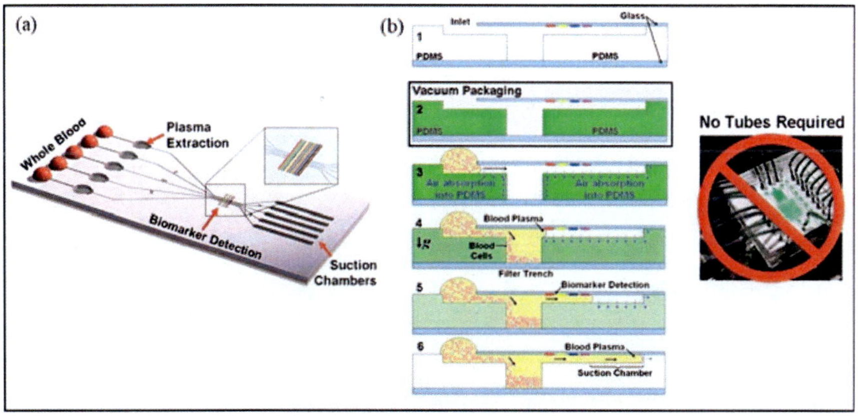

Fig. 3 SIMBAS system: **a** Schematic of the integrated microdevice showing the volume metering, plasma separation, biomarker detection zone, and suction chamber for blood flow. **b** Working of the device, showing steps vertically (1) cross section of microdevice, (2) vacuum packaging of microdevice, (3) placement of blood sample at inlet, (4) plasma separation into the microdevice, (5) biomarker detection from the separated plasma, and (6) regulation of flow by providing suction chamber. Adapted from [27] with permission from *The Royal Society of Chemistry*

the inlet port by whole blood and absorption of air due to low-pressure environment. Blood passes over the filter trench and owing to the sedimentation process plasma is extracted. Subsequently, plasma protein can be detected in the detection zone having pre-immobilized proteins capture. The device was used for successful detection of fluorescently labeled biotin.

Vella et al. [33] demonstrated a paper-based microdevice for liver function test using blood from a fingerstick. Using this device, the enzymatic markers of liver function such as alkaline phosphatase (ALP), aspartate aminotransferase (AST), and total serum protein were measured. The vertical flow microdevice consists of patterned piece of paper with wax (hydrophobic barriers and three hydrophilic zones for colorimetric tests), plastic sheath, and a plasma separating membrane. A drop of blood is injected in the device, plasma is separated in the membrane and distributed in the three reaction zones having pre-stored reagents as shown in Fig. 4A. Cali-

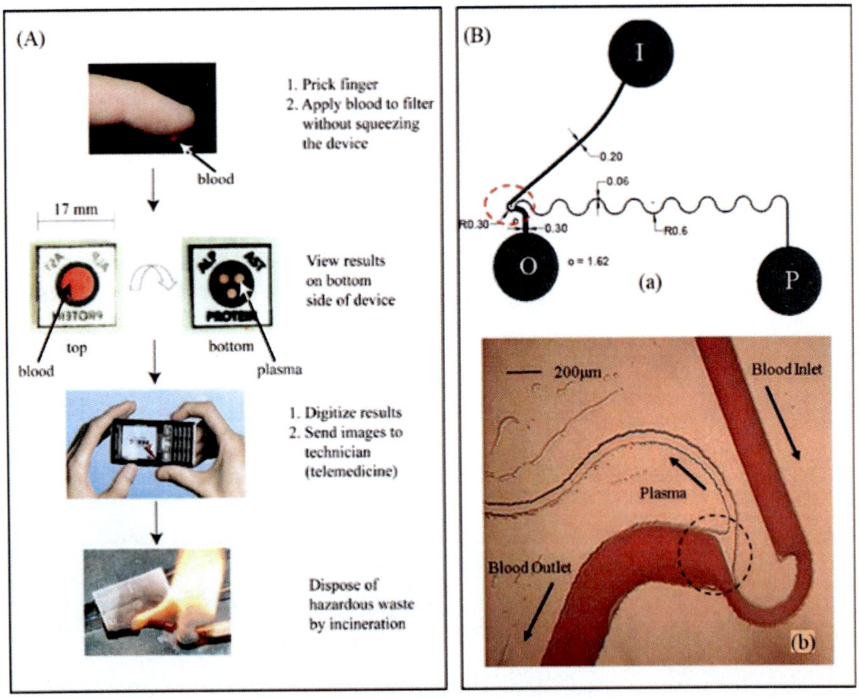

Fig. 4 Plasma-based microdevice: **A** Paper-based microfluidic device for liver function tests by measurement of ALT, AST, and proteins. Finger-prick blood is applied to the paper having pre-stored reagents; the filter separates the plasma and use cell phone for digitizing the results, and disposal of the hazardous waste. Adapted from [33] with permission from *American Chemical Society*. **B** Microdevice for plasma separation using geometrical and biophysical effects (a) schematic of the device (b) experimental photograph of plasma separation using whole blood shows the use of combination of effects such as constriction–expansion, bifurcation law, and centrifugal effect. Adapted from [7] with permission from *Springer Nature*

bration curves were generated by measuring the color intensity and intensity versus concentration graphs were generated for each analyte (AST, ALP, and protein).

Researchers claim that using these calibration curves color intensity charts can be prepared for easy readouts of the liver tests. This device is low in cost and easy to operate in resource poor setting, and it also satisfies the ASSURED criteria developed by WHO.

Tripathi et al. [7] developed a hydrodynamic-based microfluidic device utilizing biophysical and geometrical effects for plasma extraction. They utilized combination of Fahraeus effect [34–36], centrifugal effect, bifurcation law, and constriction–expansion zone to separate high purity plasma for wide range of hematocrits. As shown in Fig. 4B, as the blood flows through a bend the centrifugal forces push the cells toward the outer wall of the channel and subsequently presence of an expansion region enhances the cell-free layer. Cell-free plasma flows into the plasma line after the bifurcation. The separated plasma was subsequently tested (off-chip) to demonstrate the device capability to detect presence of glucose, hcG hormone (human chorionic gonadotropin, a pregnancy indicator), and proteins. The detection of the biomarkers agreed well with the commercially available test strips. Recently, Vazquez-Guardado et al. [37] reported the detection of neurotransmitter dopamine (responsible for functioning of neural system) by combining the plasma separator microdevice invented by Tripathi et al. [7] with an enzyme-free biosensor, Fig. 5. The biosensor is composed of nanostructured plasmonic substrate (NPS) functionalized with oxygen-deficient cerium oxide nanoparticles (CNP). Detection of dopamine at concentration of 1 nM was achieved from blood without requirement of any sample preparation steps. Additionally, the enzyme-free biosensor does not require sample preparation, handling, and storage of reagents; this further simplifies its use as a point-of-care device.

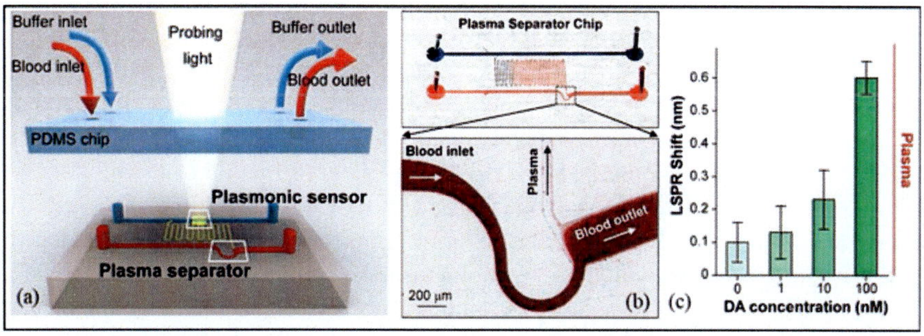

Fig. 5 Dopamine detection from blood: **a** Schematic representation of the integrated enzyme-free dopamine testing device with plasmonic sensor coupled to plasma separation component. **b** Plasma separation chip and plasma separation. **c** Sensor response for DA detection. Adapted from [37] with permission from *American Chemical Society*

Various blood-based point-of-care devices are available in the market today, these devices use either whole blood or plasma for analyte detection and reading in an automated way using finger-prick blood sample. Common analytes test which are available via point-of-care testing are measurement of glucose, prostate-specific antigen (PSA), C-reactive protein (CRP), cardiac markers, Human chorionic gonadotropin (hCG), electrolytes, and ferritin. The *i*-STAT handheld blood analyzer (Abott point-of-care) uses different cartridges for testing separate group or panel of analytes. It can provide rapid quantitative results with accuracy comparable to conventional systems. Measurements utilize different cartridges to measure total β hCG (diagnosis of pregnancy), BnP (B-type Natriuretic Peptide), CK-MB (creatinine Kinase MB), cTnI (detecting myocardial infarction) and for electrolytes [6, 21]. Point-of-care tests for measuring cholesterol, lipids, and glucose are also available (Abbott-Alere). Recently, a point-of-care device for measuring prostate-specific antigen (PSA) has been reported by OPKO diagnostics [38].

3 Red Blood Cells (RBCs)

Red blood cells form the majority of cell volume in human blood. Healthy RBCs look like biconcave discs of diameter in the range of 6–8 μm and thickness approximately 2.5 μm. Healthy humans have a normal count of 4.5–5.5 million per microliter of blood. RBCs are deformable and they do not have any nucleus. RBCs contain several proteins such as antioxidant reagents, structural protein, and hemoglobin, in which ~97% is hemoglobin [39]. The primary role of hemoglobin is to transport oxygen throughout the human body. Hemoglobin has iron atoms which are responsible for the red color of the blood. The normal values of the hemoglobin (Hb) for male ranges between 13.5 and 17.5 g/dL whereas, for an adult female it is between 11.5 and 15.5 g/dL [40]. Another important term related to red blood cell quantification is the hematocrit (Hct) also known as the packed cell volume (PCV). Hematocrit is the proportion of blood volume that is occupied by red blood cells. Normal hematocrit for average males varies from 40 to 54% and for average females vary from 36 to 48% [41]. Clinically, complete blood count gives the information of red blood cells in terms of RBCs count, hemoglobin value, and RBCs total volume (hematocrit). Deviations in these parameters from the normal range lead to disorders related to RBCs. Anemia and Polycythemia are among the most common RBCs disorders [39].

Anemia is caused due to scarcity of red blood cells whereas polycythemia results due to relatively high population of RBCs [39]. According to WHO recommendation, hemoglobin is the key parameter to monitor anemia in human. There are several types of anemia caused either due to large blood loss, or low production of RBCs, or high destruction of RBCs in the body. Sickle cell anemia is directly linked with hemoglobin. It is caused due to the structural change in hemoglobin leading to reduced deformability in RBCs. Malaria is also related to RBCs; it is caused by plasmodium falciparum parasite. This parasite is responsible for structural changes which results in impaired deformability of the RBCs [39, 42]. Malaria-infected RBCs show different dielectric, magnetic, stiffness, and optical properties [43]. Separation of infected RBCs can help in diagnosing diseases. Several physiological tests are performed on RBCs to track hemoglobin, hematocrit, and malaria detection. Several researchers have also proposed point-of-care devices for detection of these parameters. Following paragraphs will give a brief review of these tests based on point-of-care methods.

There exist various laboratory-based methods to measure hemoglobin, namely, the cyanmethemoglobin method, Sodium Lauryl Sulfate method [44], Sahli's method, Azide-methemoglobin method, Copper-sulfate method, oxyhemoglobin method, and WHO HCS method [45]. The cyanmethemoglobin method for measuring hemoglobin is the internationally recommended method. In this method, hemoglobin is converted to cyanmethemoglobin (hemiglobincyanide, HiCN) using potassium cyanide. The absorbance of the solution follows the Beer–Lambert's law [39]. The absorbance of the solution is compared with the standard HiCN solution of known concentration at a wavelength of 540 nm in a spectrophotometer to obtain the hemoglobin concentration [40]. Hemoglobin is also measured using the automated hematology analyzers and point-of-care handheld devices.

Recently, Yang et al. [46] reported a paper-based hemoglobin measuring test for resource-limited settings. Hemoglobin measurement was performed by mixing blood with Drabkin reagent (20 μL droplet) and depositing onto the patterned chromatography paper at the center. The bloodstain was detected by the algorithm and extracted the color information. The mean color intensity was used to measure the hemoglobin. Taparia et al. [47] reported a microfluidic approach for measuring hemoglobin concentration in whole blood. Their device consists of simple optics: a CMOS sensor and lens. Image analysis was performed to determine the optical density for measuring hemoglobin using generalized Steinke and Shephard model. A few studies have also reported hemoglobin measurement using blood analysis on a cell phone platform [48, 49]. The smartphone methods utilize hemolysis and spectrophotometric principles to measure absorbance and hemoglobin concentration.

Various point-of-care devices exist for hemoglobin measurement in the market today. The Hemocue hemoglobinometer and Dia spect are the popular ones [50]. The Hemocue is a portable hemoglobinometer based on the photometric principle, it consists of a battery-operated spectrometer. The analyzer uses specially designed microcuvettes containing dried reagents and serves as pipette and a reaction vessel. Sample blood (10 μL) is introduced by the microcuvette into the instrument by capillary action; the dried reagents in the microcuvette convert the hemoglobin to azide-methaemoglobin. Two wavelengths are used (570 and 880 nm); the color intensity of the solution is compared against the pre-calibrated standard, and results are generated within 1 min [45]. Another potential POCT device is the CO-oximeter, which is essentially a spectrophotometer incorporated in the gas analyzers. Yet another potent hemoglobin testing device is the WHO hemoglobin color scale (HCS). It is a low-cost hemoglobin measuring platform for developing world, and for those who have limited or no access to the laboratory facility. The principle behind this test is the fact that blood color is dependent on the amount of hemoglobin present in blood. Herein, a blood drop gets absorbed on a paper and the color is compared with a chart having six shades of red, where each shade represents the corresponding hemoglobin. This device is capable of identifying anemia and its severity but is not capable of quantifying the result [51]. Although many methods have been reported for hemoglobin measurement, only a few provide accurate results. The HiCN is considered to be the gold standard method; however, it is time consuming, requires a spectrophotometer, and is unsafe due to the use of cyanide [45, 52]. The automated analyzers are accurate but they require a lab infrastructure and are expensive. The requirement for a portable, efficient, and low-cost microfluidic-based device therefore still exists.

Clinical methods to measure hematocrit are by means of centrifugation, the CBC (impedance), conductivity, or by indirect method using the hemoglobin value [41]. The reference method is the centrifugation technique. In this method, blood is drawn into a capillary and rotated at high speeds using a centrifuge and the hematocrit (Hct) can be directly read from the scale (the ratio of RBCs to the total volume). Most commonly, the automated analyzers (based on Coulter impedance principle) are used, and hematocrit is measured from average size and number of RBCs. Hct can also be measured using a Point-of-care device, such as i-STAT blood gas analyzer [41]. These analyzers measure Hct using the principle of electrical conductivity or by evaluating Hct from the hemoglobin value. In recent work, Berry et al. [53] reported Hct measurement using a wax patterned chromatography paper. As shown in Fig. 6A, the pretreated blood sample enters the paper-based microfluidic device and flows through the microchannel. Sample having higher hematocrit travels a shorter distance as compared to samples with a smaller hematocrit value. The device measured Hct in range of 30–55% within 30 min.

Several microfluidics devices have been proposed to separate malaria-infected RBCs (*i*RBCs) from healthy RBCs, utilizing physical properties of infected RBCs. The conventional method of malaria detection is microscopic analysis carried on blood smear [55]. In this process, thin and thick blood smears are prepared and

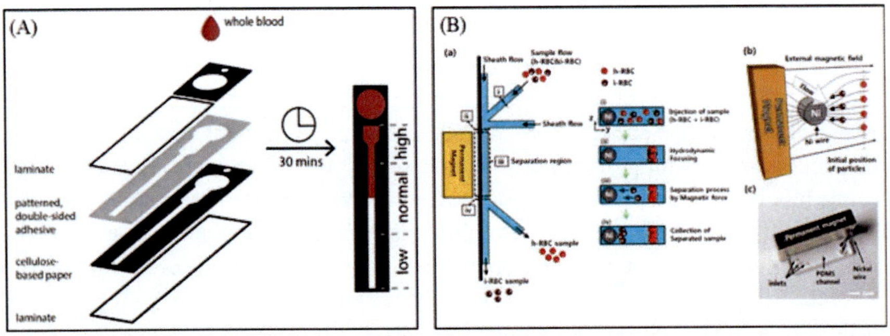

Fig. 6 Red blood cell-based detection: **A** Schematic of paper base microdevice for hematocrit measurement. Adapted from [53] with permission from *The Royal Society of Chemistry*. **B** Separation of malaria-infected RBCs using magnetic field (a) schematic of microdevice design integrated with paramagnet, (b) Schematic of working principal of microdevice, (c) Image of the PDMS microdevice attached with nickel wire and integrated with paramagnet. Adapted from [54] with permission from *American Chemical Society*

stained with Giemsa dye and analyzed by experienced personnel. This method is time consuming and needs skilled and trained personnel [55]. Nowadays, rapid diagnostic tests (RDTs) and nucleic acid amplification tests (NATT) are commonly used. RDTs also termed as point-of-care testing are fast and easy to perform and do not require any specific instrument and power supply [56]. RDTs are based on lateral flow technique using specific antibodies which bind to three types of plasmodium antigen named as *P. falciparum histidine-rich protein 2 (P fHRP2), P. falciparum lactate dehydrogenase (PfLDH), and Plasmodium Pan-specific* antigens to detect malaria [43, 57]. RDTs can detect specific antigen in 15 min from a finger-prick of blood [43]. RDTs do suffer from a few drawbacks, such as degradation of reagents in adverse environmental condition and low sensitivity in field along with high cost as compared to the conventional method [56].

Recently, several microfluidics-based lab-on-chip microdevices for detection of malaria in early stage have been proposed. These microdevices are based on detection of protein and changes in the physical properties of RBCs. Nam et al. [54] reported an active microdevice to separate early and late-stage malaria-infected RBCs based on their paramagnetic properties using magnetic field gradient. Their device consists of 100 μm wide and 50 μm deep PDMS microchannel bonded on to a glass slide and integrated with a ferromagnetic wire as shown in Fig. 6B. They successfully separated early stage and late-stage infected RBCs with recovery rate of 73% and 98.3%, respectively. Hou et al. [58] presented a passive microfluidics device for isolating *i*RBCs based on margination of stiffer *i*RBCs which get pass through an outlet, as shown in Fig. 7A. Their device comprises 100 μm wide channel at the inlet and constricted to 15 μm and expanded to 100 μm at the outlet and asymmetrically divided in three outlets. The microchannel is 10 μm deep and 3 cm long. They have utilized margination phenomenon (lateral migration of stiff *i*RBCs toward the channel walls) to separate *i*RBCs and achieved approximately 75% separation efficiency of early stage *i*RBCs and 90% late stage. Warkiani et al. [59] reported a PDMS-based inertial microfluidics device utilizing inertial lift forces acting on healthy and infected RBCs to isolate *i*RBCs for downstream polymerase chain reaction (PCR) application as shown in Fig. 7B.

They designed three constrictions and expansion array of 85 units of dimension 30/90 μm (contraction/expansion), 40/120 μm, and 50/150 μm with a depth of 100 μm. They achieved 99.99% WBC depletion and a malaria parasite collection with yield of $70.9 \pm 11.4\%$ after two-cycle run with their 30/90 constriction–expansion design.

There are many malaria POC test kits are available in market based on immunoassay lateral flow. These kits are Carestart Malaria Pf/Pan, OptiMAL, OptiMAL-IT, and Parabank, BinaxNow Malaria to name few. In past few years, world health organization (WHO 2015) has approved 86 RDTs manufactured by different companies [57].

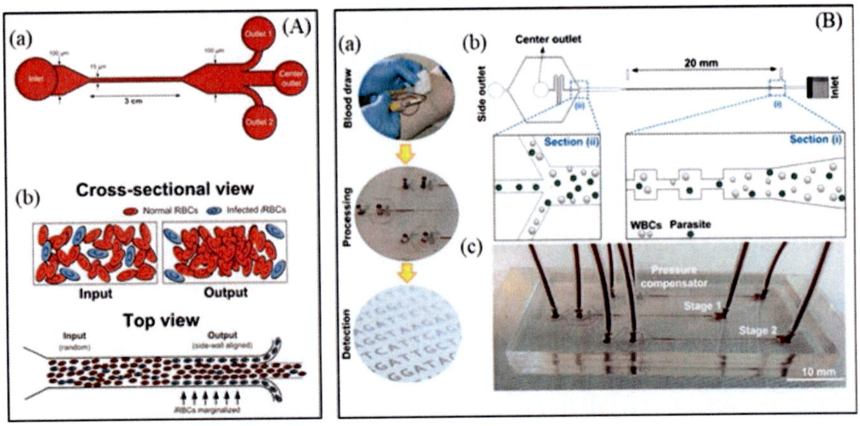

Fig. 7 Microdevices for the separation of malaria-infected RBCs: **A** Microdevice for malaria-infected RBCs separation based on cell margination (a) Schematic of microdevice design, (b) Schematic of cross-sectional view and top view of microchannel demonstrating separation phenomena. Adapted from [58] with permission from *The Royal Society of Chemistry*. **B** Malaria detection using inertial microdevice (a) Sample processing steps, (b) Schematic of contraction–expansion array microdevice and operational mechanism for enrichment of malaria-infected RBCs, (c) Experimental photograph of two stage cascade microdevice for enrichment of infected RBCs. Adapted from [59] with permission from *The Royal Society of Chemistry*

4 White Blood Cells (WBCs)

WBCs or leukocytes contain genetic materials and constitute our immune system. White blood cells defend our body against infection and participate in inflammatory processes. Leukocytes are found throughout the body, including the blood and lymphatic system [9, 60, 61]. They can be classified as granulocytes, lymphocytes, and monocytes. Granulocytes have multi-lobed nuclei whereas monocytes and lymphocytes are mononuclear. Leukocytes are potential biomarkers for blood-based diagnostics of various pathological conditions, such as heart disease [62], cancer [63], HIV [50, 64], forms of arthritis including auto-immune diseases disorders [65], and various other diseases [64]. To obtain relevant clinical and diagnostic information from WBCs their separation/isolation from other blood cells becomes imperative. As an example, diagnosis and treatment of HIV rely on the monitoring of human T-lymphocytes (CD4) cells. Also, in a clinical setting, the WBCs in a patient are often separated from other constituents of blood to get counts of individual types of WBCs. This can provide valuable information about a patient's status, especially in the diagnosis and treatment of rare diseases. Isolation of leukocytes is essential for clinical diagnostic tests, for monitoring disease progression, drug treatment analysis, and for carrying out fundamental cell studies. Separation of leucocytes is quite challenging, WBCs are present in very low quantities as compared to RBCs and platelets. The small density difference among the blood cells further complicates the separation process. Commonly, WBCs are separated using density-based centrifugation technique (Ficoll-paque) and the RBC lysis method [9]. Though frequently employed, these methods require large amount of blood sample, are labor intensive, require trained personnel, and result in compromised sample purity. The drawback associated with the conventional process of WBC isolation can be addressed using microfluidic technology.

Similar to the case of plasma separation, the microfluidic techniques for WBC separation can be classified as active and passive modes. Active forms of separation methods include Dielectrophoresis DEP [66–68], acoustophoresis [69–71], magnetophoresis [72, 73], etc. The passive techniques include on-chip microfiltration [74–77], deterministic lateral displacement [29, 78], RBC lysis [79–81], biomimetic [82–84], viscoelastic effects and hydrodynamics [85], pinched flow fractionation [86, 87], hydrophoretic filtration [88], inertial methods [89–93], and immunocapture [94, 95]. In the following section, we will discuss a few of these techniques with focus on those microfluidic devices which are specially designed toward point-of-care analysis.

The microfiltration techniques for WBC separation mainly involve membranes, pillars, and/or weirs. These designs rely on the fact that WBCs are larger in size compared to RBCs and platelets. Further, RBCs are deformable and may align and squeeze through narrow openings whereas the spherical and nucleated WBCs are retained. Sethu et al. [96] reported a microfluidic filter for leukapheresis. They devised a continuous flow diffusive filter for isolation of WBCs. The microfluidic device can operate continuously and can separate white blood cells (leukapheresis) for blood transfusion purposes. The device design constitutes a main channel which is connected to the two sides of the diffuser through sieves with filter elements, refer Fig. 8A. The deformable RBCs pass through the sieves to the side channels while the leukocytes remain in the center main channel. Almost 99% depletion of WBCs was achieved with blood flow rate of 5 μL/min using this device.

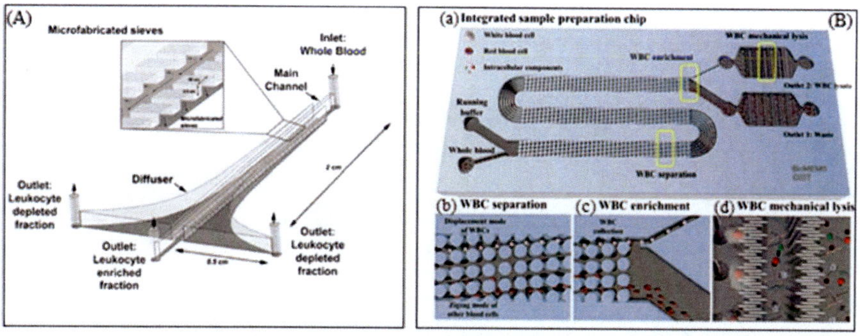

Fig. 8 Microdevice for WBC separation: **A** Diffusive filter design showing collection of leukocytes in the central channel and erythrocytes in side channels. Adopted from [96] with permission from *The Royal Society of Chemistry.* **B** Microdevice operation for extraction of intracellular molecules from WBCs (a) Schematic depiction of the integrated microchip for WBC separation and lysis, (b) lateral displacement of WBCs due to microarrays, (c) enrichment of WBCs, (d) Nano blade arrays used for rupturing WBCs. Adapted from reference [88] with permission from *Springer Nature*

Choi et al. [88] reported a microfluidic device which can separate leukocytes and extract intracellular proteins and nucleic acid for detection purposes using deterministic lateral displacement principle. The device design constitutes inlet ports, serpentine channel with microposts, and mechanical lysis structures with outlets as shown in Fig. 8B, whole blood and buffer are injected from the inlets and the lateral displacement of WBCs takes place due to the micropost arrays. Enrichment of WBCs is achieved by controlling the width ratios at the outlet and the intracellular components within the WBCs are released with the help of the mechanical nanoblade arrays having ultra-sharp edges. Quantitative analysis of the intracellular protein concentration and gDNA (Genomic deoxyribonucleic acid) levels was carried out using protein assay kit and UV-V spectrophotometer. The device was able to extract detectable HIV DNA at the cell population of $10^2/\mu l$ of the blood sample. With further upgradation and usage with microdiagnostic tool kits, the device can be used as a point-of-care microdevice in resource-limited settings.

Hassan et al. [64] reported a point -of-care microfluidic biochip to count and quantify CD64 (Cluster of Differentiation 64 is a type of integral membrane glycoprotein) expression on neutrophils. This biochip can be used for sepsis diagnostics, as increase in expression level of CD64 antigen can be correlated with infection. As shown in Fig. 9A, whole blood enters the chip along with a co-flowing buffer to lyse the RBCs. Later, the flow is halted using a quenching solution. The cells are counted at two stations, at inlet and outlet of the chamber using microfabricated electrodes. CD 64+ cells are captured in the cell capture chamber using pre-absorbed anti-CD64 antibody and the difference in cell counts was correlated with CD64 expression level. The chip requires only 10 μl of whole blood and provides the results in 30 min which were found to be in close agreement with those provided by a flow cytometer. WBC and its differential counts are important blood analysis parameters, as they relate

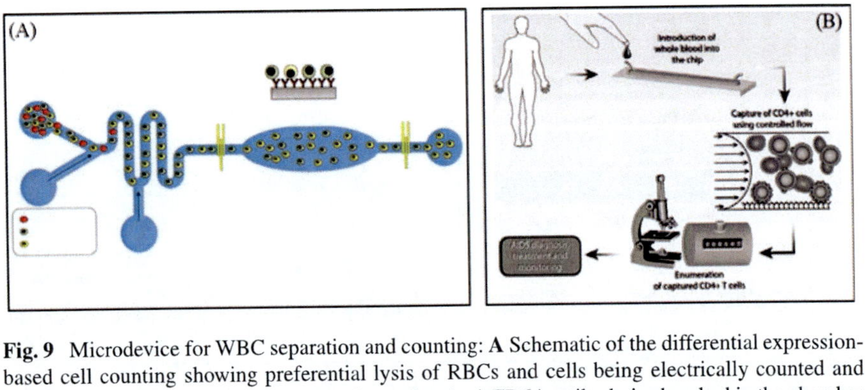

Fig. 9 Microdevice for WBC separation and counting: **A** Schematic of the differential expression-based cell counting showing preferential lysis of RBCs and cells being electrically counted and differentiated based on their size using electrodes. Anti-CD64 antibody is absorbed in the chamber and CD64+ cells are captured based on their expression level. Difference in cell counts is correlated with nCD64 expression level. Adapted from [64] with permission from *Springer Nature*. **B** Schematic showing the operational details of CD64 cell counting, CD4+ cells are captured using controlled flow rate and counted using a microscope. Adapted from [65] with permission from *The Royal Society of Chemistry*

to various disorders. WBC handheld point-of-care analyzers are also commercially available. The Hemocue WBC can provide total WBC count whereas the Chempaq XBC provides total WBC count with 3-part differential. These counters are based on the impedance cell counting method. Cell-based assays for detecting HIV depend on the CD4+ T-cell counts. A CD4 count of fewer than 350 cells/mm^3 is indicative of the disease and is one of the qualifications for a diagnosis of Acquired immune deficiency syndrome (AIDS) [4, 97, 98]. Cheng et al. [65] reported a CD4+ T-cell counting device based on microfluidic cell affinity chromatography. In this device, label-free whole blood, 10μL is directly injected into the microdevice, and CD4+ T cells are directly captured using controlled shear stress. The PDMS microdevice was functionalized by antibodies to capture CD4+ T cells, which were later counted using a simple microscope (Fig. 9B). The results were found to be in close agreement with the conventional counting using the expensive flow cytometer.

Various CD4 counting portable devices also exist in the market today. For example, CyFlow miniPOC, PointCare NOW, Daktari CD4 Counter, Visitect CD 4, BDFACS Presto, Zyomyx CD4 Test, and Pima Analyzer, to name a few [4, 98, 99]. Among these POCT devices, the Daktari CD4 counter and Zyomyx counters are quite attractive as they are very portable and low cost and easy to use systems with high sensitivity and specificity.

5 Platelets

Platelets are cytoplasmic fragments of the megakaryocytes [39]. They contain several proteins, enzymes, and platelet-derived growth factors which help in blood clotting and soft tissue healing. Blood clotting also known as hemostasis, is a multistep process including vascular constriction, platelet plug formation, and blood coagulation [100]. Although hemostasis is an essential phenomenon for sustenance of human life, the disorders of hemostasis known as thromboembolic disorders and bleeding disorders are life threatening [39]. Thromboembolic disorders arise when undesired clot forms inside the vessels or travels in microcirculation and block circulation of blood whereas bleeding disorders are those which prevent normal clot formation [39]. Platelet disorders may be linked with various diseases such as cardiovascular disease, diabetes, inflammation, haemorrage, sepsis, Von Willebrand disease, scurvy etc. [100–104]. Platelets separation have many applications such as in transfusion purpose, preparation of platelet-rich plasma (PRP), and study of hemostatic disorder. Clinically, platelets are separated using centrifugation-based methods known as PRP method, buffy coat method, and apheresis. These methods involve multistep processing, need trained personnel, are time consuming, and have adverse effect on the quality of platelets [100]. Separation of platelets on microfluidics devices is categorized as active and passive separation methods. The active separation methods involve dielectrophoresis (DEP) [105, 106] and acoustophoresis methods [107–110]. The passive separation method involves inertial methods [111], hydrophoretic techniques [101, 112], deterministic lateral displacement (DLD) method [113, 114], hydrodynamic methods [115], and filtration techniques [116], and affinity-based method [117]. A brief review of those methods which aims point-of-care analysis is provided in the following paragraph.

A passive PDMS-based hydrophoretic microdevice utilizing rotational flow and cell obstacle interaction to separate platelets was reported by Choi et al. [101]. Their device comprises two steps of photolithography, one for 1000 μm wide linear channel and other for patterns of 80° slanted grooves of 27 μm width and 33 μm pitch length as shown in Fig. 10A. Experiments were performed using diluted rat blood (1:9) at flow rate of 20 μl/min. They achieved 36.7% purity of platelets in the first round and 82.8% purity after the second round from an initial purity of 3.1%. They also parallelized the microdevice with 10 stacks of the hydrophoretic microdevice, which showed 76.8% purity from 2% of initial purity at a throughput of 2.9 million cells/s. The group has also stated that the platelets can be concentrated further by reducing the height of the grooves. Dickson et al. [116] reported a cross filtration-based macro-scalable microdevice to separate platelet-rich plasma (PRP) from whole blood. The device consists of two layers of micro-sieves of dimension 3 mm wide and 3 mm long as shown in Fig. 10B. The first layer has pore sizes of 1.2–3.5 μm for separating RBCs while the second layer with pore size of 0.45 μm is for separating platelets and plasma. They reported highest extraction fraction of 115% of platelets using 3.5 μm pore size of filter with whole blood sample. They also proposed that the macroscale device can obtain 50 ml of platelet-rich plasma (PRP) in 30 min.

Instead of separating platelets on microdevice for various purposes such as platelets transfusion, preparation of PRP, several researchers have reported on-chip coagulation test using microfluidics device. They monitored blood coagulation in terms of the *activated partial thromboplastin time (aPTT), prothrombin time (PT), thrombin time (TT), and fibrinogen assay tests* to diagnose any disorder [118, 119]. A brief discussion of these microdevices is given in the following paragraph along with commercially available point-of-care kits. In recent years, several microfluidics-based devices have been reported to measure the coagulation time of the blood. Santos et al. [120] reported a microfluidics system based on surface acoustic wave (SAW) to

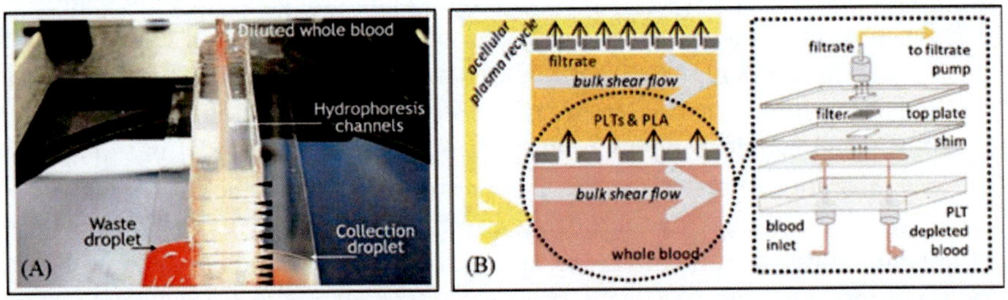

Fig. 10 Microdevice for platelets separation: **A** Experimental image of parallelized hydrophoretic microdevice demonstrating 10 layers of hydrophoretic device arranged on top of each other. Adapted from [101] with permission from The Royal Society of Chemistry. **B** Schematic of microfiltration-based microdevices for platelets extraction. Adapted from [116] with permission from Springer Nature

detect the coagulation time. They used PDMS-based arrangement and applied SAW to mix citrated blood. Thereafter, image analysis was performed to check the coagulation time. Nam et al. [121] presented Lamb wave-based acoustic streaming of the particle to measure the coagulation time of blood. They used Lamb wave for instant mixing of citrated plasma and coagulation reagents in a droplet. After that, acoustic streaming of florescent particles suspended in plasma was performed to measure the coagulation time. Coagulation tests were also performed using acoustic resonator [122].

Li et al. [123] presented paper-based measurement of coagulation time utilizing the fact that viscosity of blood change during coagulation. Their device consists of two layers, the top layer has dispensing window or sample reservoir and an opening to observe fluid flow. The bottom layer of test strip has four components named as sample pad, analytical membrane made of nitrocellulose, wicking pad, and membrane backing as shown in Fig. 11. Experiments were performed by placing citrated rabbit blood into the sample reservoir. As blood flows from sample pad to analytical membrane, images were collected from the observing window to calculate blood clotting time. They also compared the device with CoaData 2000 fibrintimer [123, 124]. Numerous microfluidics-based coagulation monitoring kits are now commercially available. These kits utilize electrochemical impedance and optical detection method to measure coagulation time. These kits are Coaguchek XS, CoaguChek S, Hemochron (ITC), ProTime (ITC), Xprecia Stride, etc. [118].

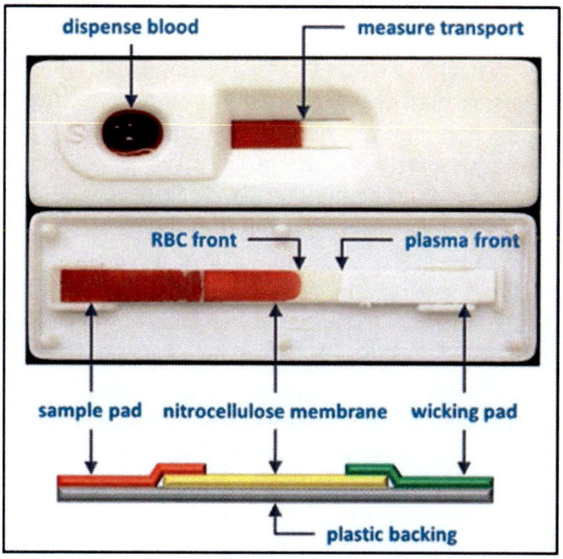

Fig. 11 Blood coagulation checking on lateral flow-based paper microdevices (**a**) (top) image of assembled device and (bottom) demonstration of lateral flow in microdevice (**b**) schematic showing component of microdevice. Adapted from [123] with permission from The Royal Society of Chemistry

6 Circulating Tumor Cells (CTCs)

Circulating tumor cells (CTCs) in human blood is a connecting link for cancer metastasis. Cancer is a fatal disease and more than 90% of cancer-related deaths are due to spread of cancer cells from the primary location to other body parts known as metastasis [125, 126]. Circulating tumor cells detach from primary tumor and start circulating in blood stream, thus known as circulating tumor cell (CTC). Size of CTCs range between 16 and 20 μm, and these cells are relatively larger than other blood cells [9]. CTCs are rare cells as they are present in very low numbers (1–1000 per ml or 1–2 per billion of normal cells) [127]. CTCs also differ from normal blood cells in biochemical and biophysical properties. CTCs are less stiff and exhibit altered electrical properties [125]. Current clinical methods of diagnosing cancer are based on detection of cancer biomarker in serum and radiological imaging of tumor tissues obtained by liquid biopsy [126]. These methods have low sensitivity and specificity. Further, repeated performance of biopsy may have adverse effect on the health of the patient. Also, performing biopsy could be challenging due to limited access to the tumor location [128].

Separation of CTCs leads to a promising outcome in early detection of cancer metastasis. Separation of CTCs is a very challenging task due to their rarity in blood. Like separation of other normal blood cells, centrifugation method is also employed to isolate CTCs. Fewcett et al. [129] separated cancer cells from peritoneal fluid using albumin as the floating media. Seal et al. [130] separated CTCs using silicon blending oil as the floatation medium. In their study, cancer cell is successfully detected in 53% gastrointestinal tract cancer cell patients sample and 33% breast cancer samples. OncoQuick is an improved centrifugation technique, in which a porous barrier is nestled within the 50 ml centrifuge tube to prevent the mixing of top and bottom layers of the blood sample [131].

Many researchers have reported CTCs separation on microfluidics device based on their size and surface markers. Size-based separation is also termed as label-free separation, whereas surface markers-based separation is known as label-dependent separation. Size-based separation involve mechanical filtration [126, 132–140], hydrodynamics separation [141–144], dielectrophoresis [145–148], and acoustophoresis [149]. Surface marker-based methods or label-dependent methods bind CTCs with cancer-specific antibodies and separate based on surface expression. Label-dependent methods include immune capturing [150, 151], immune magnetophoresis [152, 153], and deterministic cell rolling [154] to isolate CTCs. The following paragraph gives a brief discussion on label-free CTCs separation microdevices.

Zhou et al. [137] reported a 3D microfilter membrane, as shown in Fig. 12. The device comprised of two separable layers, the pores on the top layer are five times larger than the pore on the bottom membranes. Experiments were performed with healthy blood spiked with multiple cancer cell lines. A 78–83% separation efficiency with 71–74% cell viability was achieved. The main advantage of the filtration

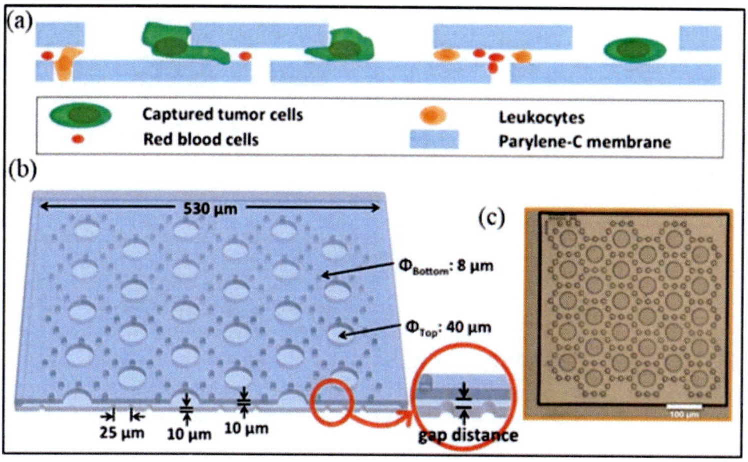

Fig. 12 Separable bilayer microdevice to separate circulating tumor cells: **a** Schematic of microdevice, **b** 3D view of bilayer filtration zone, **c** microscopic view of filtration layer. Adapted from [137] with permission from *Springer Nature*

arrangement is the simplicity in design. However, these designs are prone to clogging. They also affect the viability of the cells. To address the issue of clogging, cross-flow filtration method was reported.

Many researchers have reported hydrodynamic methods for clog-free separation of CTCs from blood sample [141–144]. Hou et al. [142] developed an inertial microdevice for CTCs separation using Dean flow-induced motion. Their device consists of spiral microchannel of 500 μm width and 160 μm height as shown in Fig. 13A. Experiments were performed with 20% hematocrit blood spiked with cancer cells, and they reported greater than 85% recovery and cell viability of more than 98%. In addition to that, this device had successfully detected CTCs in all blood samples collected from 20 patients with metastatic lung cancer ranging from 5 to 88 CTCs per ml. CTCs separation on microdevice is also attempted by active methods of separation using dielectrophoretic and acoustophoretic forces. Augustsson et al. [145] presented a microdevice to isolate cancer cells using acoustophoretic force. They have used device with trifurcation inlet and outlet to separate three different prostate cancer cell lines (DU145, PC3, and LNCaP) from blood sample spiked with mentioned cells as shown in Fig. 13B. They achieved average CTC recovery rate of DU145 cells 85.4% and 96.6% for acoustic energy of 120 J/m^3 and 188 J/m^3, respectively. Li et al. [149] reported tilted angle surface acoustic wave to separate cancer cells (MCF-7 and HeLa) from RBC lysed blood sample. It is reported in literature that the dielectrophoresis (DEP) device has been successfully employed to separate oral cancer, colon cancer [146], breast cancer [147], lung cancer, and prostate cancer cells [148], with a recovery rate of 70–90%.

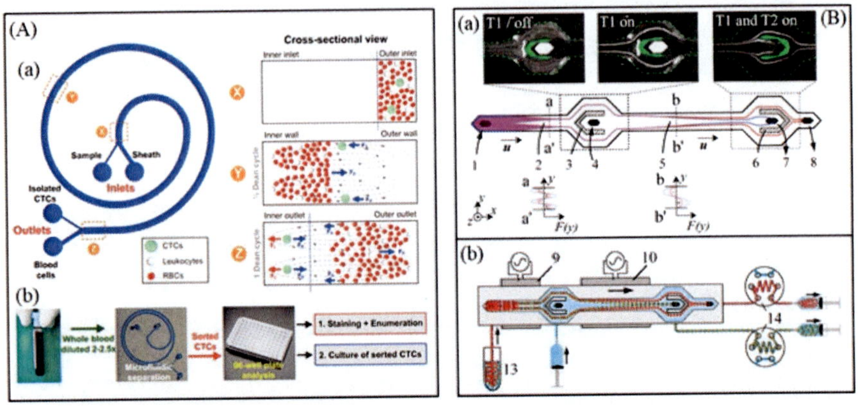

Fig. 13 Microdevices for CTCs separation: **A** Schematic of CTCs separation on a spiral microdevice using Dean flow fractionation (DFF) and demonstrating cross-sectional view of the microdevice at different Dean cycle. Adapted from [142] with permission from *Springer Nature*. **B** Schematic of CTCs separation microdevice (a) top view of microdevice showing inlet 1, followed by a channel 2 aligning cells due to acoustic field applied at (a-a′), two streams of cells bifurcated at 3 from central inlet 4 and flow along the walls of the channel. At 5 an acoustic field (b-b′) applied resulting deflection of cells based on their acoustic property and morphology. At the outlet 6, different cells guided to central outlet 7 and side outlet 8. (b) acoustic channel integrated with piezo-ceramic transducer (9 and 10), using acoustic force. Adapted from [145] with permission from *American Chemical Society*

Several CTCs separation devices are commercially available based on label-free and label-based method. Parsortix (ANGLE), Celsee PREP, ISET, ScreenCell, CellSieve, and SmartBiopsy are mechanical filtration-based label-free devices. Vortex HT and ClearCell FX system is based on hydrodynamic separation utilizing vortex formation and inertial effect. ApoStream is DEP-based device available commercially. Label-based method comprise IsoFlux, Liquid Biopsy, CellSearch, Adna Test, EPISPOT, etc. [126].

7 Conclusion

Blood is one of the most essential bodily fluids which provide information regarding the onset of various diseases that may be present in the body. Each component of the blood provides vital and profound information. Various biomarkers present in blood constituents can accurately pinpoint the associated ailment and offer appropriate diagnosis. Although many conventional methods exist and are widely used to manipulate the blood, they are neither expeditious nor economical and require substantial amount of reagents and samples. In this context, the advent of point-of-care technology becomes imminent. POCT requires minimal infrastructure, almost no prerequisite knowledge or skill by the user, is portable and provides results in a very short span of time.

The extensive literature survey conducted on the viability of this technology provides a realizable solution in the form of a microfluidic device. A number of studies have focused on the isolation of biomarkers present in the plasma, such as cardiac biomarkers, C-reactive protein, glucose level, and neurotransmitter dopamine. In addition to this, numerous studies have also proposed methods for detecting diseases such as malaria, HIV, coagulation disorders, and cancers. It was also observed that most of the commercially available products, such as glucose or blood coagulation test kits, revolve around the lateral flow of blood coming under the domain of paper microfluidics. POCT is also lacking in the field of active methods in contrast to the passive methods as they require operating sources. This requires miniaturization of operating and pumping forces to further accelerate the advancement of this technology toward revolutionizing the conventional methods. This field is an ever expanding one, offering plethoric applications, and unconventional solutions to disease diagnostics.

References

1. Fung, Y.C.: Biomechanics—Mechanical Properties of Living Tissues, 1st edn. Springer (1981)
2. Caro, C. G.: The Mechanics of the Circulation, 2nd edn. Cambridge University Press (2011)
3. Lee, J., Lee, S.H.: Lab on a chip for in situ diagnosis: from blood to point of care. Biomed. Eng. Lett. **3**(2), 59–66 (2013)
4. Jung, W., Han, J., Choi, J.W., Ahn, C.H.: Point-of-care testing (POCT) diagnostic systems using microfluidic lab-on-a-chip technologies. Microelectron. Eng. **132**, 46–57 (2015)
5. Sackmann, E.K., Fulton, A.L., Beebe, D.J.: The present and future role of microfluidics in biomedical research. Nature **507**(7491), 181–189 (2014)
6. Kersaudy-Kerhoas, M., Sollier, E.: Micro-scale blood plasma separation: from acoustophoresis to egg-beaters. Lab Chip **13**(17), 3323–3346 (2013)
7. Tripathi, S., Kumar, Y.B., Agrawal, A., Prabhakar, A., Joshi, S.S.: Microdevice for plasma separation from whole human blood using bio-physical and geometrical effects. Sci. Rep. **6**, 26749 (2016)
8. Lima, R., Ishikawa, T., Imai, Y., Yamaguchi, T.: Blood flow behavior in microchannels: past, current and future trends. In: Single and Two-Phase Flows on Chemical and Biomedical Engineering, pp. 513–547 (2012)
9. Hou, H.W., Bhagat, A.A.S., Lee, W.C., Huang, S., Han, J., Lim, C.T.: Microfluidic devices for blood fractionation. Micromachines **2**(3), 319–343 (2011)
10. Cui, F., Rhee, M., Singh, A., Tripathi, A.: Microfluidic sample preparation for medical diagnostics. Annu. Rev. Biomed. Eng. **17**, 267–286 (2015)
11. Jiang, H., Weng, X., Chon, C.H., Wu, X., Li, D.: A microfluidic chip for blood plasma separation using electro-osmotic flow control. J. Micromech. Microeng. **21**(8), 085019 (2011)
12. Nakashima, Y., Hata, S., Yasuda, T.: Blood plasma separation and extraction from a minute amount of blood using dielectrophoretic and capillary forces. Sens. Actuators B. **145**, 561–569 (2010)
13. Mohammadi, M., Madadi, H., Casals-Terré, J.: Microfluidic point-of-care blood panel based on a novel technique: reversible electroosmotic flow. Biomicrofluidics **9**(5), 054106 (2015)
14. Vemulapati, S., Erickson, D.: HERMES: rapid blood-plasma separation at the point-of-need. Lab Chip **18**(21), 3285–3292 (2018)

15. Laurell, T., Petersson, F., Nilsson, A.: Chip integrated strategies for acoustic separation and manipulation of cells and particles. Chem. Soc. Rev. **36**, 492–506 (2007)
16. Lenshof, A., Ahmad-Tajudin, A., Jarås, K., Sward-Nilsson, A.M., Åberg, L., Marko-Varga, G., Laurell, T.: Acoustic whole blood plasmapheresis chip for prostate specific antigen microarray diagnostics. Anal. Chem. **81**(15), 6030–6037 (2009)
17. MacDonald, M.P., Spalding, G.C., Dholakia, K.: Microfluidic sorting in an optical lattice. Nature **426**, 421–424 (2003)
18. Huh, D., Bahng, J.H., Ling, Y., Wei, H.H., Kripfgans, O.D., Fowlkes, J.B., Grotberg, J.B., Takayama, S.: Gravity-driven microfluidic particle sorting device with hydrodynamic separation amplification. Anal. Chem. **79**, 1369–1376 (2007)
19. Moorthy, J., Beebe, D.J.: In situ fabricated porous filters for microsystems. Lab Chip **3**(2), 62–66 (2003)
20. Thorslund, S., Klett, O., Nikolajeff, F., Markides, K., Bergquist, J.: A hybrid poly (dimethylsiloxane) microsystem for on-chip whole blood filtration optimized for steroid screening. Biomed. Microdevice **8**(1), 73–79 (2006)
21. Wang, S., Sarenac, D., Chen, M.H., Huang, S.H., Giguel, F.F., Kuritzkes, D.R., Demirci, U.: Simple filter microchip for rapid separation of plasma and viruses from whole blood. Int. J. Nanomed. **7**, 5019 (2012)
22. Shim, J.S., Ahn, C.H.: An on-chip whole blood/plasma separator using heteropacked beads at the inlet of a microchannel. Lab Chip **12**(5), 863–866 (2012)
23. Aran, K., Fok, A., Sasso, L.A., Kamdar, N., Guan, Y., Sun, Q., Ündar, A., Zahn, J.D.: Microfiltration platform for continuous blood plasma protein extraction from whole blood during cardiac surgery. Lab Chip **11**(17), 2858–2868 (2011)
24. Sun, M., Khan, Z.S., Vanapalli, S.A.: Blood plasma separation in a long two-phase plug flowing through disposable tubing. Lab Chip **12**(24), 5225–5230 (2012)
25. Zhang, X.B., Wu, Z.Q., Wang, K., Zhu, J., Xu, J.J., Xia, X.H., Chen, H.Y.: Gravitational sedimentation induced blood delamination for continuous plasma separation on a microfluidics chip. Anal. Chem. **84**(8), 3780–3786 (2012)
26. Tachi, T., Kaji, N., Tokeshi, M., Baba, Y.: Simultaneous separation, metering, and dilution of plasma from human whole blood in a microfluidic system. Anal. Chem. **81**(8), 3194–3198 (2009)
27. Dimov, I.K., Basabe-Desmonts, L., Garcia-Cordero, J.L., Ross, B.M., Ricco, A.J., Lee, L.P.: Stand-alone self-powered integrated microfluidic blood analysis system (SIMBAS). Lab Chip **11**(5), 845–850 (2011)
28. Wu, Z., Hjort, K.: Microfluidic hydrodynamic cell separation: a review. Micro Nanosyst. **1**(3), 181–192 (2009)
29. Davis, J.A., Inglis, D.W., Morton, K.J., Lawrence, D.A., Huang, L.R., Chou, S.Y., Sturm, J.C., Austin, R.H.: Deterministic hydrodynamics: taking blood apart. Proc. Natl. Acad. Sci. **103**(40), 14779–14784 (2006)
30. Kersaudy-Kerhoas, M., Kavanagh, D.M., Dhariwal, R.S., Campbell, C.J., Desmulliez, M.P.: Validation of a blood plasma separation system by biomarker detection. Lab Chip **10**(12), 1587–1595 (2010)
31. Maria, M.S., Rakesh, P.E., Chandra, T.S., Sen, A.K.: Capillary flow-driven microfluidic device with wettability gradient and sedimentation effects for blood plasma separation. Sci. Rep. **7**, 43457 (2017)
32. Yang, S., Ündar, A., Zahn, J.D.: A microfluidic device for continuous, real time blood plasma separation. Lab Chip **6**(7), 871–880 (2006)
33. Vella, S.J., Beattie, P., Cademartiri, R., Laromaine, A., Martinez, A.W., Phillips, S.T., Whitesides, G.M.: Measuring markers of liver function using a micropatterned paper device designed for blood from a fingerstick. Anal. Chem. **84**(6), 2883–2891 (2012)
34. Fahraeus, R.: The suspension stability of the blood. Physiol. Rev. **9**, 241–274 (1929)
35. Fahraus, R., Lindqvist, T.: The viscosity of the blood in narrow capillary tubes. Am. J. Physiol. (legacy content) **96**, 562–568 (1931)
36. Barbee, J.H., Cokelet, G.R.: The Fahraeus effect. Microvasc. Res. **3**, 6–16 (1971)

37. Vázquez-Guardado, A., Barkam, S., Peppler, M., Biswas, A., Dennis, W., Das, S., Chanda, D.: Enzyme-free plasmonic biosensor for direct detection of neurotransmitter dopamine from whole blood. Nano Lett. **19**(1), 449–454 (2018)
38. Taylor, N.P.: FDA approves OPKO's point-of-care PSA test, 1 Feb 2019. https://www.medtechdive.com/news/fda-approves-opkos-point-of-care-psa-test/547409/
39. Marieb, E.N., Hoehn, K.: Human Anatomy & Physiology. Pearson Education (2007)
40. Higgins, C.: Hemoglobin and its Measurement. Hemoglobin (2005)
41. Wennecke, G.: Hematocrit—A Review of Different Analytical Methods. Radiometer Medical ApS (2004)
42. Laposata M., Laboratory Medicine Diagnosis of Disease in Clinical Laboratory 2/E. McGraw-Hill Education; 2014
43. Kolluri, N., Klapperich, C.M., Cabodi, M.: Towards lab-on-a-chip diagnostics for malaria elimination. Lab Chip **18**(1), 75–94 (2018)
44. Karsan, A., Maclaren, I., Conn, D., Wadsworth, L.: An evaluation of hemoglobin determination using sodium lauryl sulfate. Am. J. Clin. Pathol. **100**(2), 123–126 (1993)
45. Srivastava, T., Negandhi, H., Neogi, S.B., Sharma, J., Saxena, R.: Methods for hemoglobin estimation: a review of "What Works". J. Hematol. Transfus. **2**(3), 1028 (2014)
46. Yang, X., Piety, N.Z., Vignes, S.M., Benton, M.S., Kanter, J., Shevkoplyas, S.S.: Simple paper-based test for measuring blood hemoglobin concentration in resource-limited settings. Clin. Chem. **59**(10), 1506–1513 (2013)
47. Taparia, N., Platten, K.C., Anderson, K.B., Sniadecki, N.J.: A microfluidic approach for hemoglobin detection in whole blood. AIP Adv. **7**(10), 105102 (2017)
48. Zhu, H., Sencan, I., Wong, J., Dimitrov, S., Tseng, D., Nagashima, K., Ozcan, A.: Cost-effective and rapid blood analysis on a cell-phone. Lab Chip **13**(7), 1282–1288 (2013)
49. Guo, T., Patnaik, R., Kuhlmann, K., Rai, A.J., Sia, S.K.: Smartphone dongle for simultaneous measurement of hemoglobin concentration and detection of HIV antibodies. Lab Chip **15**(17), 3514–3520 (2015)
50. Briggs, C., Kimber, S., Green, L.: Where are we at with point-of-care testing in haematology? Br. J. Haematol. **158**(6), 679–690 (2012)
51. Stott, G.J., Lewis, S.M.: A simple and reliable method for estimating haemoglobin. Bull. World Health Organ. **73**(3), 369 (1995)
52. Singh, A., Dubey, A., Sonker, A., Chaudhary, R.: Evaluation of various methods of point-of-care testing of haemoglobin concentration in blood donors. Blood Transfusion **13**(2), 233 (2015)
53. Berry, S.B., Fernandes, S.C., Rajaratnam, A., DeChiara, N.S., Mace, C.R.: Measurement of the hematocrit using paper-based microfluidic devices. Lab Chip **16**, 3689–3694 (2016)
54. Nam, J., Huang, H., Lim, H., Lim, C., Shin, S.: Magnetic separation of malaria-infected red blood cells in various developmental stages. Anal. Chem. **85**(15), 7316–7323 (2013)
55. Abdalla, S.H., Pasvol, G.: Malaria: A Hematological Perspective. Imperial College Press (2004)
56. Reboud, J., Xu, G., Garrett, A., Adriko, M., Yang, Z., Tukahebwa, E.M., Rowell, C., Cooper, J.M.: Paper-based microfluidics for DNA diagnostics of malaria in low resource underserved rural communities. Proc. Natl. Acad. Sci. **116**(11), 4834–4842 (2019)
57. Point-of-care tests for malaria, Horizon Scan Report 0040 (2015)
58. Hou, H.W., Bhagat, A.A., Chong, A.G., Mao, P., Tan, K.S., Han, J., Lim, C.T.: Deformability based cell margination-a simple microfluidic design for malaria-infected erythrocyte separation. Lab Chip **10**(19), 2605–2613 (2010)
59. Warkiani, M.E., Tay, A.K., Khoo, B.L., Xiaofeng, X., Han, J., Lim, C.T.: Malaria detection using inertial microfluidics. Lab Chip **15**(4), 1101–1109 (2015)
60. Janeway, C.A., Travers, P., Walport, M., Shlomchik, M.: Immunobiology: the immune system in health and disease. Lon.: Curr. Biol. **7**, 26 (1996)
61. Wu, Z., Chen, Y., Wang, M., Chung, A.J.: Continuous inertial microparticle and blood cell separation in straight channels with local microstructures. Lab Chip **16**(3), 532–542 (2016)

62. Brown, D.W., Giles, W.H., Croft, J.B.: White blood cell count: an independent predictor of coronary heart disease mortality among a national cohort. J. Clin. Epidemiol. **54**(3), 316–322 (2001)
63. Hou, L., Wang, H., Sartori, S., Gawron, A., Lissowska, J., Bollati, V., Baccarelli, A.: Blood leukocyte DNA hypomethylation and gastric cancer risk in a high-risk Polish population. Int. J. Cancer **127**(8), 1866–1874 (2010)
64. Hassan, U., Ghonge, T., Reddy Jr., B., Patel, M., Rappleye, M., Taneja, I., Jensen, T.: A point-of-care microfluidic biochip for quantification of CD64 expression from whole blood for sepsis stratification. Nat. Commun. **8**, 15949 (2017)
65. Cheng, X., Irimia, D., Dixon, M., Sekine, K., Demirci, U., Zamir, L., Toner, M.: A microfluidic device for practical label-free CD4+ T cell counting of HIV-infected subjects. Lab Chip **7**(2), 170–178 (2007)
66. Han, K.H., Frazier, A.B.: Lateral-driven continuous dielectrophoretic microseparators for blood cells suspended in a highly conductive medium. Lab Chip **8**(7), 1079–1086 (2008)
67. Yang, J., Huang, Y., Wang, X.B., Becker, F.F., Gascoyne, P.R.: Differential analysis of human leukocytes by dielectrophoretic field-flow-fractionation. Biophys. J. **78**(5), 2680–2689 (2000)
68. Yang, J., Huang, Y., Wang, X., Wang, X.B., Becker, F.F., Gascoyne, P.R.: Dielectric properties of human leukocyte subpopulations determined by electrorotation as a cell separation criterion. Biophys. J. **76**(6), 3307–3314 (1999)
69. Urbansky, A., Olm, F., Scheding, S., Laurell, T., Lenshof, A.: Label-free separation of leukocyte subpopulations using high throughput multiplex acoustophoresis. Lab Chip **19**(8), 1406–1416 (2019)
70. Grenvall, C., Magnusson, C., Lilja, H., Laurell, T.: Concurrent isolation of lymphocytes and granulocytes using prefocused free flow acoustophoresis. Anal. Chem. **87**(11), 5596–5604 (2015)
71. Petersson, F., Åberg, L., Swärd-Nilsson, A.M., Laurell, T.: Free flow acoustophoresis: microfluidic-based mode of particle and cell separation. Anal. Chem. **79**(14), 5117–5123 (2007)
72. Han, K.H., Frazier, A.B.: Paramagnetic capture mode magnetophoretic microseparator for high efficiency blood cell separations. Lab Chip **6**(2), 265–273 (2006)
73. Qu, B.Y., Wu, Z.Y., Fang, F., Bai, Z.M., Yang, D.Z., Xu, S.K.: A glass microfluidic chip for continuous blood cell sorting by a magnetic gradient without labeling. Anal. Bioanal. Chem. **392**(7–8), 1317 (2008)
74. Ji, H.M., Samper, V., Chen, Y., Heng, C.K., Lim, T.M., Yobas, L.: Silicon-based microfilters for whole blood cell separation. Biomed. Microdevice **10**(2), 251–257 (2008)
75. Tripathi, A., Riddell IV, J., Chronis, N.: A Biochip with a 3D microfluidic architecture for trapping white blood cells. Sens. Actuators B: Chem. **186**, 244–251 (2013)
76. VanDelinder, V., Groisman, A.: Perfusion in microfluidic cross-flow: separation of white blood cells from whole blood and exchange of medium in a continuous flow. Anal. Chem. **79**(5), 2023–2030 (2007)
77. Alvankarian, J., Bahadorimehr, A., Yeop Majlis, B.: A pillar-based microfilter for isolation of white blood cells on elastomeric substrate. Biomicrofluidics **7**(1), 014102 (2013)
78. Zheng, S., Tai, Y.C., Kasdan, H.: A micro device for separation of erythrocytes and leukocytes in human blood. In: 27th Annual Conference of the 2005 IEEE Engineering in Medicine and Biology, pp. 1024–1027 (2006)
79. Sethu, P., Anahtar, M., Moldawer, L.L., Tompkins, R.G., Toner, M.: Continuous flow microfluidic device for rapid erythrocyte lysis. Anal. Chem. **76**(21), 6247–6253 (2004)
80. Chen, J., Chen, D., Yuan, T., Xie, Y., Chen, X.: A microfluidic chip for direct and rapid trapping of white blood cells from whole blood. Biomicrofluidics **7**(3), 034106 (2013)
81. Kim, M., Mo Jung, S., Lee, K.H., Jun Kang, Y., Yang, S.: A microfluidic device for continuous white blood cell separation and lysis from whole blood. Artif. Organs **34**(11), 996–1002 (2010)
82. Shevkoplyas, S.S., Yoshida, T., Munn, L.L., Bitensky, M.W.: Biomimetic autoseparation of leukocytes from whole blood in a microfluidic device. Anal. Chem. **77**(3), 933–937 (2005)

83. Jain, A., Munn, L.L.: Biomimetic postcapillary expansions for enhancing rare blood cell separation on a microfluidic chip. Lab Chip **11**(17), 2941–2947 (2011)
84. Kuan, D.H., Wu, C.C., Su, W.Y., Huang, N.T.: A microfluidic device for simultaneous extraction of plasma, red blood cells, and on-chip white blood cell trapping. Sci. Rep. **8**(1), 15345 (2018)
85. Tan, J.K., Park, S.Y., Leo, H.L., Kim, S.: Continuous separation of white blood cells from whole blood using viscoelastic effects. IEEE Trans. Biomed. Circuits Syst. **11**(6), 1431–1437 (2017)
86. Yamada, M., Seki, M.: Hydrodynamic filtration for on-chip particle concentration and classification utilizing microfluidics. Lab Chip **5**(11), 1233–1239 (2005)
87. Cupelli, C., Borchardt, T., Steiner, T., Paust, N., Zengerle, R., Santer, M.: Leukocyte enrichment based on a modified pinched flow fractionation approach. Microfluid. Nanofluid. **14**(3–4), 551–563 (2013)
88. Choi, J., Hyun, J.C., Yang, S.: On-chip extraction of intracellular molecules in white blood cells from whole blood. Sci. Rep. **14**(5), 15167 (2015)
89. Nivedita, N., Papautsky, I.: Continuous separation of blood cells in spiral microfluidic devices. Biomicrofluidics **7**(5), 054101 (2013)
90. Sollier, E., Amini, H., Go, D.E., Sandoz, P.A., Owsley, K., Di Carlo, D.: Inertial microfluidic programming of microparticle-laden flows for solution transfer around cells and particles. Microfluid. Nanofluid. **19**(1), 53–65 (2015)
91. Wu, L., Guan, G., Hou, H.W., Bhagat, A.A.S., Han, J.: Separation of leukocytes from blood using spiral channel with trapezoid cross-section. Anal. Chem. **84**(21), 9324–9331 (2012)
92. Zhang, J., Yuan, D., Sluyter, R., Yan, S., Zhao, Q., Xia, H., Li, W.: High-throughput separation of white blood cells from whole blood using inertial microfluidics. IEEE Trans. Biomed. Circuits Syst. **11**(6), 1422–1430 (2017)
93. Tasadduq, B., Lam, W., Alexeev, A., Sarioglu, A.F., Sulchek, T.: Enhancing size based size separation through vertical focus microfluidics using secondary flow in a ridged microchannel. Sci. Rep. **7**(1), 17375 (2017)
94. Agrawal, N., Toner, M., Irimia, D.: Neutrophil migration assay from a drop of blood. Lab Chip **8**(12), 2054–2061 (2008)
95. Hamza, B., Irimia, D.: Whole blood human neutrophil trafficking in a microfluidic model of infection and inflammation. Lab Chip **15**(12), 2625–2633 (2015)
96. Sethu, P., Sin, A., Toner, M.: Microfluidic diffusive filter for apheresis (leukapheresis). Lab Chip **6**(1), 83–89 (2006)
97. Mauk, M., Song, J., Bau, H.H., Gross, R., Bushman, F.D., Collman, R.G., Liu, C.: Miniaturized devices for point of care molecular detection of HIV. Lab Chip **17**(3), 382–394 (2017)
98. Manoto, S., Lugongolo, M., Govender, U., Mthunzi-Kufa, P.: Point of care diagnostics for HIV in resource limited settings: an overview. Medicina **54**(1), 3 (2018)
99. Arora, D.R., Maheshwari, M., Arora, B.: Rapid point-of-care testing for detection of HIV and clinical monitoring. ISRN AIDS (2013)
100. Laxmi, V., Tripathi, S., Joshi, S.S., Agrawal, A.: Microfluidic techniques for platelet separation and enrichment. J. Indian Inst. Sci. **98**(2), 185–200 (2018)
101. Choi, S., Ku, T., Song, S., Choi, C., Park, J.K.: Hydrophoretic high-throughput selection of platelets in physiological shear-stress range. Lab Chip **11**(3), 413–418 (2011)
102. Leslie, M.: Beyond clotting: the powers of platelets. Science **328**, 562–564 (2010)
103. Von Hundelshausen, P., Weber, C.: Platelets as immune cells: bridging inflammation and cardiovascular disease. Circ. Res. **100**(1), 27–40 (2007)
104. Huebsch, L.B., Harker, L.A.: Disorders of platelet function: mechanisms, diagnosis and management. West. J. Med. **134**(2), 109 (1981)
105. Pommer, M.S., Zhang, Y., Keerthi, N., Chen, D., Thomson, J.A., Meinhart, C.D., Soh, H.T.: Dielectrophoretic separation of platelets from diluted whole blood in microfluidic channels. Electrophoresis **29**(6), 1213–1218 (2008)
106. Piacentini, N., Mernier, G., Tornay, R., Renaud, P.: Separation of platelets from other blood cells in continuous-flow by dielectrophoresis field-flow-fractionation. Biomicrofluidics **5**(3), 034122 (2011)

107. Nam, J., Lim, H., Kim, D., Shin, S.: Separation of platelets from whole blood using standing surface acoustic waves in a microchannel. Lab Chip **11**(19), 3361–3364 (2011)
108. Chen, Y., Wu, M., Ren, L., Liu, J., Whitley, P.H., Wang, L., Huang, T.J.: High-throughput acoustic separation of platelets from whole blood. Lab Chip **16**(18), 3466–3472 (2016)
109. Dykes, J., Lenshof, A., Astrand-Grundstrom, B., Laurell, T., Scheding, S.: Efficient removal of platelets from peripheral blood progenitor cell products using a novel micro-chip based acoustophoretic platform. PLoS One **6**(8), e23074 (2011)
110. Gu, Y., Chen, C., Wang, Z., Huang, P.H., Fu, H., Wang, L., Wu, M., Chen, Y., Gao, T., Gong, J., Kwun, J.: Plastic-based acoustofluidic devices for high-throughput, biocompatible platelet separation. Lab Chip **19**(3), 394–402 (2019)
111. Di Carlo, D., Edd, J.F., Irimia, D., Tompkins, R.G., Toner, M.: Equilibrium separation and filtration of particles using differential inertial focusing. Anal. Chem. **80**(6), 2204–2211 (2008)
112. Choi, S., Song, S., Choi, C., Park, J.K.: Continuous blood cell separation by hydrophoretic filtration. Lab Chip **7**(11), 1532–1538 (2007)
113. Li, N., Kamei, D.T., Ho, C.M.: On-chip continuous blood cell subtype separation by deterministic lateral displacement, In: 2nd IEEE International Conference on Nano/Micro Engineered and Molecular Systems, 2007. NEMS'07, pp. 932–936 (2007)
114. Inglis, D.W., Morton, K.J., Davis, J.A., Zieziulewicz, T.J., Lawrence, D.A., Austin, R.H., Sturm, J.C.: Microfluidic device for label-free measurement of platelet activation. Lab Chip **8**(6), 925–931 (2008)
115. Geislinger, T.M., Eggart, B., Braunmüller, S., Schmid, L., Franke, T.: Separation of blood cells using hydrodynamic lift. Appl. Phys. Lett. **100**(18), 183701 (2012)
116. Dickson, M.N., Amar, L., Hill, M., Schwartz, J., Leonard, E.F.: A scalable, micropore, platelet rich plasma separation device. Biomed. Microdevices **14**(6), 1095–1102 (2012)
117. Basabe-Desmonts, L., Ramstrom, S., Meade, G., O'neill, S., Riaz, A., Lee, L.P., Kenny, D.: Single-step separation of platelets from whole blood coupled with digital quantification by interfacial platelet cytometry (iPC). Langmuir **26**(18), 14700–14706 (2010)
118. Harris, L.F., Castro-López, V., Killard, A.J.: Coagulation monitoring devices: past, present, and future at the point of care. TrAC Trends Anal. Chem. **50**, 85–95 (2013)
119. Li, H., Steckl, A.J.: Paper microfluidics for point-of-care blood-based analysis and diagnostics. Anal. Chem. **91**(1), 352–371 (2018)
120. Santos, S.M., Zorn, A., Guttenberg, Z., Willems, B.P., Klaffling, C., Nelson, K., Klinkhardt, U., Harder, S.: A novel l-fluidic whole blood coagulation assay based on Rayleigh surface-acoustic waves as a point-of-care method to detect anticoagulants. Biomicrofluidics **7**, 56502 (2013)
121. Nam, J., Choi, H., Kim, J.Y., Jang, W., Lim, C.S.: Lamb wave-based blood coagulation test. Sens. Actuators B: Chem. **263**, 190–195 (2018)
122. Xu, W., Appel, J., Chae, J.: Real-time monitoring of whole blood coagulation using a microfabricated contour-mode film bulk acoustic resonator. J. Microelectromech. Syst. **21**(2), 302–307 (2012)
123. Li, H., Han, D., Pauletti, G.M., Steckl, A.J.: Blood coagulation screening using a paper-based microfluidic lateral flow device. Lab Chip **14**(20), 4035–4041 (2014)
124. Li, H., Han, D., Hegener, M.A., Pauletti, G.M., Steckl, A.J.: Flow reproducibility of whole blood and other bodily fluids in simplified no reaction lateral flow assay devices. Biomicrofluidics **11**(2), 024116 (2017)
125. Low, W.S., Abas, W., Bakar, W.A.: Benchtop technologies for circulating tumor cells separation based on biophysical properties. Biomed. Res. Int. **2015**, 1–22 (2015)
126. Cho, H., Kim, J., Song, H., Sohn, K.Y., Jeon, M., Han, K.H.: Microfluidic technologies for circulating tumor cell isolation. Analyst **143**(13), 2936–2970 (2018)
127. Gwak, H., Kim, J., Kashefi-Kheyrabadi, L., Kwak, B., Hyun, K.A., Jung, H.I.: Progress in circulating tumor cell research using microfluidic devices. Micromachines **9**(7), 353 (2018)
128. Joosse, S.A., Gorges, T.M., Pantel, K.: Biology, detection, and clinical implications of circulating tumor cells. EMBO Mol. Med. **7**(1), 1–11 (2015)

129. Fawcett, D.W., Vallee, B.L., Soule, M.H.: A method for concentration and segregation of malignant cells from bloody, pleural, and peritoneal fluids. Science **111**(2872), 34–36 (1950)
130. Seal, S.H.: Silicone flotation: a simple quantitative method for the isolation of free-floating cancer cells from the blood. Cancer **12**(3), 590–595 (1959)
131. Johnson, G.R.: The use of discontinuous percoll gradients to separate populations of cells from human bone marrow and peripheral blood. J. Immunol. Methods **66**(1), 9–16 (1984)
132. Lim, L.S., Hu, M., Huang, M.C., Cheong, W.C., Gan, A.T., Looi, X.L., Leong, S.M., Koay, E.S., Li, M.H.: Microsieve lab-chip device for rapid enumeration and fluorescence in situ hybridization of circulating tumor cells. Lab Chip **12**(21), 4388–4396 (2012)
133. Adams, D.L., Zhu, P., Makarova, O.V., Martin, S.S., Charpentier, M., Chumsri, S., Li, S., Amstutz, P., Tang, C.M.: The systematic study of circulating tumor cell isolation using lithographic microfilters. RSC Adv. **4**(9), 4334–4342 (2014)
134. Kim, E.H., Lee, J.K., Kim, B.C., Rhim, S.H., Kim, J.W., Kim, K.H., Jung, S.M., Park, P.S., Park, H.C., Lee, J., Jeon, B.H.: Enrichment of cancer cells from whole blood using a microfabricated porous filter. Anal. Biochem. **440**(1), 114–116 (2013)
135. Kim, T.H., Lim, M., Park, J., Oh, J.M., Kim, H., Jeong, H., Lee, S.J., Park, H.C., Jung, S., Kim, B.C., Lee, K.: FAST: size-selective, clog-free isolation of rare cancer cells from whole blood at a liquid–liquid interface. Anal. Chem. **89**(2), 1155–1162 (2016)
136. Coumans, F.A., van Dalum, G., Beck, M., Terstappen, L.W.: Filter characteristics influencing circulating tumor cell enrichment from whole blood. PLoS One **8**(4), e61770 (2013)
137. Zhou, M.D., Hao, S., Williams, A.J., Harouaka, R.A., Schrand, B., Rawal, S., Ao, Z., Brenneman, R., Gilboa, E., Lu, B., Wang, S.: Separable bilayer microfiltration device for viable label-free enrichment of circulating tumour cells. Sci. Rep. **4**(7392), 1–10 (2014)
138. Chung, J., Shao, H., Reiner, T., Issadore, D., Weissleder, R., Lee, H.: Microfluidic cell sorter (FCS) for on-chip capture and analysis of single cells. Adv. Healthc. Mater. **4**, 432–436 (2012)
139. Hvichia, G.E., Parveen, Z., Wagner, C., Janning, M., Quidde, J., Stein, A., Müller, V., Loges, S., Neves, R.P., Stoecklein, N.H., Wikman, H.: A novel microfluidic platform for size and deformability based separation and the subsequent molecular characterization of viable circulating tumor cells. Int. J. Cancer **138**(12), 2894–2904 (2016)
140. Xu, L., Mao, X., Imrali, A., Syed, F., Mutsvangwa, K., Berney, D., Cathcart, P., Hines, J., Shamash, J., Lu, Y.J.: Optimization and evaluation of a novel size based circulating tumor cell isolation system. PLoS One **10**(9), e0138032 (2015)
141. Loutherback, K., D'Silva, J., Liu, L., Wu, A., Austin, R.H., Sturm, J.C.: Deterministic separation of cancer cells from blood at 10 mL/min, AIP Adv. **2**(4), 1–7
142. Hou, H.W., Warkiani, M.E., Khoo, B.L., Li, Z.R., Soo, R.A., Tan, D.S., Lim, W.T., Han, J., Bhagat, A.A., Lim, C.T.: Isolation and retrieval of circulating tumor cells using centrifugal forces. Sci. Rep. **3**, 1259 (2013)
143. Warkiani, M.E., Guan, G., Luan, K.B., Lee, W.C., Bhagat, A.A., Chaudhuri, P.K., Tan, D.S., Lim, W.T., Lee, S.C., Chen, P.C., Lim, C.T.: Slanted spiral microfluidics for the ultra-fast, label-free isolation of circulating tumor cells. Lab Chip **14**(1), 128–137 (2014)
144. Geislinger, T.M., Franke, T.: Sorting of circulating tumor cells (MV3-melanoma) and red blood cells using non-inertial lift. Biomicrofluidics **7**(4), 1–9 (2013)
145. Augustsson, P., Magnusson, C., Nordin, M., Lilja, H., Laurell, T.: Microfluidic, label-free enrichment of prostate cancer cells in blood based on acoustophoresis. Anal. Chem. **84**(18), 7954–7962 (2012)
146. Fabbri, F., Carloni, S., Zoli, W., Ulivi, P., Gallerani, G., Fici, P., Chiadini, E., Passardi, A., Frassineti, G.L., Ragazzini, A., Amadori, D.: Detection and recovery of circulating colon cancer cells using a dielectrophoresis-based device: KRAS mutation status in pure CTCs. Cancer Lett. **335**(1), 225–231 (2013)
147. Moon, H.S., Kwon, K., Kime, S.I., Han, H., Sohn, J., Lee, S., Jung, H.I.: Continuous separation of breast cancer cells from blood samples using multi-orifice flow fractionation (MOFF) and dielectrophoresis (DEP). Lab Chip **11**(6), 1118–1125 (2011)
148. Huang, C., Liu, H., Bander, N.H., Kirby, B.J.: Enrichment of prostate cancer cells from blood cells with a hybrid dielectrophoresis and immunocapture microfluidic system. Biomed. Microdevice **15**(6), 941–948 (2013)

149. Li, P., Mao, Z., Peng, Z., Zhou, L., Chen, Y., Huang, P.H., Truica, C.I., Drabick, J.J., El-Deiry, W.S., Dao, M., Suresh, S.: Acoustic separation of circulating tumor cells. Proc. Natl. Acad. Sci. **112**(16), 4970–4975 (2015)
150. Farokhzad, O.C., Jon, S., Khademhosseini, A., Tran, T.N., LaVan, D.A., Langer, R.: Nanoparticle-aptamer bioconjugates: a new approach for targeting prostate cancer cells. Cancer Res. **64**(21), 7668–7672 (2004)
151. Paterlini-Brechot, P., Benali, N.L.: Circulating tumor cells (CTC) detection: clinical impact and future directions. Cancer Lett. **253**(2), 180–204 (2007)
152. Riethdorf, S., Fritsche, H., Müller, V., Rau, T., Schindlbeck, C., Rack, B., Janni, W., Coith, C., Beck, K., Jänicke, F., Jackson, S.: Detection of circulating tumor cells in peripheral blood of patients with metastatic breast cancer: a validation study of the Cell Search system. Clin. Cancer Res. **13**(3), 920–928 (2007)
153. Hoshino, K., Huang, Y.Y., Lane, N., Huebschman, M., Uhr, J.W., Frenkel, E.P., Zhang, X.: Microchip-based immunomagnetic detection of circulating tumor cells. Lab Chip **11**(20), 3449–3457 (2011)
154. Choi, S., Karp, J.M., Karnik, R.: Cell sorting by deterministic cell rolling. Lab Chip **12**(8), 1427–1430 (2012)

Origami MEMS

Derosh George and Marc J. Madou

1 Introduction

Origami, an ancient Japanese art of folding a single uncut paper to form complex shapes, dates back to the sixth century A. D. It was not very popular back in those days due to the high price of paper. Nevertheless, the art sustained through the generations as a Japanese tradition. There are also some evidences that similar art forms existed in Europe as early as the eighth century. It is not clear, however, if origami in Europe was independently developed or originated also in Japan and found its way to Europe via the famous *silk road*. One of the first written instructions for origami is found in Akisato Rito's 1797 book called "Sembazuru Orikata" (meaning "thousand crane folding"). Although the Origami artwork was practiced in various parts of the world, it was not until the twentieth century, during *Akira Yoshizawa*'s time, that this art form became globalized. *Yoshizawa* published a system of patterns and developed a variant of origami called *wet origami. Wet origami* uses a dampened paper instead of a dry sheet of paper to construct more accurate non-geometric origami rather than through mere folding (Fig. 1). In another variation, creation of more complex shapes were made possible through the simultaneous use of multiple sheets of papers in a technique known as modular origami [165]. Later, Demaine and Tachi developed an algorithm to make almost any conceivable polyhedron shapes from a sheet of paper through just folding (Fig. 1) [29, 165]. Collectively, traditional origami, *wet origami,* and *modular origami* can transform simple two-dimensional shapes into three-dimensional shapes of different degrees of complexities [126]. Such origami techniques, from an engineer's perspective, offer some key advantages

D. George · M. J. Madou (✉)
The Henry Samueli School of Engineering, University of California, Irvine, Irvine, CA 92697, USA
e-mail: mmadou@uci.edu

D. George
e-mail: deroshg@uci.edu

© Springer Nature Singapore Pte Ltd. 2021

U. S. Dixit and S. K. Dwivedy (eds.), *Mechanical Sciences*,
https://doi.org/10.1007/978-981-15-5712-5_9

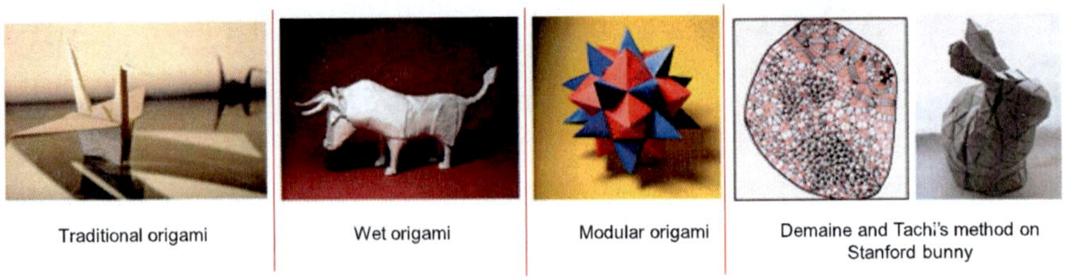

Fig. 1 Traditional origami (Picture credit: Curt Smith, https://bit.ly/353bCb4), wet origami (Designed by Stephan Weber, https://bit.ly/2MB460z), modular origami (Designed by Tomoko Fuse, https://bit.ly/2t5IsuC), and Demaine and Tachi's method [29]

(i) easy packaging and transportation due to the flat configuration, (ii) high specific strength due to minimal material usage, and (iii) simplicity of construction due to the absence of mechanical components. These advantages inspired engineers of the modern age to translate the origami technique to various engineering platforms, either using paper itself as the material (*paper-based origami*) or with the help of other materials.

A sheet of paper is the raw material, and manual-folding is the construction method in the *paper-based origami* technique. The original purpose of this type of origami was recreational. When adopted for engineering purposes, raw materials and folding strategies had to be rationally adapted for the specific application at hand [65, 118]. Before going into materials other than paper, let us briefly discuss the engineering aspects of *paper-based origami*. For manufacturing *paper-based origami*, folding methods that do not involve manual-folding is necessary. Hand-free folding of paper-based origami shapes is possible via actuation mechanisms that can be implemented on the paper through printing. For instance, folding of a paper can be realized by printing a water-based ink on the paper followed by drying. Drying shrinks the paper locally and results in its folding. The dried paper is stiff enough to sustain a stable folded configuration. Meanwhile, the field of *paper-electronics*, where paper-based electronic systems are studied, offers a host of techniques for incorporating basic electrical elements such as transistors, batteries, and actuators on a *paper-based origami* [33, 55, 83, 127, 131]. *Paper-based origami* has also found application as an inexpensive diagnostic platform. In the year 2007, Whitesides and co-workers from Harvard University introduced a paper-based microfluidic analytic platform using a single piece of patterned paper [107]. Later, the same group presented a more sophisticated form of the paper-based diagnostic platform with a stack of patterned paper layers. Eventually, integration of multilayered paper microfluidics with the origami technique enabled single-step patterning of complex microfluidic systems [171]. This simplification also facilitated easy integration of various components, e.g., a battery, into the system by bringing paper-electronics and paper microfluidics together. In this Chapter, however, we will not be going further into details on paper-based origami structure since these systems are often confined to bigger length scales that are beyond the scope of this chapter. On the other end of the spectrum, molecular methods are used, for example, for making DNA origami. Those techniques are also excluded here (details on the technique can be found in Hong et al. [53]). What we do focus on here are sheets other than paper for micro-sized origami.

2 Fabrication of Precursor Thin Sheets for Submillimeter Origami

Fabrication methods, including *photolithography, direct writing, and sheet cutting,* are employed (see Table 1) to produce origami precursors at smaller length scales [102], i.e., two-dimensional patterned shapes with submillimeter characteristic lengths. *Photolithography* is one of the most common techniques for microfabrication of planar shapes. In this fabrication paradigm, the desired locations of a planar photopolymer/photoresist layer are chemically modified by irradiating them with light. The locally altered material resulting from the exposure is either more dissolvable in developer or less dissolvable, depending on whether the polymer photoresist is positive (e.g., AZ 1400 series) or negative tone (e.g., SU8 2000 series), respectively (Fig. 2). This technique allows for the fabrication of planar patterns of photopolymers with resolutions well below 100 nm. The patterns obtained can be utilized as a transfer mold/stamp. Alternatively, these photopolymer patterns can act as a mask for selective etching or deposition (Fig. 2). In all these photolithography/photolithography-derived processes, the product is mostly limited to two-dimensional shapes.

While photolithography produces patterned surfaces by removing unwanted parts of the material (a top-down method), the *direct writing* technique creates shapes by adding material to selected locations (Fig. 3) (a bottom-up method). Generally, the dispersion of ink to the selected locations is driven through hydrodynamic forces, electrostatic forces, or mechanical pulling [24, 100, 164]. The resolution of the writing is determined by the size of the line produced by the dispenser, and this size is, in turn, a function of the dispersion technique and other working parameters. The

Table 1 General fabrication methods for thin patterned sheets

Fabrication methods for thin sheets	Length scale	Specific techniques
Photolithography	100 nm–10 mm	Optical lithography, UV nanoimprint lithography, etc. Other derived techniques include micro-molding, stamping, micro-contact printing, deposition or growth, etching, etc.
Micro/nano writing	10 nm–10 μm	Electrospinning, melt spinning, jet spinning, dry spinning, dip-pen lithography, etc.
Cutting	>10 nm	Laser cutting, focused ion beam, CNC machines, etc.

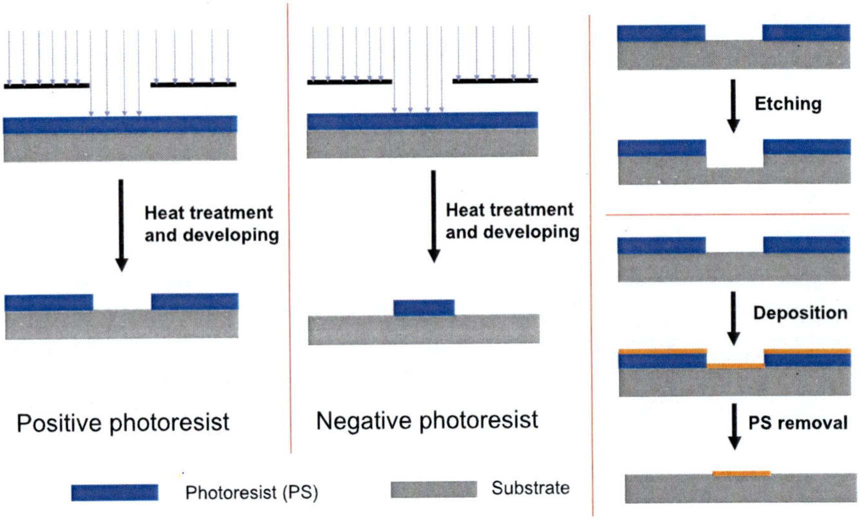

Fig. 2 Photolithography-based fabrication

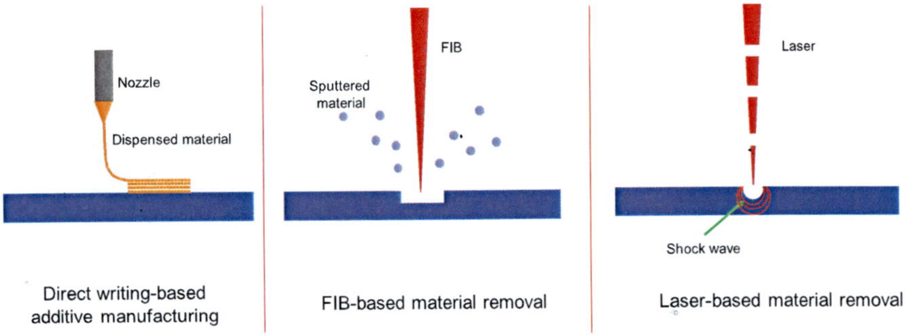

Fig. 3 Additive manufacturing, and focused ion beam (FIB) and laser-based material removal methods

writing technique can deliver complex three-dimensional shapes by implementing a layer-by-layer writing approach [149]. With the recent advances in the field of direct writing methods, many complex multi-material three-dimensional shapes with high resolution are possible. However, achieving high aspect ratio structures with high fidelity is still a challenge.

Another precursor film fabrication approach is through material removal, such as by *cutting* of thin sheets. Traditional machine-based cutting is widely adopted for the fabrication of various shapes owing to its centuries-old usage. However, this traditional method lacks the capacity to machine tiny two-dimensional shapes. Recently, numerous material removal techniques that utilize non-traditional methods have emerged. For instance, thin sheets can be directly cut using lasers or focused ion beams (FIB) for forming patterned two-dimensional sheets (Fig. 3). These new techniques made possible resolutions as small as 10 nm.

All three of these fabrication pathways described here—*photolithography, direct writing, and cutting*—have the capability to produce small patterned thin sheets. These patterned sheets, if folded, could create tiny three-dimensional shapes. Such three-dimensional structures and systems of submillimeter scales are generally referred to as *Origami MEMS (Microelectromechanical systems).* Folding in such a small length scale demands creative folding strategies since manual-folding used in traditional paper-based origami is not a feasible choice for submillimeter origami shapes. In the following section, different approaches to folding sheets in the submillimeter regime are discussed.

3 Folding Strategies for Origami MEMS

Here, we examine various strategies for folding submillimeter and sub-micron scale patterned thin sheets for origami shapes. Logical approaches to accomplish folding in the submillimeter regime include (i) borrowing existing folding strategies from a larger length scale, or (ii) mimicking folding mechanisms found in nature. Bending and buckling tactics are adopted to transform simple two-dimensional shapes to complex three-dimensional shapes in the submillimeter length scale. The bending of a sheet takes place when it experiences an out-of-the plane moment (Bending moment). Buckling, on the other hand, is a result of a compressive force on a slender object. Both bending and buckling on a sheet material is generally realized via two different methods: (1) by introducing a non-uniform material property across the sheet thickness followed by a trigger, or (2) by applying an external force to materials having a uniform property (Fig. 4). Generally, for the first case, a mismatch in the deformations at different locations of cross-sections across the thickness (a strain mismatch) causes bending or buckling. The required strain may be induced via

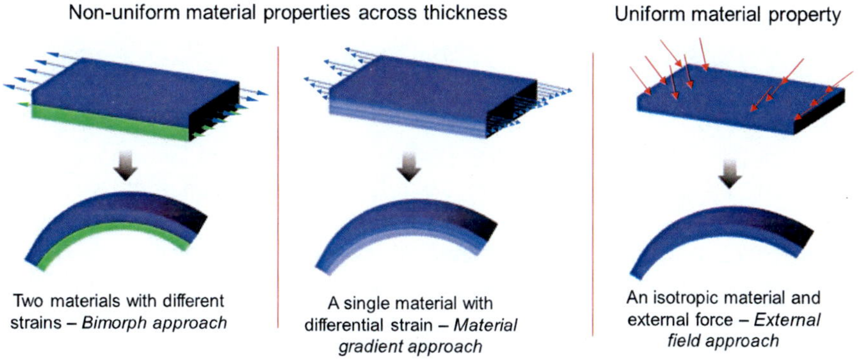

Fig. 4 Bimorph approach, material gradient approach, and external field approach

swelling, shrinkage, thermal expansion, piezoelectric effect, configuration change in a liquid crystal elastomer, or shape memory effect (Table 2). In the second case, an external force prompts bending or buckling of sheets that are isotropic in nature. Driving forces of such geometrical transformation include surface tension-based actuation, magnetic actuation, mechanical actuation, or electrical actuation (Table 2).

Bending and buckling are achieved in the submillimeter length scales using the following strategies (Fig. 4).

(1) By introducing different strain values on two attached sheet materials—***Bimorph approach***.
(2) By introducing differential strain across the thickness of a single material by designing different material properties across its thickness—***Material gradient approach***.
(3) By using external fields on a material with uniform properties across the thickness—***External field approach***. (Note: External fields may be required for the activation of both #1 and #2, but the material property across the thickness is not uniform in either of them.)

3.1 Bimorph Approach

Materials often display deformation when heated. Generally, dimensional changes generated in a material due to changes in temperature are homogeneous, leading to simple stretching or compression. However, when two such sheets (Generally, metals in the case of larger length scales) having distinct thermal expansions are bonded together (e.g., *bimetallic strip*), the resulting structure bends upon heating because of their strain mismatch [32]. The bending is such that it forms a convex curvature on the side having material with relatively higher thermal expansion material. Curvature formation by heating of bimetallic strips is one of the conventional approaches for bending thin constructs at scales larger than a millimeter (Fig. 5). The curvature, κ, is expressed based on Timoshenko's plate theory [146] and is given as follows.

$$\kappa = \frac{6\Delta\alpha(1+m)^2}{h\left\{3(1+m)^2 + (1+mn)\left(m^2 + \frac{1}{mn}\right)\right\}} \tag{1}$$

where $\Delta\alpha$ is the difference in thermal expansion coefficients of the two materials, m is the thickness ratio between them, and n is the stiffness ratio of the two materials. Although the same concept of bending driven by distinct strains of two materials (bimorphs) can be implemented for the microfabrication of bent structures, the execution of the idea at that length scale must adopt various crafty new ways. Driving forces relevant for bending sheets of micron length scales include *residual stress, swelling, liquid crystal alignment, and shape memory effect*. The general strategy is to attach these active materials to another inactive substance, followed by actuation/trigger (*Unimorph*). Alternatively, both the sheets could be active materials (*Bimorphs).*

Table 2 Different actuation mechanism for origami fabrication and their details

Actuation mechanisms	Size of origami	Advantages and disadvantages	Reversible/Irreversible
Capillary force	100 nm–1000 μm	Advantages – Inexpensive fabrication Disadvantages – Pairing of droplets and sheets demands multilayer fabrication – Difficulty making complex shapes – Elastomers return to their original shape	Reversible or irreversible
e.g.			
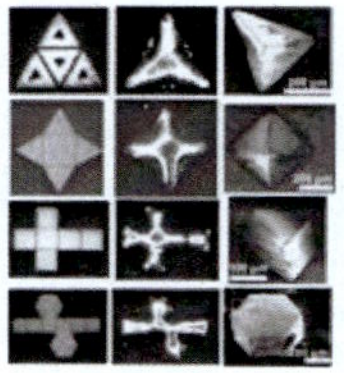 [45]	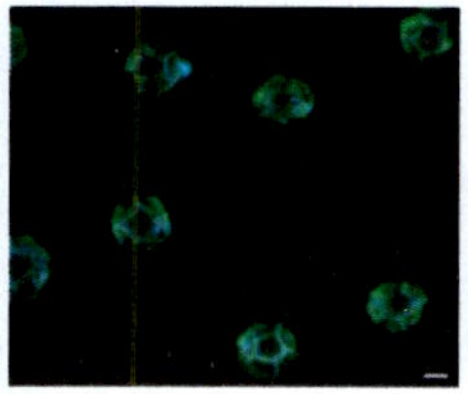 [77]	 [153]	 [20]
[45] (Copyright 2002 by John Wiley & Sons, Inc. Reproduced by permission of John Wiley & Sons, Inc.) [77, 153] (Reprinted with the permission of AIP Publishing) [20] (Copyright 2009 American Chemical Society) [38, 44, 88, 93, 117, 121]			
Electric field/magnetic field	100 μm–10 mm	Advantages – Excellent temporal control over the folding – Quick response – Easy manipulation – Wireless control – Nanorobot fabrication (Magnetic) Disadvantages – External system to control the device	Reversible

(continued)

Table 2 (continued)

Actuation mechanisms	Size of origami	Advantages and disadvantages	Reversible/Irreversible
e.g.			
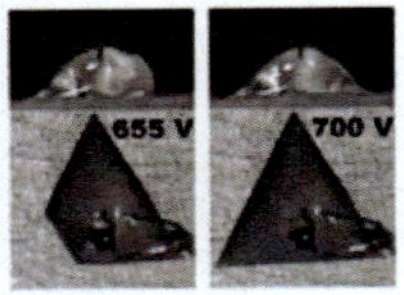 [119]	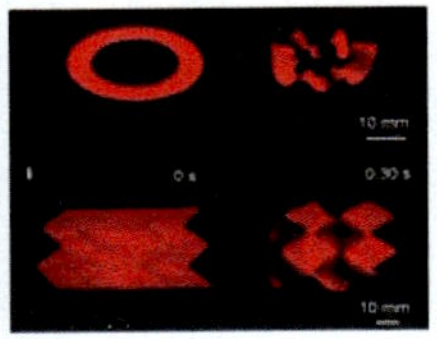 [71]	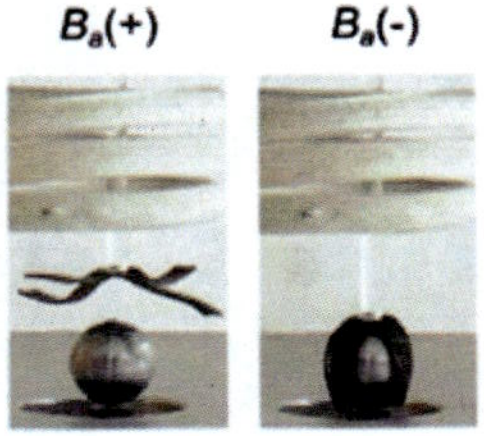 [170]	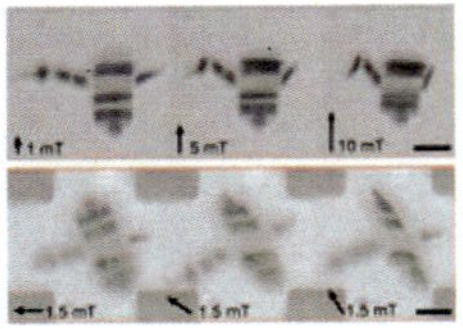 [25]
[119] (Reprinted with permission from the Royal Society of Chemistry) [71] (Reproduced by permission from Macmillan Publishers Ltd: Nature, copyright 2018) [170] (Copyright 2019 by John Wiley & Sons, Inc. Reproduced by permission of John Wiley & Sons, Inc.) [25] (Reproduced by permission from Macmillan Publishers Ltd: Nature, copyright 2018) [141, 161]			
Residual stress	>10 nm	Advantages – High accuracy – High resolution – Bending can be controlled by the thickness of the layer. – Microfabrication is possible Disadvantages – Complex multilayer fabrication	Reversible or irreversible
e.g.			

(continued)

Table 2 (continued)

Actuation mechanisms	Size of origami	Advantages and disadvantages	Reversible/Irreversible
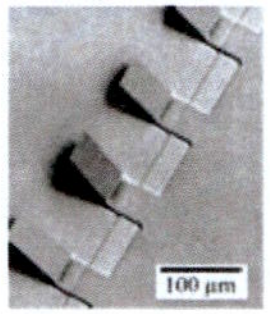 [152]	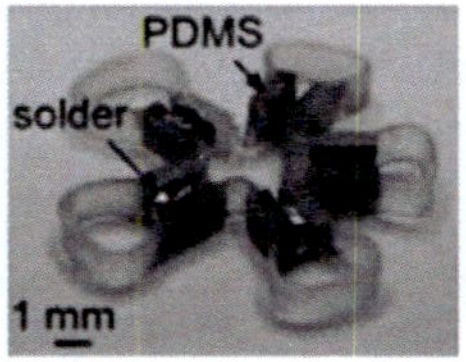 [13]	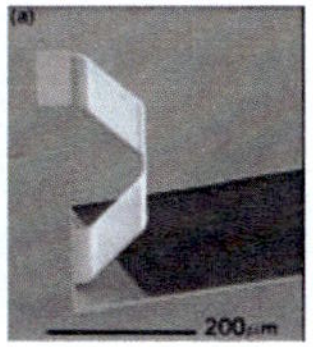 [5]	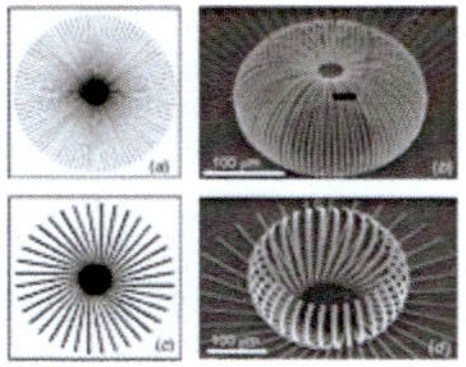 [112]
[152] (Reprinted with the permission of AIP Publishing) [13] (Copyright 2005 by John Wiley & Sons, Inc. Reproduced by permission of John Wiley & Sons, Inc.) [5] (Reprinted with the permission of AIP Publishing) [112] (Reproduced with permission. © IOP Publishing. All rights reserved) [9, 21, 23, 76, 97, 98, 112, 133, 151, 176]			
Swelling (solvent transfer, heat-assisted, pH-responsive, etc.	~100 µm	Advantages – Large deflection – Various triggered actuation – Biocompatibility – Biodegradability Disadvantages – Reduced stiffness – Slow response	Reversible
e.g.			

(continued)

Table 2 (continued)

Actuation mechanisms	Size of origami	Advantages and disadvantages	Reversible/Irreversible
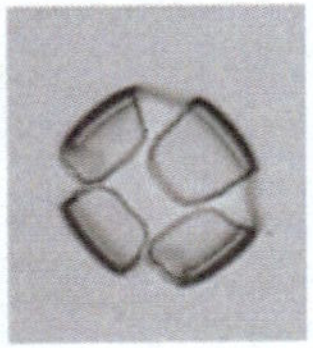 [47]	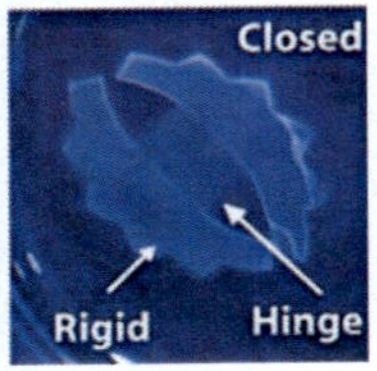 [7]	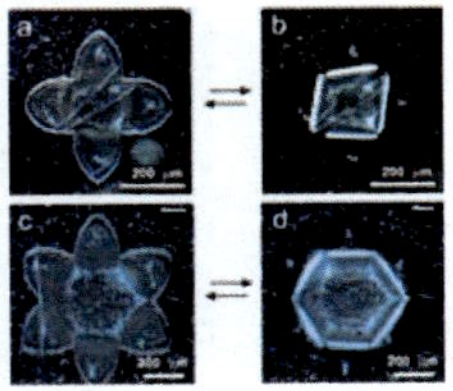 [136]	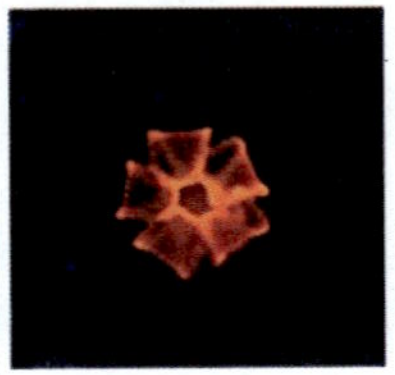 [140]
[47] (Copyright 2009 American Chemical Society) [7] (Copyright 2010, reprinted with permission from Elsevier) [136] (Reprinted with permission from the Royal Society of Chemistry) [140] (Reproduced by permission from Macmillan Publishers Ltd: Nature Materials, copyright 2016) [69, 74, 85, 138, 166, 175]			
Shape memory effect	~1 mm	Advantages – Good mechanical properties – Relatively faster response Disadvantages – Material response depends on the original configuration – Low efficiency – Difficulty to microfabricate – Difficulty to make complex structures	Reversible or irreversible
e.g.			

(continued)

Table 2 (continued)

Actuation mechanisms	Size of origami	Advantages and disadvantages	Reversible/Irreversible
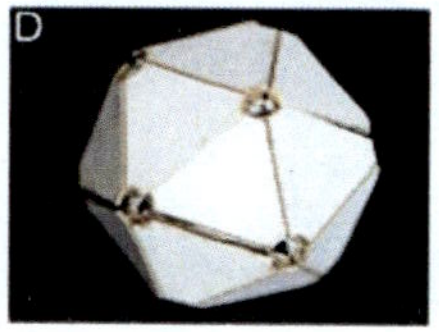 [147]	 [106]	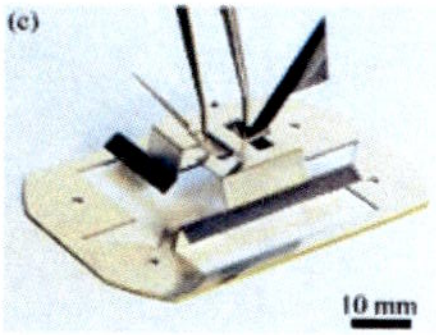 [35]	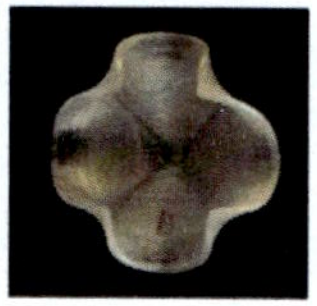 [54]
[147] (Reproduced with permission. © IOP Publishing. All rights reserved) [35, 106] (Reproduced with permission. © IOP Publishing. All rights reserved) [54] (Reproduced with permission. © IOP Publishing. All rights reserved) [31, 142, 144, 157]			
Liquid crystal	~1 mm	Advantages – Fast response – Different triggers Disadvantages – Difficulty to microfabricate	Reversible
e.g.			
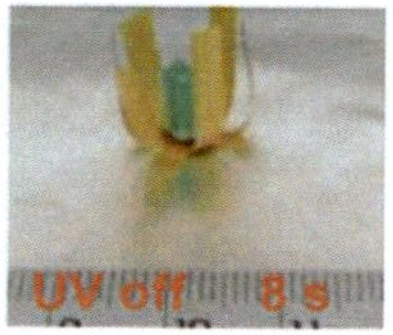 [99]	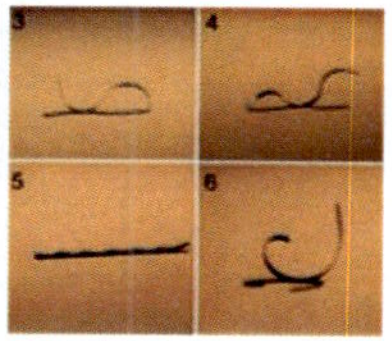 [81]	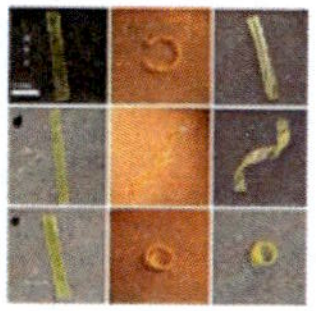 [14]	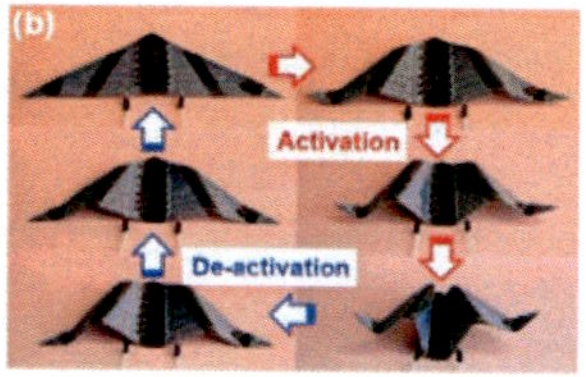 [168]
[99] (Copyright 2018 by John Wiley & Sons, Inc. Reproduced by permission of John Wiley & Sons, Inc.) [14, 81] (Reprinted with permission from the Royal Society of Chemistry) [168] (Reprinted with permission from the Royal Society of Chemistry) [86, 158]			

(continued)

Table 2 (continued)

Actuation mechanisms	Size of origami	Advantages and disadvantages	Reversible/Irreversible
Buckling	<1 mm	Advantages – Fabrication of complex configurations Disadvantages – Less explored – Difficult to control (Multiple stable configurations)	Irreversible
e.g.			
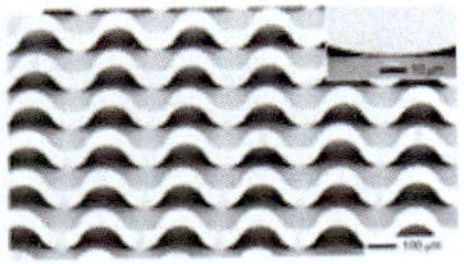 [139]	 [162]	 [40]	

[139] (Reproduced by permission from Macmillan Publishers Ltd: Nature Nanotechnology, copyright 2006) [162] (Copyright 2016 by John Wiley & Sons, Inc. Reproduced by permission of John Wiley & Sons, Inc.) [40] (Reproduced by permission from Macmillan Publishers Ltd: Nature Materials, copyright 2016) [130, 160]

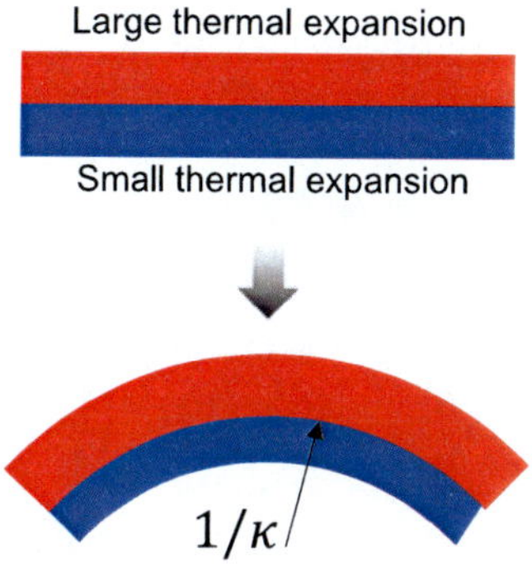

Fig. 5 Bimetallic strip

The term *bimorph* is often used as a more generalized term encompassing both *uni-* and *bimorphs*. We will be following the same convention from here onwards.

3.1.1 Fabrication Strategies for Curved Shapes Using a Bimorph Approach

Rational exploitation of *residual stress, swelling, configuration change in liquid crystal elastomer, or shape memory effect* results in bent shapes (Fig. 6). In microfabrication, material deposition often entails compressive or tensile *residual stress* within the layers of materials. This *residual stress* is, generally, considered detrimental for microfabrication since it acts as a bottleneck for some of the traditional fabrication processes since a large amount of such pre-stress leads to cracking of the films or even of the substrate. Conversely, the same pre-stress may be able to generate a bending moment capable of forming origami shapes if adjoined with another material having different internal stress. For instance, one of the ways to grow SiO_2 on silicon is by thermal oxidation. This thermal growth produces a SiO_2 layer with compressive stress. When a metal with lesser compressive stress is deposited on top of this pre-stressed SiO_2 layer, and then the bilayer is released from the silicon, the strain mismatch results in the bending of this *bimorph* [112]. Similarly, chromium deposition also introduces *residual stress*. Consequently, a bilayer of chromium (Cr) and copper (Cu) exhibits bending as soon as it is released from the supporting structure. As an extension of this bimorph approach, bending in both directions can be achieved by depositing copper and chromium in a Cr/Cu/Cr arrangement [9]. The bending direction, in this case, is dictated by the thicknesses of the patterns of the chromium layer on the top and on the bottom.

Nature presents a variety of movement mechanisms in plants by exploiting *swelling* caused by water. Generally, a gradient in the solvent concentration across a cell wall leads to water intake by the cell through osmotic pressure build-up. Intake of water, in turn, leads to their *swelling*, and this can trigger movement in

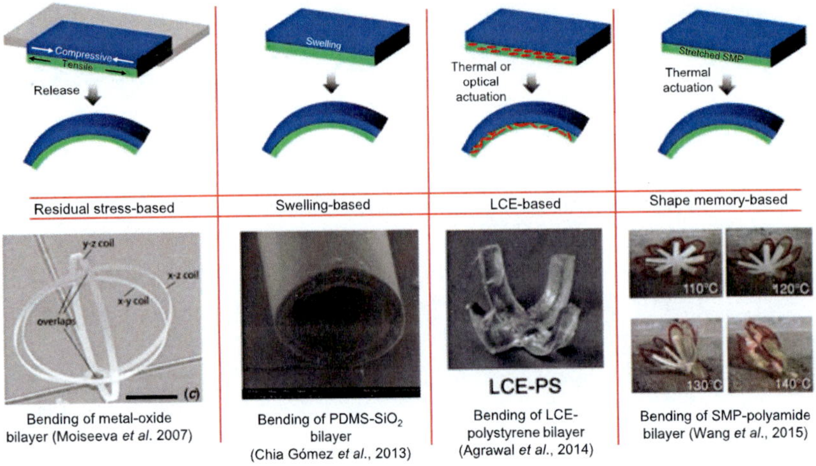

Fig. 6 Implementation of bimorph approach to obtain *bending* using residual stress (Reproduced with permission. © IOP Publishing. All rights reserved) [112], swelling/shrinking (Republished with permission of the Royal Society of Chemistry; permission conveyed through Copyright Clearance Center, Inc.) [19], liquid crystal elastomer (Republished with permission of the Royal Society of Chemistry) [1], and shape memory polymers [157]

plants [68]. Among various examples from nature, one of the most studied is the bilayer design that causes the release of ripe seed from *conifer cones* [27]. A technique emulating such bending can be adopted for the fabrication of small origami shapes [3]. Hydrogels and certain other polymers are capable of absorbing solvents. Temporal and spatial control of this absorption and the resulting swelling of polymers enable bending and folding in those precursors (Fig. 6) [59]. Additional control over the bending or folding may be achieved by designing materials that respond to stimulations such as a change in pH, temperature, enzyme concentration, light intensity, electric field, and solvent concentration [7, 91, 96, 132, 148, 156].

A *liquid crystalline elastomer* (LCE), another type of active material, is synthesized by aligning liquid crystals (LCs) inside a cross-linked polymer matrix by means of mechanical, surface, or chemical forces [115]. The aligned LCs enhance the stiffness along the length of their orientation. On the other hand, when transformed into a randomly oriented configuration, the misalignment is manifested at the macroscale as a contraction of the LCE along the initial alignment direction (Fig. 7). This material deformation characteristic of an LCE makes it possible to fabricate three-dimensional shapes that are activated through different mechanisms. One such method utilizes a light-sensitive molecule called Azobenzene (AZ). AZ exhibits *cis-trans* isomerization under the influence of UV light, where *trans*-AZ has a rod shape and lets the LCE keep its ordered shape, *cis*-AZ has a bent configuration, and its presence adversely affects the ordered structure of the LCE [87]. Disorder in an LCE, as mentioned earlier, leads to the deformation of the material. Therefore, attaching this kind of LCE to an inactive material and then shining UV light invokes the bending of

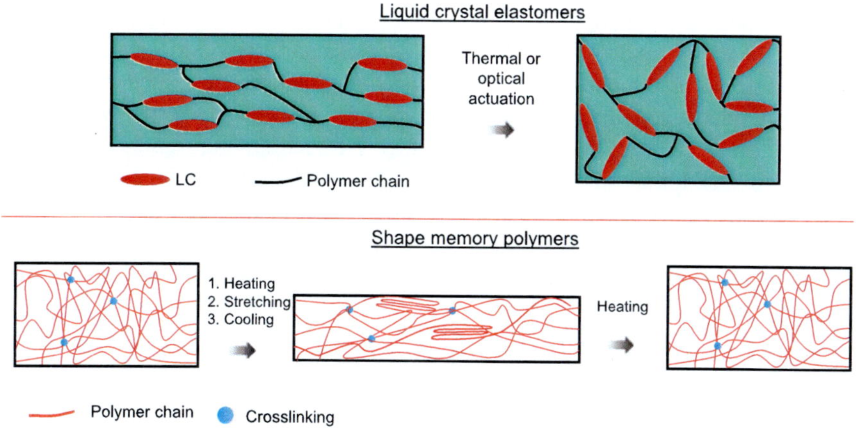

Fig. 7 Working principle of liquid crystal elastomers and shape memory polymers

this assembly (Fig. 6). Similar to this chemically engineered degree of crystallinity, mechanical forces can also induce alignment inside an LCE [28]. In the event of heating beyond its LC–isotropic phase transition temperature, the material loses its alignment, leading to mechanical deformation.

Shape memory is another actuation mechanism that can be employed for bending. *The shape memory effect* is a material's ability to remember a configuration and return to that state upon actuation. Among various materials showing shape memory effects, shape memory alloys (SMAs) and shape memory polymers (SMPs) are the two major categories used in origami fabrication. SMA displays the shape memory effects because of its microstructure. For instance, NiTi, an SMA, has two stable crystal structures: *austenite* (stable at high temperatures) and *martensite* (stable at low temperatures). A deformation applied in the SMA's *martensitic* state is reversed upon converting it to *austenite*. Generally, SMAs are stretched in the *martensitic* state. The stretched SMAs contract when heated. This contraction results in bending in a bilayer, as depicted in Fig. 6. SMA increases the flexibility associated with system design since heating required for the actuation can be induced simply by applying a Joule heating current. Wireless powering with the help of an electromagnetic field is sufficient to attain the folding of centimeter-scale origami [15]. Despite these advantages, SMA is rarely used in smaller length scales due to the difficulty associated with its micropatterning.

SMPs function based on an entirely different principle [11, 89, 124]. SMPs are generally amorphous polymers (polymer chains are randomly oriented) with restricted polymer chain mobility at room temperature. When they are heated above a specific transition temperature, the polymer chains attain more flexibility and are able to twist and rotate. At this stage, deformation is induced easily due to the enhanced mobility of the chains (Fig. 7). The polymer is then cooled down, and the deformation is temporarily locked down through physical or chemical interactions of the chains. When reheated, the original shape is recovered. Therefore, bending can be induced with such a deformed SMP that abuts an undeformed SMP sheet (Fig. 6).

Although we mentioned various techniques where strain mismatch is used for bending, it must be noted that not any combinations of two materials that are having different strains offer a uniform bending toward one direction. A thin and stiff material on top of soft and thick material can result in wrinkled shapes on the thin material [18]. This kind of deformation is not of interest at the moment for designing origami fabrication processes. Another peculiar case is when the bilayer is longitudinal rather than along thickness. The differential strain along the lateral dimension can lead to curled shapes [17, 51, 69].

3.1.2 From Bending to Folding Using Bimorph Approach

A *fold* features a large curvature over a length that is considerably smaller than the overall size to the sheet [84]. Two typical design strategies are followed for fabricating a single fold shape and are depicted in Fig. 8. In the first strategy, the actuator is present only at the fold, and its deformation induces folding on the sheet assembly. The same technique can also be utilized for designing the timely folding of a bimorph. For instance, the application of a phase transforming or stiffness changing polymer at the folds of the bimorph can prevent the spontaneous folding of residual stress-based bimorphs and instead enables a controlled bending of the film driven by softening of the polymer with an appropriate trigger [8, 80]. In the second case, although the actuator is present along the whole length of the sheet, the bending is majorly focused on the fold region. Such local bending is possible by increasing the bending stiffness

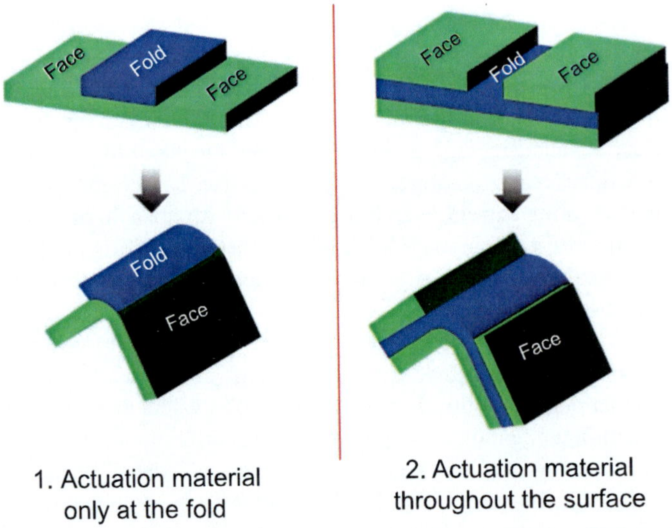

Fig. 8 General *folding* strategies

of the faces (shown in Fig. 8). A typical example is a strip of bimorph with two rigid panels at the face regions. The bending predominantly happens at the folds since the faces are reinforced with an additional layer of material. As an extension of this folding technique, bi-directional folding (mountains and valleys) can be achieved if a soft material is sandwiched between two sheets having patterned openings on the top and bottom layers. This ability to fold in both directions allows for the fabrication of complex shapes. In one of the demonstrations by Na et al., such a tri-layer design was used to make a "crane" shape at micron length scale [114] (Fig. 9). The soft hydrogel sandwiched between two rigid layers had an elastic modulus two orders of magnitude less than the ones that were sandwiching it. This difference in their moduli ensured that the deformation led to *folding*, but not *bending*. It must be noted that, alternatively, a *local actuation* of the bimorph at the folds can also result in folding.

We mentioned earlier that bimorphs, without any other modifications, results in *bending*. An exception to such uniform bending is the hierarchical *bending* of polymer bilayers, which leads to *folding* [137]. Initially, when the absorption of water is activated at the edges of the bimorph, the bending occurs only at the borders of the pattern. This localized absorption and swelling lead to a tube formation at the edges. High rigidity of such tubes restricts the subsequent bending only to the

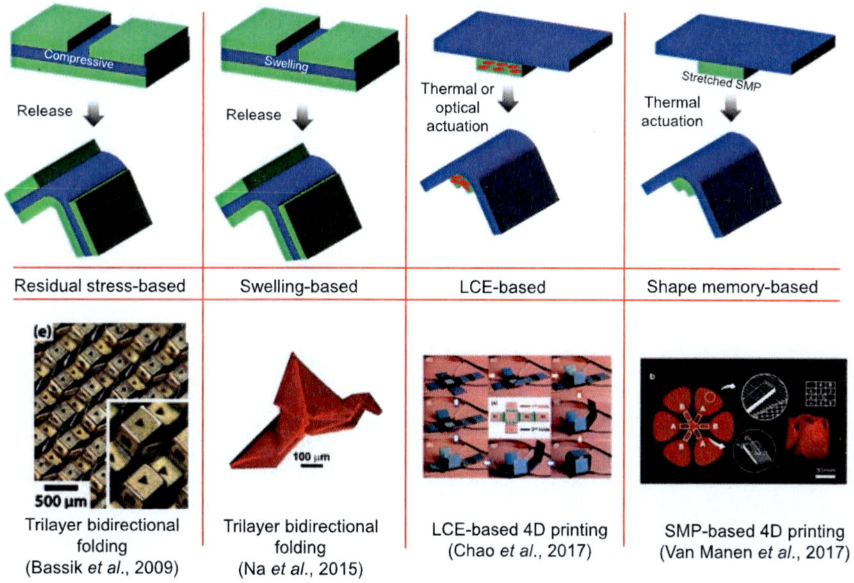

Fig. 9 Implementation of bimorph approach to obtain *folding* using residual stress (Reprinted from Bassik et al. [9], with the permission of AIP Publishing), swelling/shrinking [114] (Copyright 2014 by John Wiley & Sons, Inc. Reproduced by permission of John Wiley & Sons, Inc.), liquid crystal elastomer [168] (Reproduced with permission from The Royal Society of Chemistry), and shape memory polymers [145]

intersections of those tubes. This localized bending enables folding without any additional patterning [136] (Table 2).

As far as the execution of the bimorph-based folding strategies is concerned, most of the examples discussed so far were based on *lithography*. For patterning some of the materials, we may have to rely on other techniques such as *direct writing*. Let us revisit the direct writing-based additive manufacturing method that was mentioned earlier in this chapter. It is a process of making complex three-dimensional shapes by *layer-by-layer* writing. In this technique, the volume of the target shape determines the time required for the fabrication, irrespective of whether a structure is hollow or solid. Consequently, fabrication time for a two-dimensional patterned origami precursor sheet is an order of magnitude less than the time required for the fabrication of corresponding three-dimensional shape. Therefore, folding of two-dimensional sheets is preferred over the direct writing of the corresponding three-dimensional shapes, as far as the throughput of fabrication is concerned. A further improvement could be made by introducing 4D printing, where the morphing of the shape with respect to time is the fourth degree of freedom (Fig. 9). This kind of morphing can reduce the production time and enables an extra degree of freedom, giving rise to *free-form* manufacturing. This kind of manufacturing is generally realized by incorporating active materials such as LCE and SMP in additive manufacturing (multi-material printing), followed by actuation [42].

Incorporation of the LCE into direct writing technologies was made possible with a relatively less viscous LCE ink. Low viscosity simplified the writing process and helped introducing the required degree of molecular alignment in the resulting precursor pattern. Shear-thinning during direct writing causes the needed alignment of liquid crystals. The folding action can be initiated through Joule heating of an embedded wire inside a bimorph containing LCE [168].

Multi-materials polymer 3D printers that are compatible with SMPs can fabricate foldable multi-material precursors. SMP is printed only on one side of the fold region to introduce a strain difference favorable for folding [41]. At a temperature higher than the transition temperature, the printed sheets are stretched and brought to a temperature lower than the transition temperature. When cooled, the SMP retains its original shape, but the elastomer to which it is attached does not. This difference in strain results in the bending of the sheet. When this bent shape is heated again, SMP goes back to the original shape leading to the flattening of the sheet. Printing at a temperature above the material's transition temperature can simplify the overall process even further as shown by Van Manen et al. [154]. During the printing, the ink is extruded. This stretching is memorized by the polymer and leads to the morphing when reheated (Fig. 9).

3.2 *Material Gradient Approach*

Fabrication of bimorphs requires either multilayer photolithography or a multi-material writing, increasing the complexity of the process. Moreover, the strain

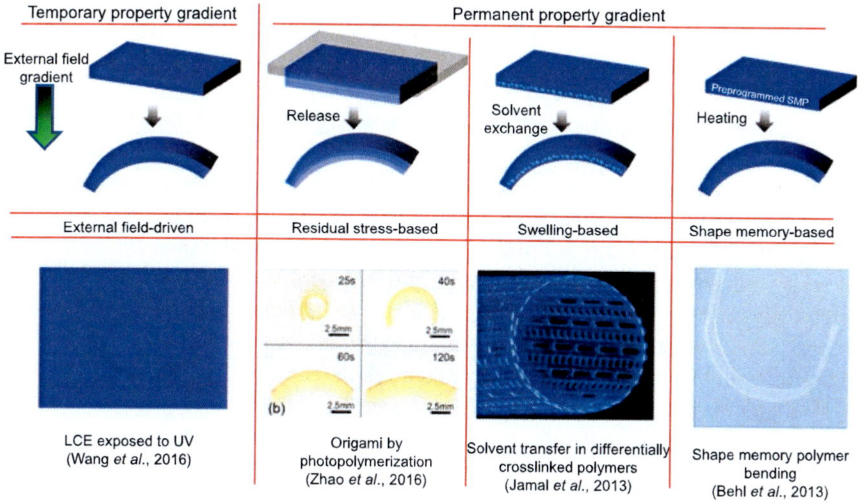

Fig. 10 Implementation of material gradient approach to obtain *bending* using a gradient in the external field [158], residual stress [174] (Copyright 2017 by John Wiley & Sons, Inc. Reproduced by permission of John Wiley & Sons, Inc.), swelling/shrinking [64] (Reproduced by permission from Macmillan Publishers Ltd: Nature Communications, copyright 2013), and shape memory polymers [10] (Copyright 2013 by John Wiley & Sons, Inc. Reproduced by permission of John Wiley & Sons, Inc.)

mismatch that drives the morphing of the bimorph may cause occasional delamination at the interface of the two materials. Having a smooth property gradient is the remedy for such problems that arise from a large strain mismatch (Fig. 10).

3.2.1 Fabrication Strategies for Curved Shapes Using Material Gradient Approach

Material property gradient within a sheet material can be categorized into two types: (i) *transient property gradient* and (ii) *permanent property gradient*. A *transient property gradient* is attained via reversible molecular rearrangement. An LCE sheet containing AZ, when irradiated, experiences different degrees of rearrangements at its polymer matrix across its thickness, with the side closer to the light source experiencing more changes. The difference in molecular rearrangement results in a gradient in deformation with more contraction on the side closer to the light source, leading to a bent shape. This type of tactic may be applied for swelling-based bending too [49, 52, 123]. A concentration gradient of absorbing molecule in the medium results in a difference in the absorption across the cross-section of the sheet. However, these techniques are rarely utilized for the fabrication of complex origami shapes due to its unidirectional bending (Note: here, external field changes the material property across the thickness unlike *external field approach* that we are going to discuss later, where field induces a bending moment without interfering with the material properties). Therefore, a property gradient that is permanently embedded in the material is preferred to a method that purely relies on an external field to fold origami. A permanent property gradient across the thickness of a sheet is fulfilled during its synthesis [75, 120]. Fabrication of photopolymer sheets with varying *crosslinking density* is one of such synthesis methods. Photopolymers crosslinks when exposed to light. The *crosslinking* occurs first in the region that is closer to the light source. For explaining the process, this region is referred as the first layer, even though the *crosslinking* is not really a layer-by-layer process. The polymer undergoes shrinkage as a result of the crosslinking. As far as the first layer is concerned, the polymer is free to shrink in every direction. However, once a solidified layer is formed, the contraction of the subsequent layers would be constrained by the first layer. This constraint results in residual stress inside the material. The *residual stress* this region experiences is analogous to a stretched band. If this residual tension is released, the sheet will bend as shown in Fig. 10. This photopolymer-based execution of the gradient approach, *frontal photopolymerization*, can create complex shapes having millimeter length scale [173]. A similar type of bending can be achieved using direct writing techniques too. Like the *frontal photopolymerization*-driven bending, the residual strain build-up during the writing of the second layer can lead to the bending of the film.

The degree of crosslinking across the thickness of a photopolymer reduces with the distance from the light source. This crosslinking density gradient directly correlated to the porosity density inside its polymer matrix. Specifically, a less cross-linked SU8 sheet possesses a relatively more porosity compared to a more cross-linked sheet. Porosity in a material allows for the absorption of various solvents by the substance. The more porosity the material has, the more solvent it absorbs. Consider a thin sheet with a crosslinking density gradient varying from a high level of crosslinking on the top to a low level of crosslinking on the bottom. If one fabricates such flat sheets with absorbed liquid (e.g., developer solution) inside them, the bending of the faces occurs toward the bottom surface when the solvent is removed. The cause of the bending is attributed to the shrinkage resulting from the removal of the large quantity of the solvent from the highly porous bottom surface of the sheet [63]. Please keep in mind that the reported fabrication based on this technique showed only a reversible bending of the SU8 film since nothing prevents the folded structures from going back to the original shape when placed inside the same solvent again. We will discuss a strategy to make permanent folding later in this chapter.

In the case of an SMP with a preprogrammed bending trained on it, the sheet bends to the "memorized" shape upon heating. Unlike in the case of bimorph, where the linear deformation of the SMP bends a bilayer sheet consisting of the polymer and another material, bending, in this case, is not assisted by a second material.

3.2.2 From Bending to Folding Using Material Gradient Approach

Localized modification of material properties is easy to achieve with photopolymers by selectively illuminating the polymer sheet. In one of the methods, selective

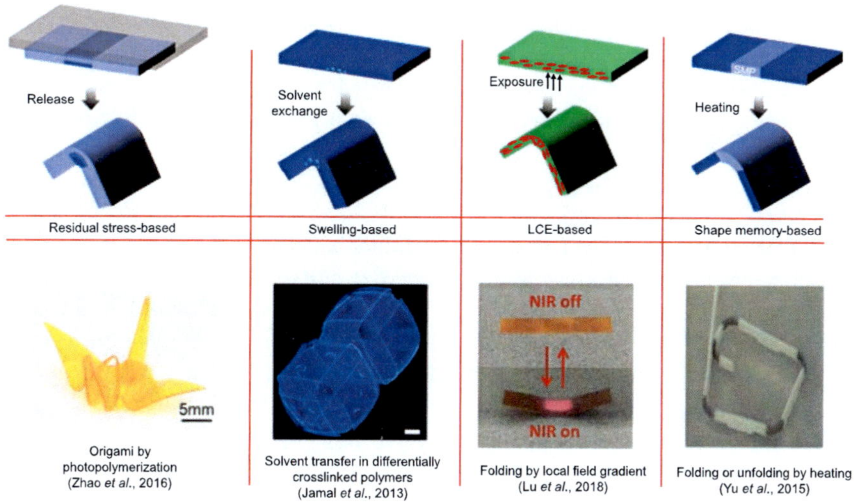

Fig. 11 Implementation of material gradient approach to obtain *folding* using residual stress [174] (Copyright 2017 by John Wiley & Sons, Inc. Reproduced by permission of John Wiley & Sons, Inc.), swelling/shrinking [64] (Reproduced by permission from Macmillan Publishers Ltd: Nature Communications, copyright 2013), liquid crystal elastomer [99] (Copyright 2017 by John Wiley & Sons, Inc. Reproduced by permission of John Wiley & Sons, Inc.), and shape memory polymers [167] (Copyright 2015, with permission from Elsevier)

exposure was performed such a way that it concentrates the residual stress from the aforementioned sequential crosslinking to the fold region, leading to folding (Fig. 11). Selective irradiation, in certain other photopolymer material, results in stress relaxation inside the material due to a polymer network rearrangement [90]. Mechanical and optical stimuli have been combined in a technique called *photo-origami* to exploit this stress relaxation to accomplish folding [128]. In this method, the sheets are stretched perpendicular to the length of the fold, followed by a UV exposure at the fold region. By exposing the fold region of a stretched film, localized stress relaxation can be controlled. Exposure energy density varies across the thickness of the sheet. The side closer to the light source experiences a higher energy density. As a result, the irradiated side of the film experiences relatively more stress relaxation, causing local bending. The porosity density gradient that emerges from a differential crosslinking can also be focused on the fold region to achieve folding at submillimeter scales (Fig. 11).

In another related light-based folding, sequential folding in mesoscale is achieved with *Shrinky dinks*, by printing the folds with inks of different colors followed by exposure to high-intensity light [97]. Unlike the previously mentioned photosensitive materials, here, the bending occurs due to the local heating of the material. The absorption wavelength by the ink depends on its color/absorption property. Absorbed light heats up the fold. Therefore, folds having different colors can be locally heated when they are exposed to the complimentary spectrum that they can absorb. *Shrinky dinks* contract considerably upon heating because of strain release, and as a result, folding of the sheet happens. Instead of colored inks on the hinges, a graphene ink may be deposited to bend *Shrinky dink* material to actuate the folding with microwave [26]. When exposed to microwave, since graphene is not transparent to microwave, the sheet is locally heated. Like the previous case, this local heating results in bending. However, the orientation of the fold with respect to the wave affects the folding of the material, limiting the versatility of the technique. In nanoscale, a focused ion beam (FIB) is used for the fabrication of folded shapes. FIB, when bombarded with the nanofilm, induces tensile or compressive stress on the film depending on the operating parameters. Gallium ion from FIB when colliding with the film can remove atoms from the gold film. The vacancies formed via this bombardment lead to grain coalescence and as a result, residual tensile stress on the top layer of the film, as illustrated in Fig. 12. This leads to the bending of the thin film [133].

Heating-based folding can be induced in an LCE by locally heating the sheet beyond its transition temperature with IR irradiation (Fig. 11). The side that is heated experiences a shrinkage due to the molecular rearrangement, leading to the bending of the LCE sheet. While folding in LCE is achieved through IR illumination of an isotropic material, an SMP-based folding requires local preprogramming of an SMP. A direct writing technique may be employed to fabricate folds with SMP and the faces with an inactive material. By training this SMP-made fold region, folded origami can be obtained (Fig. 11).

3.3 External Field Approach

Here we discuss a new approach where bending of a sheet is carried out without changing its properties across the cross-section, contrasting previously discussed cases where the material properties of the sheets are tuned to induce bending. Various forces, including capillary, magnetic, and external compressive forces, can be the cause of bending in an external field approach. Since the external field approach involves different forces, it is essential to discuss them briefly before delving into the fabrication aspects of it.

Capillary effects scale favorably for the micron-scale actuation. A comparison of the body forces ($\rho g l^3$) with the surface tension forces (γl), where l is the length scale, γ is the surface tension, ρ is the density, and g is the gravitational constant, reveals that the surface tension effects dominate in a smaller length scale. Therefore, surface tension can act as an actuation mechanism for micro-origami fabrications. Nature takes advantage of this fact, as observed in the case of the drinking mechanisms of both phalaropes and hummingbirds. For a phalarope, this heightened effect of surface tension at small scales is a way to accomplish, a gravity-defying intake of prey-laden water toward its mouth. A Hummingbird's tongue exploits the surface tension to close lamellae on its tongue to entrap honey at each of its dips into honey-filled flowers. Here, the tongue closes because of the interaction between the soft

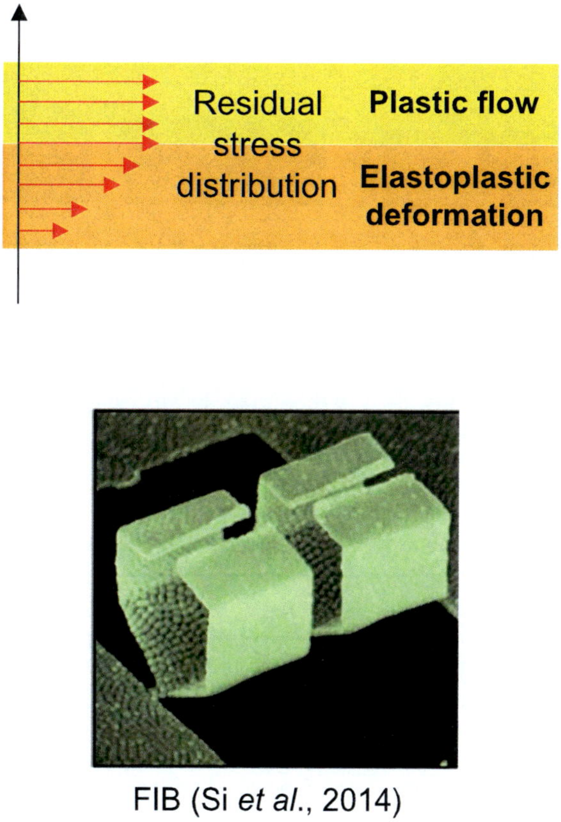

Fig. 12 Folding induced on nanofilms using focused ion beam. Reprinted with permission from Si et al. [133]. Copyright 2014 American Chemical Society

lamellae and the honey. This bending phenomenon, surface tension-driven bending of the soft materials, is known as *elastocapillary* bending.

Droplets most often assume a spherical cap shape on a solid or a liquid surface to minimize the total surface energy. A droplet on a hard surface has three different surface energies associated with it: surface energy due to the air–liquid, solid–liquid, and solid–gas interfaces. For a droplet-surface-gas system, if the energy of liquid–gas and solid–liquid interfaces is very high as compared to the other interface, the droplet forms a complete sphere to minimize the total energy. On the other hand, if the surface energy of the solid–liquid and liquid–gas is sufficiently small, the liquid spreads on the surface to avoid having the high energy offered by the solid–gas interface. In a case that is midway between these two cases, the droplet forms part of a sphere. Now imagine having a droplet on a thin sheet. The system has an extra option to wrap the sheet around the droplet to minimize the liquid–gas interfacial energy at the expense of the bending energy of the sheet. For a sufficiently thin sheet, this increase in energy is less than the reduction in surface energy. As a result, bending occurs (Fig. 13). Length scale below which the effect of capillary forces becomes significant is found by comparing the Laplace pressure, γ/l with the hydrostatic pressure, $\rho g l$, where l

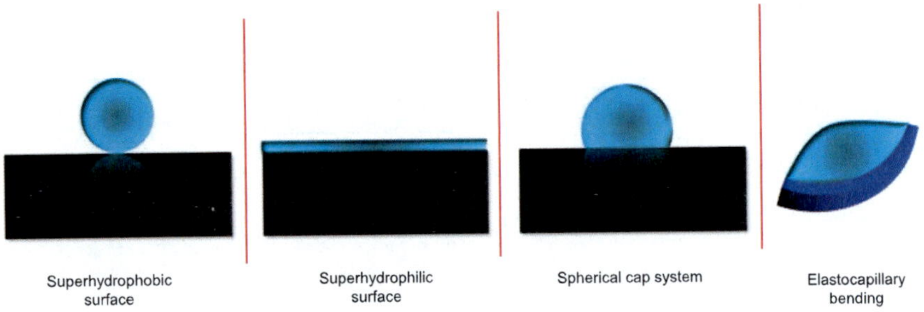

Fig. 13 Different configurations resulting from droplet-surface interactions

is the length scale, γ is the surface tension, ρ is the density, and g is the gravitational constant and is given by the capillary length, $l_c = \sqrt{\frac{\gamma}{\rho g}}$. For a significant bending, the curvature can be assumed to be of the order of $1/l$. Therefore, the bending energy density ($E_B = \frac{1}{2}B\kappa^2$) scales as $Eh^3/24(1-\nu^2)/l^2$. The length scale at which the surface energy density of a droplet, γ, becomes significant is then obtained by equating the bending energy density (E_B) with it and is given by *elastocapillary* length $L_{EC} = \left(\frac{Eh^3}{24(1-\nu^2)\gamma}\right)^{\frac{1}{2}}$.

Controlled bending of a thin sheet using *elastocapillary effect* is achieved either with different droplet sizes or by adjusting the interfacial tensions of the sheet substrate. Droplets with a characteristic length too small or too big compared to the thin sheet do not bend the sheet significantly. In the case of a small droplet, energy gain would be negligible, and for a big droplet, bending of the sheet would require a significant deformation of the droplet, an energetically expensive process. Applying an electric field to the droplet changes the surface energy of the droplet. This change is manifested as change in the angle that the droplet makes with the substrate. When an electric field is applied to the droplet system, the energy associated with it is modified with additional electrostatic energy which is of the order of $\frac{\epsilon L^2 V^2}{2(d+h)}$, where ϵ is the dielectric constant, d is the insulating layer thickness, V is the voltage, h is the thickness of the sheet, and L is the length of the sheet. Comparing the electrical energy with the surface energy would give an idea about the order of magnitude of the voltage at which the contact angle is significantly affected by it and is given by $V \sim \sqrt{\frac{\gamma(d+h)}{\epsilon}}$. Alternatively, the energy of the solid surface can be changed for controlling the folding angle [105, 119]. Oxygen plasma treatment is usually used for controlling the surface energy of the solid.

The magnetic effect can be utilized for causing deformation in materials. Elastomeric films embedded with magnetic particles can be magnetically controlled. An external magnetic field, B forces the particles to orient along its direction [56].

Buckling occurs when a thin sheet is subjected to a compressive force. Energy density E_s of a compressed flat plate of a thickness t and a length L is proportional to $t\left(\frac{\delta}{L}\right)^2$, where 2δ is the displacement by which it is compressed. Pure bending energy E_B is, on the other hand proportional to $t^3\kappa^2$, where κ is the curvature of the flat plate. Thicker films experience a uniform compression across the cross-section of the plate. However, as the thickness reduces, the bending of the sheet becomes energetically more favorable. Therefore, thinner plates experience out of the plane buckling when compressed.

3.3.1 Fabrication Strategies for Curved Shapes Using External Field Approach

Researchers from École Polytechnique, France, made complex three-dimensional shapes by applying capillary force by placing a drop of water on top of a precut patterned two-dimensional polydimethylsiloxane (PDMS) sheet (Fig. 14) [121]. They used a droplet bigger than the volume of the target shape, and then the surplus water is evaporated to form completely closed 3D shapes. Until then, the capillary effect in microfabrication was considered purely detrimental, the one that leads to undesirable stiction in multilayer lithography and coalescence in the case of high aspect ratio structures. The three-dimensional shapes formed using PDMS, however, was temporary. The sheets returned to their original shapes, as soon as the water droplets holding the three-dimensional shape together evaporated. Patterning of the

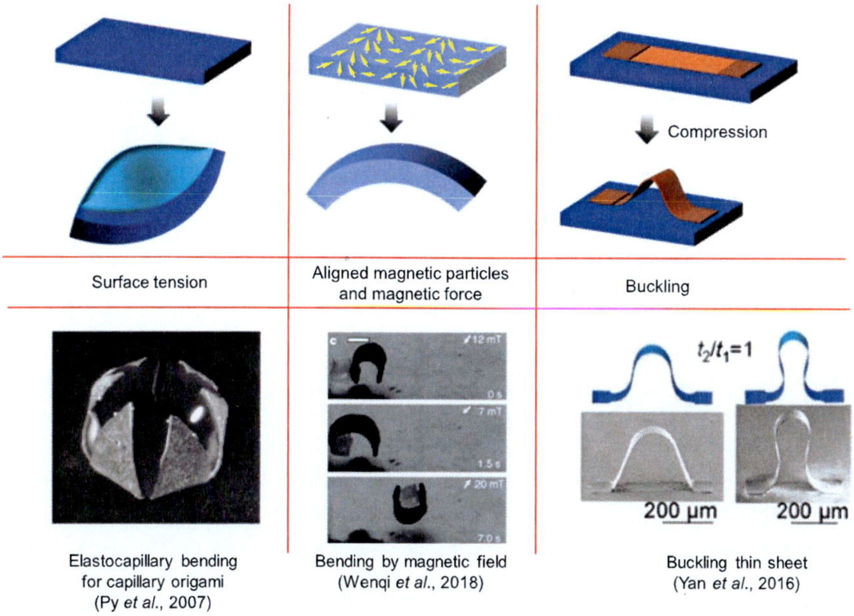

Fig. 14 Implementation of external field approach to obtain *bending* using surface tension [121] (Reproduced with permission, Copyright 2007 by the American Physical Society. https://doi.org/10.1103/PhysRevLett.98.156103), magnetic force [56] (Reproduced by permission from Macmillan Publishers Ltd: Nature, copyright 2018), and compressive force [163] (Copyright 2017 by John Wiley & Sons, Inc. Reproduced by permission of John Wiley & Sons, Inc.)

PDMS sheets to make two-dimensional shapes was challenging too. Moreover, the PDMS sheets had the same material property throughout their planform. Despite these shortcomings, the capillary origami method gave a new facade to one of the lithography limitations and opened a new door to three-dimensional fabrication. High elastic modulus may pose a limitation to this approach when other materials are used. A thinner sheet can be used to overcome the constraints posed by the high elastic modulus values of the materials. For instance, a silicon nitride with a thickness of 100 nm can be bent by the surface tension of water. Another approach to bend a high elastic modulus material is by softening it temporarily, followed by subjecting the material to capillary bending. Polymer sheets, for example, can be softened by heating them above its glass transition temperature.

Magnetic force is exploited for the reversible bending of the elastomers. The fabrication of sheets that can be bent using an external magnet is achieved either using photolithography or using direct writing techniques. The orientation of magnetic particles determines the final folding shape of the sheet. Differently oriented magnetic particles on different faces are initially achieved by orienting the particles inside a curable silicone material using an external magnetic field followed by locking them in place by curing the silicone. A precision of 100 μm is possible with photocurable silicone materials, as shown by Xu et al. [161]. A sinusoidally arranged magnetic particles inside the sheet induce uniform bending on the sheet in a magnetic field, as shown in Fig. 14.

Releasable multilayered 2D precursors are buckled to form complex three-dimensional shapes. These shapes are fabricated using SOI wafers and sacrificial layers. A bonding location is designed, and multiple layers are transferred using polyvinyl alcohol onto a stretched elastomer. When the strain of the elastomer is released, the transferred structure buckles (Fig. 14). Numerous configurations may be formed if the strain releasing sequence is controlled.

3.3.2 From Bending to Folding Using External Field Approach

Capillary-based folding of two-dimensional sheets made of rigid panels and flexible hinges can result in folded three-dimensional shape rather than a bent shape (Fig. 15). The shape obtained represents the configuration corresponding to the energy-optimized state, as explained earlier. (Therefore, the angle to which the sheet folds can be estimated by minimizing both bending and surface energies by assuming that the volume of the droplet placed on the structure remains constant during the folding process [153]). In a negative photoresist, the exposure energy density in a location determines the local degree of crosslinking. The higher the crosslinking density, the stiffer the sheet becomes [36, 43, 44]. Therefore, an origami precursor sheet with stiff faces and compliant folds can be made easily by controlling the exposure. Polymer softens when they are heated to a temperature close to or beyond its glass transition temperature. The softening temperature of a polymer is a function of the degree of crosslinking. As a result, when heated, the precursors soften more at the fold region, leading to folding of the shape upon capillary actuation [43].

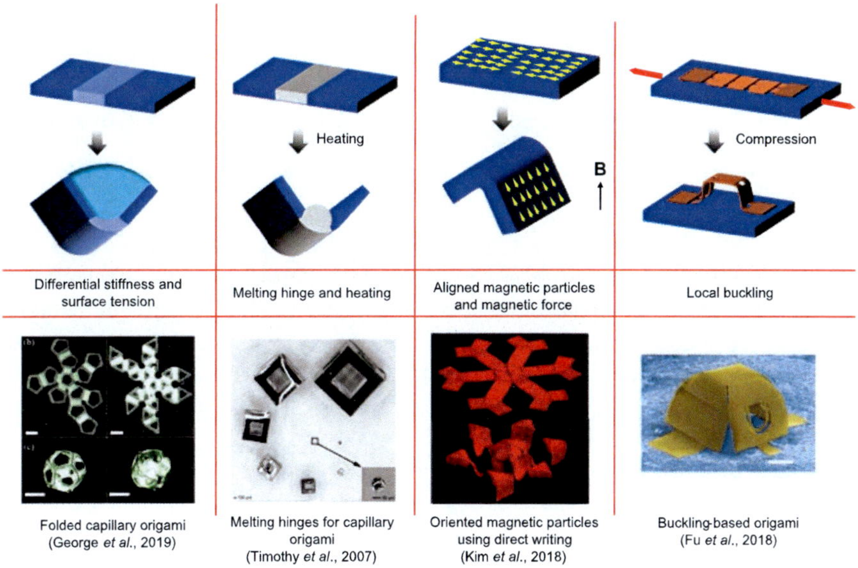

Fig. 15 Implementation of external field approach to obtain *folding* using surface tension [43, 93] (Copyright 2007, American Chemical Society), magnetic force [71] (Reproduced by permission from Macmillan Publishers Ltd: Nature, copyright 2018) and compressive force [40] (Reproduced by permission from Macmillan Publishers Ltd: Nature Materials, copyright 2018)

Multilayer microfabrication technique can be used to make two-dimensional shapes having different materials at different locations, along the length and the thickness. Precursors can be made with high melting point rigid panels and low melting point folds. After releasing from the supporting structure, if these patterned structures are heated, the low melting point materials melt. Consequently, the panels made of high melting point materials is folded because of the surface tension of the melted hinge. Folding by a locally applied solder is an example of such a technique [12, 45]. Photolithography made local deposition of solder possible. This deposition of solder can be combined with other lithography techniques such as e-beam lithography and nanoimprint lithography to make panels with hinges in micro- and nanoscales [79]. These fabrication techniques enable the creation of smaller shapes. Moreover, the absence of any droplets at room temperature makes the fabrication process simpler. However, the materials used in this method are not ideal for most of the biological applications. Implementation of a folding procedure on polymers can overcome this shortcoming [6].

The electrostatic force can drive folding action in millimeter-sized electro-origami robots [141]. Here, the folding is achieved by introducing opposite charges on each face. However, merely applying voltage is not enough to induce folding. The force which drives the folding can be improved dramatically by introducing a liquid droplet having high permittivity and high breakdown strength at the folds. The placement of liquid bead at the fold region enhances the driving force (Maxwells pressure). Moreover, an application of a high electric field is possible in this technique since the breakdown electric field increases considerably with the introduction of the liquid. Therefore, a high actuation force can be attained here, as compared to a system without any such liquid. This liquid bead prefers to be at the folds due to the electrophoretic forces generated by the high electric field there.

The integration of capillary origami to other existing MEMS components can be realized by patterning conductive materials on top of them. Gold is considered as one of the best options for this application due to its high conductivity and low elastic modulus. The electrical connection between the faces is achieved through the gold connections that are running through the hinges [88]. The low modulus of the gold makes sure that the effect of gold on the folding is minimal. A hand-free bending of the faces is achieved by building conducting loops on them using gold, followed by the application of Lorentz force [129].

As far as the magnetically driven folding is concerned, a sheet where the particle orientation on adjacent faces of the origami is such that they are facing each gives rise to folding in the presence of a uniform magnetic field, as shown in Fig. 15. Aligned magnetic materials are achieved in a polymer matrix through a direct writing technique by sending unaligned magnetizable NdFeB particles through a printer head equipped with aligning magnetic field. In this technique, the aligning magnetic field reorients the particles, and the rheology of the printed material keeps the aligned particle in place. The orientation of the magnetic particles with respect to the printing platform is adjusted by controlling the magnetic field direction and the printing direction.

4 Permanent Folding

A strategy to make a permanently folded structure is to alter the required bending energy temporarily by softening and stiffening the material at relevant instants. A soft material folded with elastocapillary effects if converted into stiff material after the folding can retain the bent shape. This retention is due to the energy-expensive route that it must take to go back to the original flat configuration. One of the ways to achieve this is by exploiting the softening of the polymers. At a lower temperature, the mobility of the molecular chains is low, and this low mobility causes a high stiffness. Polymers, when heated, undergo softening because of the increased mobility of the polymer chains inside the material. The reduced stiffness enables easy folding at an elevated temperature. Subsequent cooling ensures retention of the folded structure. Although such a technique lets the folded structure to keep its shape, it is still not a completely permanent folding method since it can go back to the original shape at an elevated temperature. Chemically altering the polymer could help in overcoming this shortcoming. One way to achieve this change is by controlling the crosslinking of the polymer. The higher the level of crosslinking of a polymer, the higher the stiffness of the material. One of the most explored techniques where crosslinking can be

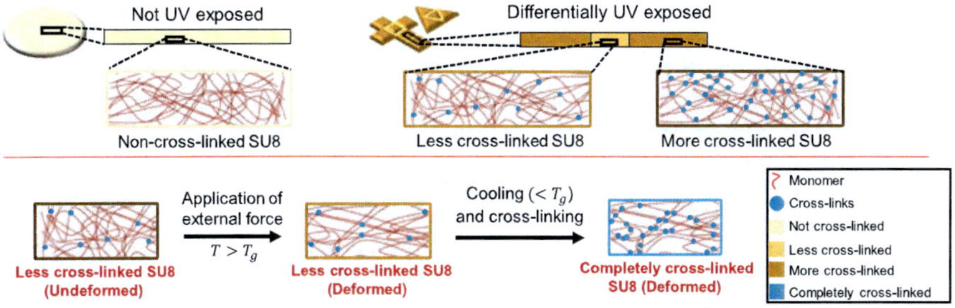

Fig. 16 A strategy to obtain permanently folded shapes using photopolymer [44]. Reprinted with permission from the Royal Society of Chemistry

manipulated with ease is by using a photopolymer. In the case of a negative photoresist that we mentioned earlier in the external field approach section, a partially cross-linked material can be folded at an elevated temperature, brought back to a lower temperature, and then can be completely cross-linked to form a three-dimensional shape which is stable in a wide range of temperature window [44] (Fig. 16).

Folding of two-dimensional sheets that are made of silicon nitride often resulted in permanently closed shapes owing to van der Waals forces, albeit with limited control. Photolithography techniques reduce the surface roughness of the resulting shapes. A sheet pattern with less roughness increases the van der Waals interaction and, therefore, the adhesion. This force can often keep the shape in its folded configuration. However, utilizing this interaction energy alone for making a permanently folded structure in the event of complete evaporation would be unreliable since the gain in energy associated with this may not be always enough to produce repeatable results. Instead of water, if a material that exhibits solid phase at room temperature is used (e.g.,. solder), the folding remains permanent at room temperature. However, the folding has to be performed at an elevated temperature to melt the solder. In one of the early attempts, researchers dipped one whole side of a two-dimensional patterned sheet on melted solder to get enough amount of solder to adhere to that side to drive the folding [45]. The amount of solder transferred to the sheet was critical in achieving the desired target shape. By controlling the surface tension of the liquid and surface energy of the sheet surface, one can control the amount of adhered solder. The energy of the surface can be adjusted with surface treatments such as plasma treatment and chemical cleaning. The surface tension of a liquid, on the other hand, can be altered by changing its temperature. The higher the temperature, the less the surface energy associated with the liquid. These two methods can be utilized to get the right amount of solder on the sheets. Later, researchers lithographically patterned the solder at the folds, allowing for more precise control of the location and quantity of solder.

Although we refrained from mentioning any manual-folding up until now, it would be worth mentioning some such techniques here that form permanently folded shapes. Micro-origami fabrication by Whitesides and group utilized electroplating to join the edges to form permanently folded three-dimensional shapes [16]. Though the folding of the two-dimensional shapes was performed manually, this welding method has the potential to be extended to smaller dimensions. In a fabrication system, which is a fusion of both direct writing and wet origami technique, origami shapes are obtained by the folding of a directly written layer [2]. The folding is facilitated by the solvent contained in ink, which controls the storage modulus of the structure. The storage modulus increases as the solvent evaporate from the structure. Therefore, a graded volatility solvent system is adopted to have better control over the mechanical properties of the sheets and, thus, a controlled folding [2]. This technique has the capability or makes TiH_2 shape, which later can be converted into TiO_2 by annealing at 1050 °C for 2 h in air. Here patterning is performed by cutting the film, and folding is carried out manually. Another method for making permanently folded shapes is by designing parylene C balloons filled with paraffin wax at the hinges [150]. Wax is melted by sending Joule current to a heater that is fabricated on balloons. The hinges become flexible as a result of this process. The bending is achieved manually, and the heater is turned off to obtain the required rigidity of the folds. The folded shape will go back to the original shape upon reheating. The major drawback of this method is the involvement of complicated fabrication routes and the difficulty associated with injecting wax into the balloons.

5 Carbon Origami MEMS

Carbon microparticles and micropatterns have found various applications, including sensors, flexible electronics, and photonics owing to their unique properties. Surface area, topology, and material properties of carbon influence the design and development of those types of system components. All these qualities can be controlled with the origami design techniques. Releasing of the two-dimensional elements from supporting structures itself increases the available surface area. Folding them and tightly packing them could further increase the available surface area in a unit volume and can create complex topologies that are otherwise difficult to achieve. Carbon also possesses excellent mechanical properties. Shapes like Miura-ori can favorably modify overall mechanical properties even further [101, 134]. However, the carbon materials that we see in day-to-day life are not foldable. That begs the question: how can we make folded carbon shapes?

There are three significant ways to fabricate carbon-based origami.

(1) By folding graphene or graphene-based materials

Although most of the carbon materials that we are accustomed to are brittle, carbon sheets made of single or multiple layers or atoms—single layer or multilayer graphene—are flexible. Researchers observed sudden changes in the profiles during atomic force microscopy (AFM) scanning of functionalized graphene sheets (FGS). Such changes in the horizontal scan lines indicate the folded FGS resulting from the lateral force exerted by the AFM tip. It proves that the folding of FGS is possible with an external force. Molecular dynamics simulations show that the folding of graphene using a droplet of liquid is possible too.

A solvent exchange folding is possible with graphene. Graphene-based paper can be designed in such a way that it folds when absorbs water [113]. Fabrication of such a material is performed by locally converting graphene oxide (GO) to GO with polydopamine. When reduced, GO is converted into reduced GO (rGO) that are hydrophobic whereas GO-poly dopamine remains hydrophilic. This difference in the affinity of both materials toward water leads to the local absorption of water to induce reversible bending.

(2) By integrating carbon onto other materials

Graphene grown by chemical vapor deposition can be transferred onto SU8 by taking advantage of the adhesion force between them [30]. Polyhedral shapes containing graphene are fabricated by combining this transfer with the self-folding of differentially cross-linked SU8 that was explained earlier. This technique offers a new pathway to exploit the attractive qualities of the graphene more effectively. Earlier, we also mentioned about SiO_2-based bimorphs that fold due to residual stress. By including graphene also into that bimorph, a graphene origami can be developed [109].

(3) By converting folded polymer shape to carbon

Polymers with carbon-rich backbones can be converted into carbon through heat treatment in an inert environment (pyrolysis). During the process, molecules other than carbon are removed, leaving the carbon behind. Interestingly, such a conversion into carbon retains the original polymer shape, but isometrically shrunken. Cellulosic paper survives pyrolysis. So does any forms that are made using cellulosic paper [60]. Therefore, structures made of carbon can be realized by pyrolyzing the paper-based origami. SU8-based polyhedral shapes that are made using capillary origami also can be converted into the corresponding carbon shapes through pyrolysis.

6 Applications of Origami MEMS

Origami at a smaller length scale has various applications in fields including optics, electronics, and biology. The major applications are listed in Table 3.

7 Conclusion

In this chapter, we discussed various strategies for microfabricating origami. Different tactics are generalized as bimorph approach, gradient approach, and

Table 3 Applications of origami

Field	Applications	Representative figures and related references
Optics	Micromirrors	 [169] (Reprinted with the permission of AIP Publishing)
	Three-dimensional optics	 [22] (Copyright 2011 by John Wiley & Sons, Inc. Reproduced by permission of John Wiley & Sons, Inc.)
	Stages	[76]
Electrical and electronic	Three-dimensional electronics	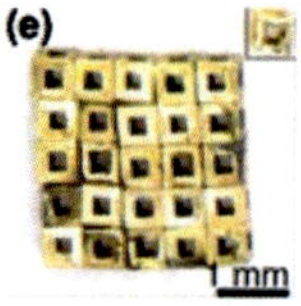 [122] (Reprinted with the permission of AIP Publishing) [46, 50, 72, 73, 172]
	Flexible electronics	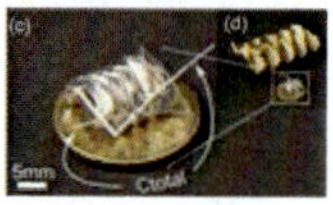 [111] (Reproduced with permission. © IOP Publishing. All rights reserved)
	Self-folding and self-degrading robots	 [159] (Reproduced with permission. © IOP Publishing. All rights reserved) [110, 116]
	Solar panels	 [82] [108, 143]

(continued)

Table 3 (continued)

Field	Applications	Representative figures and related references
	Energy storage	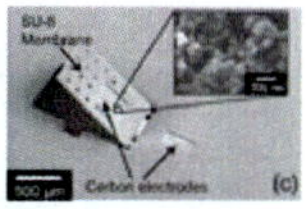 [58] (Reprinted with the permission of AIP Publishing) [135]
Biomedical	Encapsulation	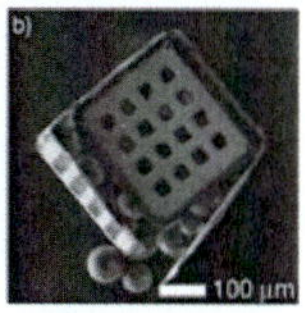 [95] (Reprinted with permission from the Royal Society of Chemistry) [48, 92, 94]
	Drug delivery	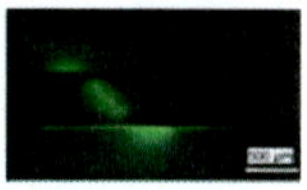 [67] (Copyright 2011 by John Wiley & Sons, Inc. Reproduced by permission of John Wiley & Sons, Inc.) [37]
	Microtube fabrication	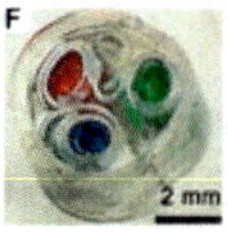 [66] (Copyright 2013 by John Wiley & Sons, Inc. Reproduced by permission of John Wiley & Sons, Inc.) [19, 57, 64]
	Scaffold for tissue engineering	 [61] (Copyright 2010, reprinted with permission from Elsevier) [62, 68]

(continued)

Table 3 (continued)

Field	Applications	Representative figures and related references
	Micro-grippers	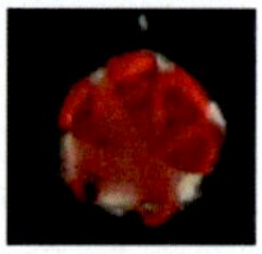 [104] (Copyright 2014 by John Wiley & Sons, Inc. Reproduced by permission of John Wiley & Sons, Inc.) [94]
	Stent to open blocked arteries	 [78] (Copyright 2010, reprinted with permission from Elsevier)
In nature	Leaves	[103]
	Hummingbird's tongue	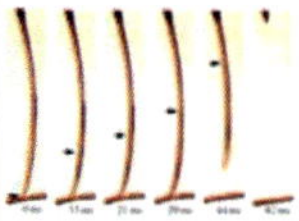 [70] (Reprinted with permission from the Royal Society, U. K.) [125]
	Mimosa	[155]
	Insect wing	[34]
	Seed pod	[4, 39]

external field approach. The folding methods explained in this chapter in conjunction with bottom-up methods such as DNA origami are hoped to span the entire submillimeter range. While there are numerous fabrication approaches for achieving folding at submillimeter scale, process of making positive and negative Gaussian curvatures within a single shape, curved folds (curved origami), and sequentially folding folds at this scale are remaining as paths less traveled. Despite having these less explored fabrication regimes, origami finds applications as micro-containers, micro-grippers, cell encapsulating device, 3D electronics platform, tissue scaffold, and many more. It is these applications that fueled the rapid growth of the origami MEMS technology in the last decade. Origami will continue to benefit electronics, photonics, energy sector, and biomedical field, if the fabrication routes are simplified, and are made more reliable. Considering that the design of origami MEMS can be inspired by other matured fields including macroscale origami, and compliant mechanism, we expect the growth in this field to sustain its pace. The development of design platforms and modeling software should follow the ongoing origami research to achieve these goals of making it simpler and more reliable, and thus to open itself to a broader audience. It should also be noted that the future of the origami MEMS is not just confined to three-dimensional shapes that it can form, rather the dynamics of the folding process and the effect the folding can have on the material properties will also be of importance. The dynamics of the folding process has already been investigated a little in the 4D printing methods that were mentioned earlier in the chapter. Such transformable structures could form reconfigurable shapes and meta-materials. Folding realized by various techniques discussed in this chapter can have different effects on the obtained origami structure and/or the material. Those side-effects are yet to be studied in detail. Such exploration is imperative to develop reliable and affordable, high-resolution multi-functional origami fabrication techniques.

References

1. Agrawal, A., Yun, T., Pesek, S.L., et al.: Shape-responsive liquid crystal elastomer bilayers. Soft Matter **10**, 1411–1415 (2014). https://doi.org/10.1039/c3sm51654g
2. Ahn, B.Y., Shoji, D., Hansen, C.J., et al.: Printed origami structures. Adv. Mater. **22**, 2251–2254 (2010). https://doi.org/10.1002/adma.200904232
3. Andres, C.M., Zhu, J., Shyu, T., et al.: Shape-morphing nanocomposite origami. Langmuir **30**, 5378–5385 (2014). https://doi.org/10.1021/la404955s
4. Armon, S., Efrati, E., Kupferman, R., Sharon, E.: Geometry and mechanics in the opening of chiral seed pods. Science (80-) **297**, 5–9 (2002)
5. Arora, W.J., Nichol, A.J., Smith, H.I., Barbastathis, G.: Membrane folding to achieve three-dimensional nanostructures: nanopatterned silicon nitride folded with stressed chromium hinges. Appl. Phys. Lett. **88**, 1–3 (2006). https://doi.org/10.1063/1.2168516
6. Azam, A., Laflin, K.E., Jamal, M., et al.: Self-folding micropatterned polymeric containers. Biomed. Microdevices **13**, 51–58 (2011). https://doi.org/10.1007/s10544-010-9470-x
7. Bassik, N., Abebe, B.T., Laflin, K.E., Gracias, D.H.: Photolithographically patterned smart hydrogel based bilayer actuators. Polymer (Guildf) **51**, 6093–6098 (2010). https://doi.org/10.1016/j.polymer.2010.10.035
8. Bassik, N., Brafman, A., Zarafshar, A.M., et al.: Enzymatically triggered actuation of miniaturized tools. J. Am. Chem. Soc. **132**, 16314–16317 (2010). https://doi.org/10.1021/ja106218s
9. Bassik, N., Stern, G.M., Gracias, D.H.: Microassembly based on hands free origami with bidirectional curvature. Appl. Phys. Lett. **95**, 1–4 (2009). https://doi.org/10.1063/1.3212896
10. Behl, M., Kratz, K., Zotzmann, J., et al.: Reversible bidirectional shape-memory polymers. Adv. Mater. **25**, 4466–4469 (2013). https://doi.org/10.1002/adma.201300880
11. Behl, M., Razzaq, M.Y., Lendlein, A.: Multifunctional shape-memory polymers. Adv. Mater. **22**, 3388–3410 (2010). https://doi.org/10.1002/adma.200904447
12. Boncheva, M., Bruzewicz, D.A., Whitesides, G.M.: Millimeter-scale self-assembly and its applications. Pure Appl. Chem. **75**, 621–630 (2003). https://doi.org/10.1351/pac200375050621
13. Boncheva, M., Whitesides, G.M.: Templated self-assembly: Formation of folded structures by relaxation of pre-stressed, planar tapes. Adv. Mater. **17**, 553–557 (2005). https://doi.org/10.1002/adma.200400940
14. Boothby, J.M., Ware, T.H.: Dual-responsive, shape-switching bilayers enabled by liquid crystal elastomers. Soft Matter **13**, 4349–4356 (2017). https://doi.org/10.1039/c7sm00541e
15. Boyvat, M., Koh, J.S., Wood, R.J.: Addressable wireless actuation for multijoint folding robots and devices. Sci. Robot. **2**, 1–10 (2017). https://doi.org/10.1126/scirobotics.aan1544

16. Brittain, S.T., Schueller, O.J.A., Wu, H.K., et al.: Microorigami: fabrication of small, three-dimensional, metallic structures. J. Phys. Chem. B **105**, 347–350 (2001). https://doi.org/10.1021/Jp002556e
17. Byun, M., Santangelo, C.D., Hayward, R.C.: Swelling-driven rolling and anisotropic expansion of striped gel sheets. Soft Matter **9**, 8264–8273 (2013). https://doi.org/10.1039/c3sm50627d
18. Chen, D., Yoon, J., Chandra, D., et al.: Stimuli-responsive buckling mechanics of polymer films. J. Polym. Sci. Part B Polym. Phys. **52**, 1441–1461 (2014). https://doi.org/10.1002/polb.23590
19. Chia Gómez, L.P., Bollgruen, P., Egunov, A.I., et al.: Vapour processed self-rolled poly(dimethylsiloxane) microcapillaries form microfluidic devices with engineered inner surface. Lab Chip **13**, 3827–3831 (2013). https://doi.org/10.1039/c3lc50542a
20. Cho, J., Gracias, D.H.: Self-assembly of lithographically patterned nanoparticles. Nano Lett. **9**, 4048–4051 (2009)
21. Cho, J.H., James, T., Gracias, D.H.: Curving nanostructures using extrinsic stress. Adv. Mater. **22**, 2320–2324 (2010). https://doi.org/10.1002/adma.200904410
22. Cho, J.H., Keung, M.D., Verellen, N., et al.: Nanoscale origami for 3D optics. Small **7**, 1943–1948 (2011). https://doi.org/10.1002/smll.201100568
23. Chua, C.L., Fork, D.K., Van, S.K., Lu, J.: Out-of-plane high-Q inductors on low-resistance silicon. J. Microelectromech. Syst. **12**, 989–995 (2003)
24. Cisquella-Serra, A., Magnani, M., Gual-Mosegui, Á., et al.: Study of the electrostatic jet initiation in near-field electrospinning. J. Colloid Interface Sci. **543**, 106–113 (2019). https://doi.org/10.1016/j.jcis.2019.02.041
25. Cui, J., Huang, T.Y., Luo, Z., et al.: Nanomagnetic encoding of shape-morphing micromachines. Nature **575**, 164–168 (2019). https://doi.org/10.1038/s41586-019-1713-2
26. Davis, D., Mailen, R., Genzer, J., Dickey, M.D.: Self-folding of polymer sheets using microwaves and graphene ink. RSC Adv. **5**, 89254–89261 (2015). https://doi.org/10.1039/c5ra16431a
27. Dawson, C., Vincent, J.F.V., Rocca, A.M.: How pine cones open. Nature **390**, 668 (1997). https://doi.org/10.1038/37745
28. De Haan, L.T., Sánchez-Somolinos, C., Bastiaansen, C.M.W., et al.: Engineering of complex order and the macroscopic deformation of liquid crystal polymer networks. Angew. Chem. Int. Ed. **51**, 12469–12472 (2012). https://doi.org/10.1002/anie.201205964
29. Demaine, E.D., Tachi, T.: Origamizer: a practical algorithm for folding any polyhedron. Leibniz Int. Proc. Inform. LIPIcs **77**, 341–3416 (2017). https://doi.org/10.4230/LIPIcs.SoCG.2017.34
30. Deng, T., Yoon, C., Jin, Q., et al.: Self-folding graphene-polymer bilayers. Appl. Phys. Lett. **106** (2015). https://doi.org/10.1063/1.4921530
31. Ding, Z., Yuan, C., Peng, X., et al.: Direct 4D printing via active composite materials. Sci. Adv. **3** (2017). https://doi.org/10.1126/sciadv.1602890
32. Dunn, M.L., Zhang, Y., Bright, V.M.: Deformation and structural stability of layered plate microstructures subjected to thermal loading. J. Microelectromech. Syst. **11**, 372–384 (2002). https://doi.org/10.1109/JMEMS.2002.800932
33. Eder, F., Klauk, H., Halik, M., et al.: Organic electronics on paper. Appl. Phys. Lett. **84**, 2673–2675 (2004). https://doi.org/10.1063/1.1690870
34. Faber, J.A., Arrieta, A.F., Studart, A.R.: Bioinspired spring origami. Science (80-) **359**, 1386–1391 (2018). https://doi.org/10.1126/science.aap7753
35. Felton, S.M., Becker, K.P., Aukes, D.M., Wood, R.J.: Self-folding with shape memory composites at the millimeter scale. J. Micromech. Microeng. **25** (2015). https://doi.org/10.1088/0960-1317/25/8/085004
36. Feng, R., Farris, R.J.: Influence of processing conditions on the thermal and mechanical properties of SU8 negative photoresist coatings. J. Micromech. Microeng. **13**, 80–88 (2003)
37. Fernandes, R., Gracias, D.H.: Self-folding polymeric containers for encapsulation and delivery of drugs. Adv. Drug Deliv. Rev. **64**, 1579–1589 (2012). https://doi.org/10.1016/j.addr.2012.02.012

38. Filipiak, D.J., Azam, A., Leong, T.G.: Hierarchical self-assembly of complex polyhedral microcontainers. J. Micromech. Microeng. **075012**, 1–6 (2009). https://doi.org/10.1088/0960-1317/19/7/075012
39. Forterre, Y., Dumais, J.: Generating helices in nature **333**, 1715–1717 (2011)
40. Fu, H., Nan, K., Bai, W., et al.: Morphable 3D mesostructures and microelectronic devices by multistable buckling mechanics. Nat. Mater. **17**, 268–276 (2018). https://doi.org/10.1038/s41563-017-0011-3
41. Ge, Q., Dunn, C.K., Qi, H.J., Dunn, M.L.: Active origami by 4D printing. Smart Mater. Struct. **23** (2014). https://doi.org/10.1088/0964-1726/23/9/094007
42. Ge, Q., Qi, H.J., Dunn, M.L.: Active materials by four-dimension printing. Appl. Phys. Lett. **103** (2013). https://doi.org/10.1063/1.4819837
43. George, D., Hernandez, E.A.P., Madou, M.: Origami fabrication using single layer photolithography and capillary folding. In: Proceedings of the 7th International Workshop on Nanotechnology and its Applications (2019a)
44. George, D., Madou, M., Hernandez, E.A.P, Lo, R.C.: Fabrication of polymer and carbon polyhedra through controlled cross-linking and capillary deformations 9171–9177 (2019b). https://doi.org/10.1039/c9sm01410a
45. Gracias, D.H., Kavthekar, V., Love, J.C., et al.: Fabrication of micrometer-scale, patterned polyhedra by self-assembly. Adv. Mater. **14**, 235–238 (2002). https://doi.org/10.1002/1521-4095(20020205)14:3%3c235:AID-ADMA235%3e3.0.CO;2-B
46. Gracias, D.H., Tien, J., Breen, T.L., et al.: Forming electrical networks in three dimensions by self-assembly. Science (80-) **289**, 1170–1172 (2000). https://doi.org/10.1126/science.289.5482.1170
47. Guan, J., He, H., Hansford, D.J., Lee, L.J.: Self-folding of three-dimensional hydrogel microstructures. J. Phys. Chem. B **109**, 23134–23137 (2005). https://doi.org/10.1021/jp054341g
48. Gultepe, E., Randhawa, J.S., Kadam, S., et al.: Biopsy with thermally-responsive untethered microtools. Adv. Mater. **25**, 514–519 (2013). https://doi.org/10.1002/adma.201203348
49. Guo, L., Anderson, D.G., Langer, R.: Bio-Inspired polymer composite actuator and generator driven by water gradients. Science (80-) **339**, 9–12 (2012)
50. Guo, X., Li, H., Yeop Ahn, B., et al.: Two- and three-dimensional folding of thin film single-crystalline silicon for photovoltaic power applications. Proc. Natl. Acad. Sci. **106**, 20149–20154 (2009). https://doi.org/10.1073/pnas.0907390106
51. Hauser, A.W., Evans, A.A., Na, J.H., Hayward, R.C.: Photothermally reprogrammable buckling of nanocomposite gel sheets. Angew. Chem. Int. Ed. **54**, 5434–5437 (2015). https://doi.org/10.1002/anie.201412160
52. Holmes, D.P., Roché, M., Sinha, T., Stone, H.A.: Bending and twisting of soft materials by non-homogenous swelling. Soft Matter **7**, 5188–5193 (2011). https://doi.org/10.1039/c0sm01492c
53. Hong, F., Zhang, F., Liu, Y., Yan, H.: DNA Origami: Scaffolds for Creating Higher Order Structures (2017)
54. Hu, G.F., Damanpack, A.R., Bodaghi, M., Liao, W.H.: (2017) Increasing dimension of structures by 4D printing shape memory polymers via fused deposition modeling. Smart Mater. Struct. **26**. https://doi.org/10.1088/1361-665X/aa95ec
55. Hu, L., Choi, J.W., Yang, Y., et al.: Highly conductive paper for energy-storage devices. Proc. Natl. Acad. Sci. U.S.A. **106**, 21490–21494 (2009). https://doi.org/10.1073/pnas.0908858106
56. Hu, W., Lum, G.Z., Mastrangeli, M., Sitti, M.: Small-scale soft-bodied robot with multimodal locomotion. Nature **554**, 81–85 (2018). https://doi.org/10.1038/nature25443
57. Huang, W., Koric, S., Yu, X., et al.: Precision structural engineering of self-rolled-up 3D nanomembranes guided by transient quasi-static FEM modeling. Nano Lett. **14**, 6293–6297 (2014). https://doi.org/10.1021/nl5026369
58. In, H.J., Kumar, S., Shao-Horn, Y., Barbastathis, G.: Origami fabrication of nanostructured, three-dimensional devices: electrochemical capacitors with carbon electrodes. Appl. Phys. Lett. **88**, 6–9 (2006). https://doi.org/10.1063/1.2177639

59. Ionov, L.: Soft microorigami: self-folding polymer films. Soft Matter **7**, 6786–6791 (2011). https://doi.org/10.1039/c1sm05476g
60. Islam, M., Flach, J., Martinez-Duarte, R.: Carbon origami: a method to fabricate lightweight carbon cellular materials. Carbon N Y **133**, 140–149 (2018). https://doi.org/10.1016/j.carbon.2018.03.033
61. Jamal, M., Bassik, N., Cho, J.H., et al.: Directed growth of fibroblasts into three dimensional micropatterned geometries via self-assembling scaffolds. Biomaterials **31**, 1683–1690 (2010). https://doi.org/10.1016/j.biomaterials.2009.11.056
62. Jamal, M., Kadam, S.S., Xiao, R., et al.: Bio-origami hydrogel scaffolds composed of photocrosslinked PEG bilayers. Adv. Healthc. Mater. **2**, 1142–1150 (2013). https://doi.org/10.1002/adhm.201200458
63. Jamal, M., Zarafshar, A.M., Gracias, D.H.: Differentially photo-crosslinked polymers enable self-assembling microfluidics. Nat. Commun. **2**, 526–527 (2011). https://doi.org/10.1038/ncomms1531
64. Jamal, M., Zarafshar, A.M., Gracias, D.H.: Differentially photo-crosslinked polymers enable self-assembling microfluidics. Nat. Commun. **2**, 527 (2013). https://doi.org/10.1038/ncomms1531.Differentially
65. Janbaz, S., Hedayati, R., Zadpoor, A.A.: Programming the shape-shifting of flat soft matter: from self-rolling/self-twisting materials to self-folding origami. Mater. Horiz. **3**, 536–547 (2016). https://doi.org/10.1039/c6mh00195e
66. Jin, Y., Wang, N., Yuan, B., et al.: Stress-induced self-assembly of complex three dimensional structures by elastic membranes. Small **9**, 2410–2414 (2013). https://doi.org/10.1002/smll.201300929
67. Kalinin, Y.V., Randhawa, J.S., Gracias, D.H.: Three-dimensional chemical patterns for cellular self-organization. Angew. Chem. Int. Ed. **50**, 2549–2553 (2011). https://doi.org/10.1002/anie.201007107
68. Kempaiah, R., Nie, Z.: From nature to synthetic systems: Shape transformation in soft materials. J. Mater. Chem. B **2**, 2357–2368 (2014). https://doi.org/10.1039/c3tb21462a
69. Kim, J., Hanna, J.A., Hayward, R.C., Santangelo, C.D.: Thermally responsive rolling of thin gel strips with discrete variations in swelling. Soft Matter **8**, 2375–2381 (2012). https://doi.org/10.1039/c2sm06681e
70. Kim, W., Peaudecerf, F., Baldwin, M.W., Bush, J.W.M.: The hummingbird's tongue: a self-assembling capillary syphon. Proc. R. Soc. B Biol. Sci. **279**, 4990–4996 (2012). https://doi.org/10.1098/rspb.2012.1837
71. Kim, Y., Yuk, H., Zhao, R., et al.: Printing ferromagnetic domains for untethered fast-transforming soft materials. Nature **558**, 274 (2018). https://doi.org/10.1038/s41586-018-0185-0
72. Kimionis, J., Georgiadis, A., Isakov, M., et al.: 3D/inkjet-printed origami antennas for multi-direction RF harvesting. In: 2015 IEEE MTT-S International Microwave Symposium, IMS 2015, pp.1–4 (2015a). https://doi.org/10.1109/MWSYM.2015.7166878
73. Kimionis, J., Isakov, M., Koh, B.S., et al.: 3D-printed origami packaging with inkjet-printed antennas for RF harvesting sensors. IEEE Trans. Microw. Theory Tech. **63**, 4521–4532 (2015). https://doi.org/10.1109/TMTT.2015.2494580
74. Kobayashi, K., Oh, S.H., Yoon, C., Gracias, D.H.: Multitemperature responsive self-folding soft biomimetic structures. Macromol. Rapid Commun. **1700692**, 1–7 (2018). https://doi.org/10.1002/marc.201700692
75. Konotop, I.Y., Nasimova, I.R., Tamm, M.V., et al.: Novel pH-responsive hydrogels with gradient charge distribution. Soft Matter **6**, 1632–1634 (2010). https://doi.org/10.1039/b923804b
76. Kubota, K., Fleischmann, T., Saravanan, S., et al.: Self-assembly of micro-stage using micro-origami technique on GaAs. In: 2002 International Microprocesses and Nanotechnology Conference, MNC 2002, p. 326 (2002). https://doi.org/10.1109/IMNC.2002.1178675
77. Kuribayashi-Shigetomi, K., Onoe, H., Takeuchi, S.: Cell Origami: Self-folding of three-dimensional cell-laden microstructures driven by cell traction force. PLoS One **7**, 1–8 (2012). https://doi.org/10.1371/journal.pone.0051085

78. Kuribayashi, K., Tsuchiya, K., You, Z., et al.: Self-deployable origami stent grafts as a biomedical application of Ni-rich TiNi shape memory alloy foil. Mater. Sci. Eng. A **419**, 131–137 (2006). https://doi.org/10.1016/j.msea.2005.12.016
79. Kwag, H.R., Cho, J.-H., Park, S.-Y., et al.: Self-folding nanostructures with imprint patterned surfaces (SNIPS). Faraday Discuss. **191**, 61–71 (2016). https://doi.org/10.1039/c6fd00021e
80. Laflin, K.E., Morris, C.J., Muqeem, T., Gracias, D.H.: Laser triggered sequential folding of microstructures. Appl. Phys. Lett. **101** (2012). https://doi.org/10.1063/1.4754607
81. Lahikainen, M., Zeng, H., Priimagi, A.: Reconfigurable photoactuator through synergistic use of photochemical and photothermal effects. Nat. Commun. **9**, 1–8 (2018). https://doi.org/10.1038/s41467-018-06647-7
82. Lamoureux, A., Lee, K., Shlian, M., et al.: Dynamic kirigami structures for integrated solar tracking. Nat. Commun. **6**, 1–6 (2015). https://doi.org/10.1038/ncomms9092
83. Lamprecht, B., Thünauer, R., Ostermann, M., et al.: Organic photodiodes on newspaper. Phys. Status Solidi Appl. Mater. Sci. **202**, 50–52 (2005). https://doi.org/10.1002/pssa.200510010
84. Lauff, C., Simpson, T.W., Frecker, M., et al.: Differentiating bending from folding in origami engineering using active materials. In: International Design Engineering Technical Conferences & Computers and Information in Engineering Conference, pp 1–12 (2014)
85. Lee, B.P., Konst, S.: Novel hydrogel actuator inspired by reversible mussel adhesive protein chemistry. Adv. Mater. **26**, 3415–3419 (2014). https://doi.org/10.1002/adma.201306137
86. Lee, D.W., Phadikar, J., Shankar, M.R.: Multiplicity of shape selection in functionally graded liquid crystalline polymers. RSC Adv. **7**, 23046–23054 (2017). https://doi.org/10.1039/c7ra03465b
87. Lee, K.M., Smith, M.L., Koerner, H., et al.: Photodriven, flexural-torsional oscillation of glassy azobenzene liquid crystal polymer networks. Adv. Funct. Mater. **21**, 2913–2918 (2011). https://doi.org/10.1002/adfm.201100333
88. Legrain, A., Berenschot, J.W., Tas, N.R., Abelmann, L.: Capillary origami of micro-machined micro-objects: Bi-layer conductive hinges. Microelectron. Eng. **140**, 60–66 (2015). https://doi.org/10.1016/j.mee.2015.06.004
89. Lendlein, A., Gould, O.E.C.: Reprogrammable recovery and actuation behaviour of shape-memory polymers. Nat. Rev. Mater. **4**, 116–133 (2019). https://doi.org/10.1038/s41578-018-0078-8
90. Lendlein, A., Jiang, H., Junger, O., Langer, R.: Light-induced shape-memory polymers. Nature **434**, 695–697 (2005). https://doi.org/10.1038/nature03438.1
91. Lendlein, A., Shastri, V.P.: Stimuli-sensitive polymers. Adv. Mater. **22**, 3344–3347 (2010). https://doi.org/10.1002/adma.201002520
92. Leong, T.G., Benson, B.R., Call, E.K., Gracias, D.H.: Thin film stress driven self-folding of microstructured containers. Small **4**, 1605–1609 (2008). https://doi.org/10.1002/smll.200800280
93. Leong, T.G., Lester, P.A., Koh, T.L., et al.: Surface tension-driven self-folding polyhedra. Langmuir **23**, 8747–8751 (2007). https://doi.org/10.1021/la700913m
94. Leong, T.G., Randall, C.L., Benson, B.R., et al.: Tetherless thermobiochemically actuated microgrippers. Proc. Natl. Acad. Sci. U.S.A. **106**, 703–708 (2009). https://doi.org/10.1073/pnas.0807698106
95. Leong, T.G., Randall, C.L., Benson, B.R., et al.: Self-loading lithographically structured microcontainers: 3D patterned, mobile microwells. Lab Chip **8**, 1621–1624 (2008). https://doi.org/10.1039/b809098j
96. Liu, F., Urban, M.W.: Recent advances and challenges in designing stimuli-responsive polymers. Prog. Polym. Sci. **35**, 3–23 (2010). https://doi.org/10.1016/j.progpolymsci.2009.10.002
97. Liu, Y., Boyles, J.K., Genzer, J., Dickey, M.D.: Self-folding of polymer sheets using local light absorption. Soft Matter **8**, 1764–1769 (2012). https://doi.org/10.1039/c1sm06564e
98. Liu, Y., Miskiewicz, M., Escuti, M.J., et al.: Three-dimensional folding of pre-strained polymer sheets via absorption of laser light. J. Appl. Phys. **115** (2014). https://doi.org/10.1063/1.4880160

99. Lu, X., Zhang, H., Fei, G., et al.: Liquid-crystalline dynamic networks doped with gold nanorods showing enhanced photocontrol of actuation. Adv. Mater. **30**, 1–8 (2018). https://doi.org/10.1002/adma.201706597
100. Luo, C.J., Stoyanov, S.D., Stride, E., et al.: Electrospinning versus fibre production methods: from specifics to technological convergence. Chem. Soc. Rev. **41**, 4708 (2012). https://doi.org/10.1039/c2cs35083a
101. Lv, C., Krishnaraju, D., Konjevod, G., et al.: Origami based mechanical metamaterials. Sci. Rep. **4** (2014). https://doi.org/10.1038/srep05979
102. Madou, M.J.: Fundamentals of Microfabrication and Nanotechnology-Volume II (2011)
103. Mahadevan, L., Rica, S.: Self-organized origami. Science (80-) **307**, 1740 (2005). https://doi.org/10.1126/science.1105169
104. Malachowski, K., Breger, J., Kwag, H.R., et al.: Stimuli-responsive theragrippers for chemomechanical controlled release. Angew. Chem. Int. Ed. **53**, 8045–8049 (2014). https://doi.org/10.1002/anie.201311047
105. Manakasettharn, S., Ashley Taylor, J., Krupenkin, T.N.: Bio-inspired artificial iridophores based on capillary origami: fabrication and device characterization. Appl. Phys. Lett. **99**, 2012–2015 (2011). https://doi.org/10.1063/1.3646394
106. Mao, Y., Yu, K., Isakov, M.S., et al.: Sequential self-folding structures by 3D printed digital shape memory polymers. Sci. Rep. **13616**, 1–12 (2015). https://doi.org/10.1038/srep13616
107. Martinez, A.W., Phillips, S.T., Butte, M.J., Whitesides, G.M.: Patterned paper as a platform for inexpensive, low-volume, portable bioassays. Angew. Chem. Int. Ed. **46**, 1318–1320 (2007). https://doi.org/10.1002/anie.200603817
108. Mikulas, M.M., Pappa, R.S., Warren, .J, Rose, G.: Telescoping Solar Array Concept for Achieving High Packaging Efficiency. In: 2nd AIAA SciTech Forum, pp. 1–21 (2015). https://doi.org/10.2514/6.2015-1398
109. Miskin, M.Z., Dorsey, K.J., Bircan, B., et al.: Graphene-based bimorphs for micron-sized, tautonomous origami machines. Proc. Natl. Acad. Sci. U.S.A. **115**, 466–470 (2018). https://doi.org/10.1073/pnas.1712889115
110. Miyashita, S., Guitron, S., Ludersdorfer, M., et al.: An untethered miniature origami robot that self-folds, walks, swims, and degrades. In: Proceedings of the IEEE International Conference on Robotics and Automation, pp. 1490–1496 (2015). https://doi.org/10.1109/ICRA.2015.7139386
111. Miyashita, S., Meeker, L., Tolley, M.T., et al.: Self-folding miniature elastic electric devices. Smart Mater. Struct. **23** (2014). https://doi.org/10.1088/0964-1726/23/9/094005
112. Moiseeva, E., Senousy, Y.M., McNamara, S., Harnett, C.K.: Single-mask microfabrication of three-dimensional objects from strained bimorphs. J Micromech. Microeng. **17** (2007). https://doi.org/10.1088/0960-1317/17/9/N01
113. Mu, J., Hou, C., Wang, H., et al.: Origami-inspired active graphene-based paper for programmable instant self-folding walking devices. Sci. Adv. **1**, 1–9 (2015). https://doi.org/10.1126/sciadv.1500533
114. Na, J.H., Evans, A.A., Bae, J., et al.: Programming reversibly self-folding origami with micropatterned photo-crosslinkable polymer trilayers. Adv. Mater. **27**, 79–85 (2015). https://doi.org/10.1002/adma.201403510
115. Ohm, C., Brehmer, M., Zentel, R.: Liquid crystalline elastomers as actuators and sensors. Adv. Mater. **22**, 3366–3387 (2010). https://doi.org/10.1002/adma.200904059
116. Paik, J.K., Kramer, R.K., Wood, R.J.: Stretchable circuits and sensors for robotic origami. In: IEEE International Conference on Intelligent Robots and Systems, pp. 414–420 (2011). https://doi.org/10.1109/IROS.2011.6048353
117. Pandey, S., Ewing, M., Kunas, A., et al.: Algorithmic design of self-folding polyhedra. Proc. Natl. Acad. Sci. **108**, 19885–19890 (2011). https://doi.org/10.1073/pnas.1110857108
118. Peraza-Hernandez, E.A., Hartl, D.J., Malak, R.J., Lagoudas, D.C.: Origami-inspired active structures: a synthesis and review. Smart Mater. Struct. **23** (2014). https://doi.org/10.1088/0964-1726/23/9/094001

119. Piñeirua, M., Bico, J., Roman, B.: Capillary origami controlled by an electric field. Soft Matter **6**, 4491 (2010). https://doi.org/10.1039/c0sm00004c
120. Piqué, A., Mathews, S.A., Charipar, N.A., Birnbaum, A.J.: Laser origami: a new technique for assembling 3D microstructures. In: Laser-Based Micro- and Nanopackaging and Assembly VI, vol. 8244, p. 82440B (2012). https://doi.org/10.1117/12.909187
121. Py, C., Reverdy, P., Doppler, L., et al.: Capillary origami: spontaneous wrapping of a droplet with an elastic sheet. Phys. Rev. Lett. **98**, 2–5 (2007). https://doi.org/10.1103/PhysRevLett.98.156103
122. Randhawa, J.S., Gurbani, S.S., Keung, M.D., et al.: Three-dimensional surface current loops in terahertz responsive microarrays. Appl. Phys. Lett. **96**, 1–4 (2010). https://doi.org/10.1063/1.3428657
123. Rath, A., Mathesan, S., Ghosh, P.: Folding behavior and molecular mechanism of cross-linked biopolymer film in response to water. Soft Matter **12**, 9210–9222 (2016). https://doi.org/10.1039/C6SM01994C
124. Ratna, D., Karger-Kocsis, J.: Recent advances in shape memory polymers and composites: a review. J. Mater. Sci. **43**, 254–269 (2008). https://doi.org/10.1007/s10853-007-2176-7
125. Rico-guevara, A., Rubega, M.A.: The hummingbird tongue is a fluid trap, not a capillary tube (2011) **108**, 1–5. https://doi.org/10.1073/pnas.1016944108
126. Lang, R.J.: Origami Design Secrets: Mathematical Methods for an Ancient Art (2011)
127. Rojas, J.P., Conchouso, D., Arevalo, A., et al.: Paper-based origami flexible and foldable thermoelectric nanogenerator. Nano Energy **31**, 296–301 (2017). https://doi.org/10.1016/j.nanoen.2016.11.012
128. Ryu, J., D'Amato, M., Cui, X., et al.: Photo-origami-bending and folding polymers with light. Appl. Phys. Lett. **100** (2012). https://doi.org/10.1063/1.3700719
129. Shaar, N.S., Barbastathis, G., Livermore, C.: Integrated folding, alignment, and latching for reconfigurable origami microelectromechanical systems. J. Microelectromech. Syst. **24**, 1043–1051 (2015). https://doi.org/10.1109/JMEMS.2014.2379432
130. Shankland, K., Mccusker, L.B., Palatinus, L., et al.: Shaping of elastic sheets by prescription of non-Euclidean metrics. Science **9**, 1116–1120 (2007). https://doi.org/10.1126/science.1137920
131. Shigemune, H., Maeda, S., Hara, Y., et al.: Origami robot: a self-folding paper robot with an electrothermal actuator created by printing. IEEE/ASME Trans. Mechatron. **21**, 2746–2754 (2016). https://doi.org/10.1109/TMECH.2016.2593912
132. Shim, T.S., Kim, S.H., Heo, C.J., et al.: Controlled origami folding of hydrogel bilayers with sustained reversibility for robust microcarriers. Angew. Chem. Int. Ed. **51**, 1420–1423 (2012). https://doi.org/10.1002/anie.201106723
133. Si, K.J., Sikdar, D., Chen, Y., et al.: Giant plasmene nanosheets, nanoribbons, and origami. ACS Nano **8**, 11086–11093 (2014)
134. Silverberg J.L., Evans, A.A., McLeod, L., et al.: Using origami design principles to fold reprogrammable mechanical metamaterials. Science (80-) **345**, 647–650 (2014)
135. Song, Z., Ma, T., Tang, R., et al.: Origami lithium-ion batteries. Nat. Commun. **5** (2014). https://doi.org/10.1038/ncomms4140
136. Stoychev, G., Puretskiy, N., Ionov, L.: Self-folding all-polymer thermoresponsive microcapsules. Soft Matter **7**, 3277–3279 (2011). https://doi.org/10.1039/c1sm05109a
137. Stoychev, G., Turcaud, S., Dunlop, J.W.C., Ionov, L.: Hierarchical multi-step folding of polymer bilayers. Adv. Funct. Mater. **23**, 2295–2300 (2013). https://doi.org/10.1002/adfm.201203245
138. Stoychev, G., Zakharchenko, S., Turcaud, S., et al.: Shape-programmed folding of stimuli-responsive polymer bilayers. ACS Nano **6**, 3925–3934 (2012). https://doi.org/10.1021/nn300079f
139. Sun, Y., Choi, W.M., Jiang, H., et al.: Controlled buckling of semiconductor nanoribbons for stretchable electronics. Nat. Nanotechnol. **1**, 201–207 (2006). https://doi.org/10.1038/nnano.2006.131

140. Sydney Gladman, A., Matsumoto, E.A., Nuzzo, R.G., et al.: Biomimetic 4D printing. Nat. Mater. **15**, 413–418 (2016). https://doi.org/10.1038/nmat4544
141. Taghavi, M., Helps, T., Rossiter, J.: Electro-ribbon actuators and electro-origami robots **9795** (2018)
142. Tang, J., Zhou, Y., Wan, L., Huang, F.: Automatically programmable shape-memory polymers based on asymmetric swelling of bilayer structures. Macromol. Rapid Commun. **39**, 1–7 (2018). https://doi.org/10.1002/marc.201800039
143. Tang, R., Huang, H., Tu, H., et al.: Origami-enabled deformable silicon solar cells. Appl. Phys. Lett. **104** (2014). https://doi.org/10.1063/1.4866145
144. Tao, Y., Gu, J., An, B., et al.: Demonstrating thermorph: democratizing 4D printing of self-folding materials and interfaces. In: Conference on Human Factors in Computing Systems—Proceedings 2018, pp. 1–12 (2018). https://doi.org/10.1145/3170427.3186478
145. Teunis, J.S., Zadpoor, A.A.: Programming 2D/3D shape-shifting with hobbyist 3D printers. Mater. Horiz. **4**, 1064–1069 (2017). https://doi.org/10.1039/c7mh00269f
146. Timoshenko, S.: Analysis of bi-metal thermostats. JOSA **11**, 233–255 (1925)
147. Tolley, M.T., Felton, S.M., Miyashita, S., et al.: Self-folding origami: Shape memory composites activated by uniform heating. Smart Mater. Struct. **23** (2014). https://doi.org/10.1088/0964-1726/23/9/094006
148. Tomatsu, I., Peng, K., Kros, A.: Photoresponsive hydrogels for biomedical applications. Adv. Drug Deliv. Rev. **63**, 1257–1266 (2011). https://doi.org/10.1016/j.addr.2011.06.009
149. Truby, R.L., Lewis, J.A.: Printing soft matter in three dimensions. Nature **540**, 371–378 (2016). https://doi.org/10.1038/nature21003
150. Tu, H., Jiang, H., Yu, H., Xu, Y.: Hybrid silicon-polymer platform for self-locking and self-deploying origami. Appl. Phys. Lett. **103** (2013). https://doi.org/10.1063/1.4842235
151. Tyagi, P., Bassik, N., Leong, T.G., et al.: Self-assembly based on chromium/copper bilayers. J. Microelectromech. Syst. **18**, 784–791 (2009). https://doi.org/10.1109/JMEMS.2009.2023841
152. Vaccaro, P.O., Kubota, K., Aida, T.: Strain-driven self-positioning of micromachined structures. Appl. Phys. Lett. **78**, 2852–2854 (2001). https://doi.org/10.1063/1.1371525
153. Van Honschoten, J.W., Berenschot, J.W., Ondaŗuhu, T., et al.: Elastocapillary fabrication of three-dimensional microstructures. Appl. Phys. Lett. **97** (2010). https://doi.org/10.1063/1.3462302
154. Van Manen, T., Janbaz, S., Zadpoor, A.A.: Programming 2D/3D shape-shifting with hobbyist 3D printers. Mater. Horiz. **4**, 1064–1069 (2017). https://doi.org/10.1039/c7mh00269f
155. Volkov, A.G., Foster, J.C., Ashby, T.A., et al.: Mimosa pudica: electrical and mechanical stimulation of plant movements. Plant Cell Environ. **33**, 163–173 (2010). https://doi.org/10.1111/j.1365-3040.2009.02066.x
156. Wang, E., Desai, M.S., Lee, S.W.: Light-controlled graphene-elastin composite hydrogel actuators. Nano Lett. **13**, 2826–2830 (2013). https://doi.org/10.1021/nl401088b
157. Wang, H., Wang, Y., Tee, B.C.K., et al.: Shape-controlled, self-wrapped carbon nanotube 3D electronics. Adv. Sci. **2** (2015). https://doi.org/10.1002/advs.201500103
158. Wang, M., Lin, B.P., Yang, H.: A plant tendril mimic soft actuator with phototunable bending and chiral twisting motion modes. Nat. Commun. **7**, 1–8 (2016). https://doi.org/10.1038/ncomms13981
159. Whitney, J.P., Sreetharan, P.S., Ma, K.Y., Wood, R.J.: Pop-up book MEMS. J. Micromech. Microeng. **21** (2011). https://doi.org/10.1088/0960-1317/21/11/115021
160. Xu, S., Yan, Z., Jang, K.I., et al.: Assembly of micro/nanomaterials into complex, three-dimensional architectures by compressive buckling. Science (80-) **347**, 154–159 (2015). https://doi.org/10.1126/science.1260960
161. Xu, T., Zhang, J., Salehizadeh, M., et al.: Millimeter-scale flexible robots with programmable three-dimensional magnetization and motions. Sci. Robot. **4** (2019). https://doi.org/10.1126/scirobotics.aav4494
162. Yan, Z., Zhang, F., Liu, F., et al.: Mechanical assembly of complex, 3D mesostructures from releasable multilayers of advanced materials. Sci. Adv. **2** (2016a). https://doi.org/10.1126/sciadv.1601014

163. Yan, Z., Zhang, F., Wang, J., et al.: Controlled mechanical buckling for origami-inspired construction of 3D microstructures in advanced materials. Adv. Funct. Mater. **26**, 2629–2639 (2016). https://doi.org/10.1002/adfm.201504901
164. Yang, Y., Song, X., Li, X., et al.: Recent progress in biomimetic additive manufacturing technology: from materials to functional structures. Adv. Mater. **30**, 1–34 (2018). https://doi.org/10.1002/adma.201706539
165. Yang, Y., You, Z.: Geometry of transformable metamaterials inspired by modular origami. J. Mech. Robot. **10**, 1–10 (2018). https://doi.org/10.1115/1.4038969
166. Yoon, C, Xiao, R., Park, J., et al.: Functional stimuli responsive hydrogel devices by self-folding. Smart Mater. Struct. **23** (2014). https://doi.org/10.1088/0964-1726/23/9/094008
167. Yu, K., Ritchie, A., Mao, Y., et al.: Controlled sequential shape changing components by 3D printing of shape memory polymer multimaterials. Procedia IUTAM **12**, 193–203 (2015). https://doi.org/10.1016/j.piutam.2014.12.021
168. Yuan, C., Roach, D.J., Dunn, C.K., et al.: 3D printed reversible shape changing soft actuators assisted by liquid crystal elastomers. Soft Matter **13**, 5558–5568 (2017). https://doi.org/10.1039/c7sm00759k
169. Zanardi, J.M., Vaccaro, P.O., Fleischmann, T., et al.: Optical actuation of micromirrors fabricated by the micro-origami technique. Appl. Phys. Lett. **83**, 3647–3649 (2003). https://doi.org/10.1063/1.1622800
170. Ze, Q., Kuang, X., Wu, S., et al.: Magnetic shape memory polymers with integrated multi-functional shape manipulations. Adv. Mater. **1906657**, 1–8 (2019). https://doi.org/10.1002/adma.201906657
171. Zhang, X., Li, J., Chen, C., et al.: A self-powered microfluidic origami electrochemiluminescence biosensing platform. Chem. Commun. **49**, 3866–3868 (2013). https://doi.org/10.1039/c3cc40905h
172. Zhao, Y., Nandra, M.S., Tai. Y.C.: A MEMS intraocular origami coil. In: 2011 16th International Solid-State Sensors, Actuators and Microsystems Conference, TRANSDUCERS'11, pp. 2172–2175 (2011). https://doi.org/10.1109/TRANSDUCERS.2011.5969378
173. Zhao, Z., Fang, D., et al.: Origami by frontal photopolymerization. Sci. Adv. **3**, e1602326 (2017a)
174. Zhao, Z., Wu, J., Mu, X., et al.: Desolvation induced origami of photocurable polymers by digit light processing. Macromol. Rapid Commun. **38**, 1–6 (2017). https://doi.org/10.1002/marc.201600625
175. Zhou, X., Li, T., Wang, J., et al.: Mechanochemical regulated origami with tough hydrogels by ion transfer printing. ACS Appl. Mater. Interfaces **10**, 9077–9084 (2018). https://doi.org/10.1021/acsami.8b01610
176. Zhu, Y., Yao, S., Huang, Q., et al.: Controlling the self-folding of a polymer sheet using a local heater: the effect of the polymer–heater interface. Soft Matter **13**, 3863–3870 (2017). https://doi.org/10.1039/c7sm00568g

Simulations of Machining Processes at Small Spatio-temporal Scales

Anish Roy, Qiang Liu, Uday S. Dixit, and Vadim V. Silberschmidt

1 Introduction

Manufacturing at small scale has become increasingly important with the advent of modern electronic devices. In this regard, microelectromechanical systems (MEMS) are a typical example, since their components range in sizes from 1 μm to 1 mm. Manufacturing such small devices is no easy task, as precision is of utmost importance. Production of small components and devices is highly restrictive with manufacturability, with its associated costs largely determining what is possible. Thus, component designers conceive parts in terms of processes and materials that were used in the past and proved to be economical. The design process is usually performed using trial and error, typically requiring numerous iterations before satisfying the performance requirements of a given device. This non-ideal design methodology combined with the length of time and high costs associated with prototyping makes the entire process inefficient for commercial product development.

Numerical modelling of the manufacturing process if done accurately can be invaluable in assessing the consequences of various techniques in manufacture, thereby opening possibilities not previously considered. However, inaccurate material information can render the component's response and performance virtually useless. Thus, there is a clear need to understand the mechanical behaviour of materials and develop advanced simulation and modelling tools to help designers to select appropriate manufacturing processes and materials for component manufacture at a

A. Roy (✉) · Q. Liu · V. V. Silberschmidt
Wolfson School of Mechanical, Electrical and Manufacturing Engineering, Loughborough University, Loughborough LE11 3TU, UK
e-mail: a.roy3@lboro.ac.uk

U. S. Dixit
Department of Mechanical Engineering, Indian Institute of Technology Guwahati, Guwahati 781039, India

© Springer Nature Singapore Pte Ltd. 2021

U. S. Dixit and S. K. Dwivedy (eds.), *Mechanical Sciences*,
https://doi.org/10.1007/978-981-15-5712-5_10

small scale. Additionally, there is a need to improve the current state of conventional machining (CM) processes as they pose growing challenges in efficient and economical cutting of newer materials [1–3]. As a result, various hybrid machining techniques were proposed to enhance and improve CM. One such hybrid technique that showed to yield tangible benefits in machining of advanced materials is ultrasonically assisted machining (UAM) [4, 5]. In UAM, high-frequency electrical energy is converted into mechanical vibration, which is superimposed on movement of a cutting tool with a specific intensity and in a specific direction during the machining process [3]. Such vibration is typically of high frequency (~20 kHz) and relatively low amplitude (12–50 μm). Due to the nature of imposed vibration, the cutting tool periodically loses contact with the workpiece in UAM, transforming a machining process into a micro-chipping one. UAM can lead to a considerable reduction of average cutting forces when compared to CM [3, 4]. For example, the cutting force in UAM was demonstrated to be reduced in excess of 50% relative to CM for nickel alloys [6] and titanium [2, 7] alloys. Additionally, better surface quality was obtained for workpieces in UAM [1].

Although UAM exhibits tractable benefits when compared to CM, there is a need to understand the main micro-mechanisms that drive plasticity at small length scales under superimposed constrained dynamic rapidly changing loading states. The temporal scale in such a process is essentially very small. UAM results in extreme deformation conditions in its process zone, where strain levels can exceed 2, accompanied by strain rates of up to 10^5 s^{-1} and temperatures in excess of 700–800 °C. Additionally, extremely high strain gradients are observed in this zone. Hence, the influence of machining parameters (e.g. depth of cut and cutting speed) in UAM on the machining-quality characteristics of workpieces needs a careful study [1, 3, 4, 8, 9]. Consequently, finite-element (FE) simulations were widely used to study the machining processes of various metallic materials [10, 11]. An increase in demand for high-precision micro-featured components requires a thorough understanding of grain-level deformation of metallic materials [9, 12]. To better understand local deformation at a tool–workpiece interface, single-crystal plasticity (SCP) is usually employed to study micromechanics at the smallest practical length scale. Zhang et al. [13] adopted an SCP model to describe the mechanical behaviour of grains at mesoscale in cutting simulations of titanium alloys. In the work of Kota and Ozdoganlar [14], a significant lattice rotation due to machining-related plastic deformation was observed in single-crystal aluminium. An initial crystal orientation of a workpiece with respect to the cutting direction was found to have a significant effect on chip formation and a cutting force of single-crystal copper in the micro-cutting simulation of Abolfazl et al. [15] and Tajalli et al. [16]. Moreover, the work of Demiral et al. [17] and Pal and Stucker [18] indicated that inhomogeneous plastic deformation could affect machinability of a workpiece based on strain-gradient crystal-plasticity simulations.

In contrast to extensive investigations of micromechanics of the CM process, similar fundamental studies for UAM are scarce. In this paper, CM and UAM at micro-scale were studied using SCP simulations based on our previous work [11]. A special attention is paid to the influence of ultrasonic vibration on the cutting force and chip formation for different nominal cutting velocities. This paper is organized as follows: a theoretical framework of the SCP theory is summarized in Sect. 2, followed by a description of the FE modelling procedure in Sect. 3. Simulation results are discussed in the subsequent section. We conclude with some remarks in Sect. 5.

2 Crystal-Plasticity Theory

In this section, the classical crystal-plasticity theory adopted in this study is reviewed. Deformation gradient $\mathbf{F}$ can be decomposed into its elastic and plastic parts:

$$\mathbf{F} = \mathbf{F_e F_p}, \tag{1}$$

where the subscripts '**e**' and '**p**' denote the elastic and plastic parameters, respectively. Often it is assumed that the plastic part of the spin tensor is zero. By applying the product rule of differentiation, one can obtain the rate of the total deformation gradient $\dot{\mathbf{F}}$:

$$\dot{\mathbf{F}} = \dot{\mathbf{F}}_\mathbf{e}\mathbf{F_p} + \mathbf{F_e}\dot{\mathbf{F}}_\mathbf{p}. \tag{2}$$

Therefore, the velocity gradient $\mathbf{L}$ can be introduced following its definition $\mathbf{L} = \dot{\mathbf{F}}\mathbf{F}^{-1}$ as

$$\mathbf{L} = \dot{\mathbf{F}}_\mathbf{e}\mathbf{F}_\mathbf{e}^{-1} + \mathbf{F_e}(\dot{\mathbf{F}}_\mathbf{p}\mathbf{F}_\mathbf{p}^{-1})\,\mathbf{F}_\mathbf{e}^{-1} = \mathbf{L_e} + \mathbf{L_p}. \tag{3}$$

It is assumed that the plastic velocity gradient, $\mathbf{L_p}$, is induced by shearing on each slip system in a crystal. Hence, $\mathbf{L_p}$ is formulated as the sum of shear rates on all the slip systems, i.e.

$$\mathbf{L_p} = \sum_{\alpha=1}^{N} \dot{\gamma}^{(\alpha)}\mathbf{s}^{(\alpha)} \otimes \mathbf{m}^{(\alpha)}, \tag{4}$$

where $\dot{\gamma}^{(\alpha)}$ is the shear slip rate on the slip system α, N is the total number of slip systems, and unit vectors $\mathbf{s}^{(\alpha)}$ and $\mathbf{m}^{(\alpha)}$ define the slip direction and the normal to the slip plane in the deformed configuration, respectively. Furthermore, the velocity gradient can be expressed in terms of a symmetric rate of stretching $\mathbf{D}$ and an antisymmetric rate of spin $\mathbf{W}$:

$$\mathbf{L} = \mathbf{D} + \mathbf{W} = (\mathbf{D_e} + \mathbf{W_e}) + (\mathbf{D_p} + \mathbf{W_p}). \tag{5}$$

Using Eqs. (3) and (4), it can be deduced

$$\mathbf{D_e} + \mathbf{W_e} = \dot{\mathbf{F}}_\mathbf{e}\mathbf{F}_\mathbf{e}^{-1}, \quad \mathbf{D_p} + \mathbf{W_p} = \sum_{\alpha=1}^{N} \dot{\gamma}^{(\alpha)}\mathbf{s}^{(\alpha)} \otimes \mathbf{m}^{(\alpha)}. \tag{6}$$

Following the work of Huang [19], a constitutive law is expressed as the relationship between the elastic part of the symmetric rate of stretching, $\mathbf{D_e}$, and the Jaumann rate of Cauchy stress, $\overset{\nabla}{\boldsymbol{\sigma}}$, i.e.

$$\overset{\nabla}{\boldsymbol{\sigma}} + \boldsymbol{\sigma}(\boldsymbol{I} : \boldsymbol{D}_\mathbf{e}) = \mathbf{C} : (\mathbf{D} - \mathbf{D_p}), \tag{7}$$

where $\mathbf{I}$ is the second-order unit tensor, $\mathbf{C}$ is the fourth-order, possibly anisotropic, elastic stiffness tensor. The Jaumann stress rate is expressed as

$$\overset{\nabla}{\boldsymbol{\sigma}} = \dot{\boldsymbol{\sigma}} - \mathbf{W_e}\boldsymbol{\sigma} + \boldsymbol{\sigma}\boldsymbol{W}_\mathbf{e}. \tag{8}$$

On each slip system, the resolved shear stress, $\tau^{(\alpha)}$, is expressed by a Schmid law:

$$\tau^{(\alpha)} = \mathbf{sym}(\mathbf{s}^{(\alpha)} \otimes \mathbf{m}^{(\alpha)}) : \boldsymbol{\sigma}. \tag{9}$$

The relationship between the shear rate, $\dot{\gamma}^{(\alpha)}$, and the resolved shear stress, $\tau^{(\alpha)}$, on the slip system α is expressed by a power law proposed by Hutchinson [20]:

$$\dot{\gamma}^{(\alpha)} = \dot{\gamma}_0 \left| \frac{\tau^{(\alpha)}}{g^{(\alpha)}} \right|^n \mathrm{sgn}\left(\tau^{(\alpha)}\right), \tag{10}$$

where $\dot{\gamma}_0$ is the reference shear rate, $g^{(\alpha)}$ is the slip resistance and n is the rate-sensitivity parameter. The evolution of $g^{(\alpha)}$ is given by

$$\dot{g}^{(\alpha)} = \sum_{\beta=1}^{N} h_{\alpha\beta} \left| \dot{\gamma}^{(\beta)} \right|, \tag{11}$$

where $h_{\alpha\beta}$ is the hardening modulus that can be calculated in the form modified from that proposed by Asaro [21]:

$$h_{\alpha\alpha} = (h_0 - h_s)\mathrm{sech}^2\left(\frac{h_0\gamma}{\tau_S - \tau_0}\right) + h_s, \quad h_{\alpha\beta} = q\, h_{\alpha\alpha}\, (\alpha \neq \beta), \quad \gamma = \sum_{\alpha} \int_0^t \left| \dot{\gamma}^{(\alpha)} \right| \mathrm{d}t. \tag{12}$$

Here, h_0 and h_s are the initial and saturated hardening moduli, respectively, q is the latent hardening ratio, τ_0 and τ_S are the shear stresses at the onset of yield and the saturation of hardening, respectively, and γ is the accumulative shear strain over all the slip systems.

3 Finite-Element Modelling Procedure

A finite-element (FE) model was developed to simulate micromachining processes (Fig. 1). The cutting tool was assumed to be rigid, with a clearance angle of 5°. In this study, the cutting depth was fixed as 5 μm. The length of the workpiece, L, was 160 μm, the height (H) and width (W) were 60 μm and 10 μm, respectively. Without loss of generality, the workpiece material was considered as single-crystal copper with FCC crystallographic structure. Its deformation behaviour was described by the SCP model introduced in Sect. 2, implemented in the commercial finite-element code ABAQUS/Explicit by employing the user subroutine VUMAT. For the FCC single-crystal copper, the slip was assumed to occur on the usual 12 [111]<110> slip systems.

The workpiece was meshed using eight-node brick elements with reduced integration (C3D8R) in ABAQUS. Reduced integration essentially reduced the element

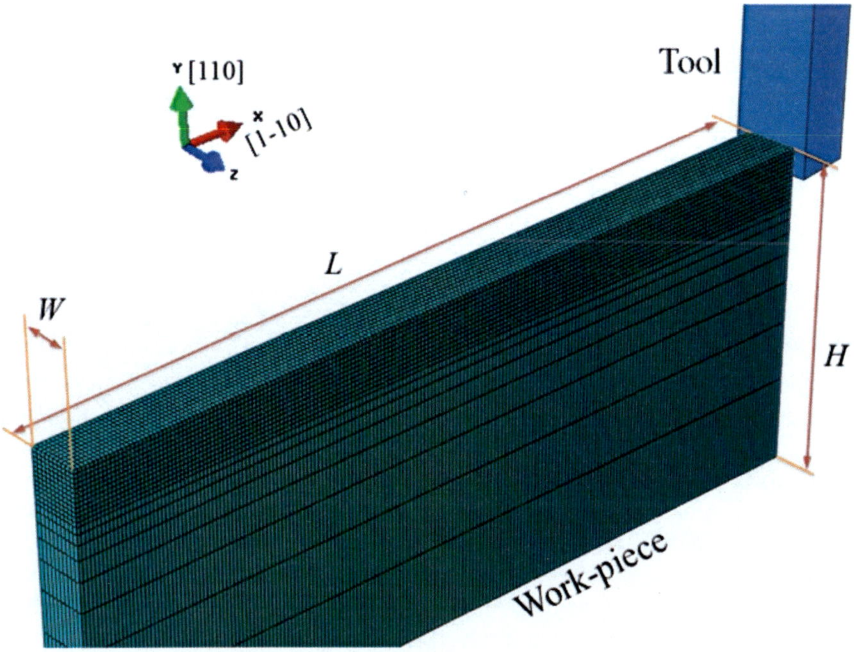

Fig. 1 Finite-element model for simulation of micromachining

to a single integration point and thereby helps in eliminating the problem of parasitic shear. In total, 60,000 brick elements were used to mesh the workpiece after a mesh-sensitivity study. To ensure the accuracy of FE simulations, a finer local mesh was used in regions near the cutting zone (corresponding to a height of 10 μm). In this study, a normal to the top surface of the workpiece coincided with the [110] crystal orientation and the cutting direction with the [1-10] crystal orientation. A process of removal of the workpiece material was simulated using an element-deletion technique in ABAQUS. Contact conditions between the cutting tool and the workpiece were assumed to be frictionless.

A constant velocity was specified for the cutting tool to simulate the CM process, whilst a harmonic motion was superimposed on the cutting tool motion to simulate the ultrasonically assisted machining process. The relationship between the displacement of the cutting tool with time (t) in UAM is expressed as

$$x(t) = V_0 t + a \sin(2\pi f t), \tag{13}$$

where V_0 is the constant nominal velocity in the CM process, a and f are the vibration amplitude and frequency, respectively. In the simulation of UAM, the motion of the cutting tool was controlled by a velocity boundary condition in ABAQUS. The velocity of the cutting tool was obtained by differentiation of Eq. (13) as

$$V(t) = V_0 + V_c \cos(2\pi f t), \quad V_c = 2\pi f a, \tag{14}$$

where V_c is the critical oscillatory speed of the cutting tool. In the current study, the values of amplitude and frequency of vibration were 15 μm and 20 kHz, respectively. The critical oscillatory speed was calculated to be 1.885 mm/ms. Three values of V_0 were considered in our simulations: 2.0, 1.5 and 1.0 mm/ms.

4 Results and Analysis

First, data obtained from experimental studies performed on single-crystal copper [22] were used to calibrate the material parameters of workpiece. As shown in Fig. 2, numerical results based on the described SCP theory show an excellent match with the experimental data for both [100] and [111] crystal orientations. The calibrated model parameters are listed in Table 1, which were used to simulate the CM and UAM process at the subsequent stage of the present investigation.

For the nominal velocity of 2.0 mm/ms, a relationship between the velocity and displacement of the cutting tool and the cutting time is shown in Fig. 3 for CM and UAM. In this case, the normal velocity, V_0, was slightly higher than the critical oscillatory speed, V_c and the ratio of V_0 to V_c was about 1.06. This implies that the cutting tool did not separate from the workpiece during the UAM process. However, the superimposed motion due to UAM does affect the overall kinematics of the machining process, showing a minor difference in the average cutting forces (and

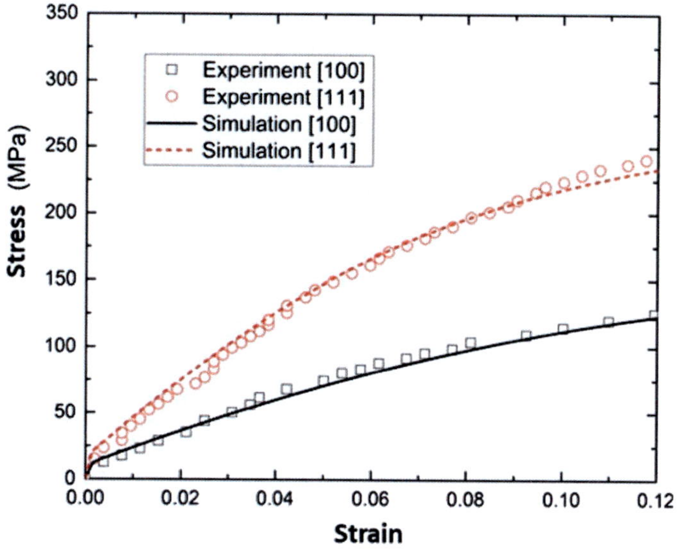

Fig. 2 Comparison of experimental and numerical stress–strain behaviour under compression for different orientations

Table 1 Material parameters of single-crystal copper

Parameter	Meaning	Value	Unit
$\dot{\gamma}_0$	Reference shear rate	0.001	s^{-1}
n	Rate-sensitivity parameter	50	–
τ_0	Initial slip resistance	4.0	MPa
τ_s	Saturated slip resistance	52	MPa
h_0	Initial hardening modulus	180	MPa
h_s	Saturated hardening modulus	24	MPa
q	Latent hardening ratio	1.2	–

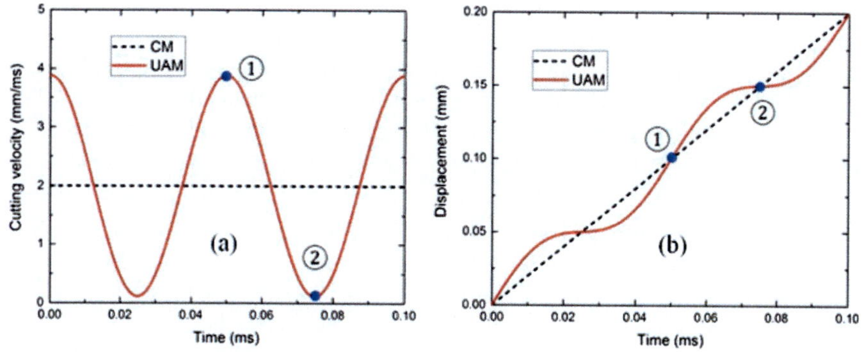

Fig. 3 Comparison of velocity (**a**) and displacement (**b**) of cutting tool for CM and UAM simulations for $V_0 = 2.0$ mm/ms

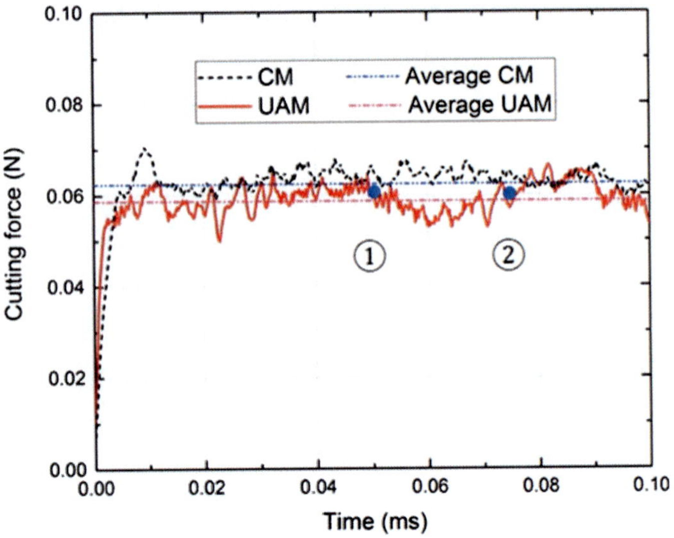

Fig. 4 Comparison of cutting force in CM and UAM for $V_0 = 2.0$ mm/ms

the overall cutting response) as presented in Fig. 4. The difference between the average cutting forces (after the initial engagement stage) in UAM relative to that in CM was only about 8%. For both CM and UAM, the chip morphology at the time corresponding to the maximum and minimum cutting velocities in UAM is shown in Figs. 5a and b, respectively (these times are indicated by ① and ② in Figs. 3 and 4). The dissipated plastic-energy density (PENER in ABAQUS) is indicated by colour contours in the workpiece and the chip. As demonstrated in Fig. 5b, vibration of the cutting tool shows a clear effect on chip morphology although it was not significant for the cutting force. The separation of the chip from the workpiece was easier in UAM than in CM. However, there was no significant difference between the distribution of PENER produced by CM and UAM.

In the next step of our analysis, a lower level of nominal velocity of $V_0 = 1.5$ mm/ms was chosen; the respective velocities and displacements of the cutting tool are shown in Fig. 6 for both CM and UAM. In this case, V_0 was about 80% of V_c; it implies that the tool separated from the chip in each cycle of vibration in the UAM process. As a result, a clear difference in the cutting force was observed for CM and UAM. The cutting force in CM was almost constant after the initial engagement process due to the constant cutting velocity (Fig. 7). In contrast, the cutting force in UAM changed with time, and its evolution could be divided into two main stages in each full cycle of vibration. In the first stage, the cutting force in UAM was comparable with that in CM although more fluctuations were observed on the cutting force–time curves for UAM as a consequence of the varying cutting velocity. In the second stage, the cutting force in UAM rapidly dropped to zero after the cutting tool started to lose contact with the workpiece. Consequently, the average cutting force in UAM was

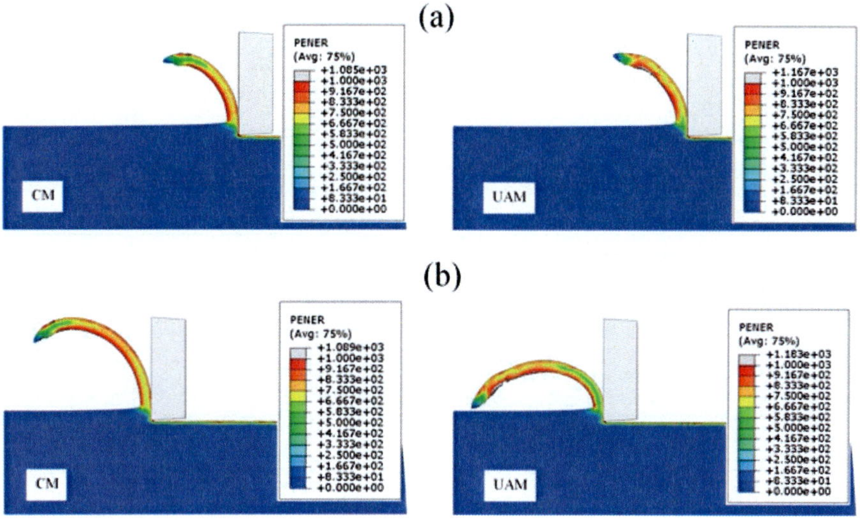

Fig. 5 Comparison of chip morphology in CM and UAM for $V_0 = 2.0$ mm/ms for moments ① (**a**) and ② (**b**) in Figs. 3 and 4

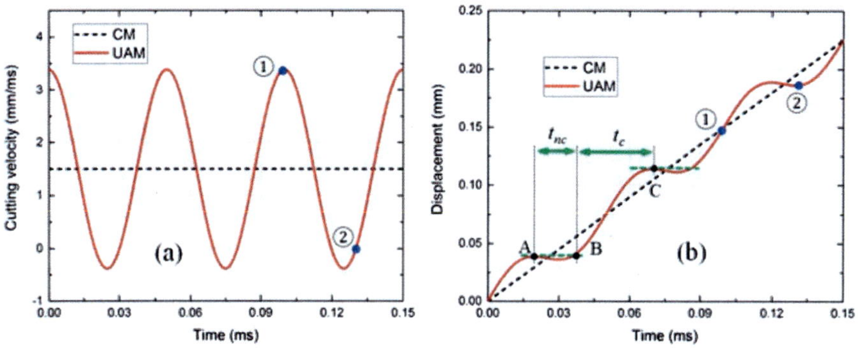

Fig. 6 Comparison of velocity (**a**) and displacement (**b**) of cutting tool for CM and UAM simulations for $V_0 = 1.5$ mm/ms

reduced significantly when compared to that in CM; the former was only 65% of the latter. As shown in Fig. 6(b), the time corresponding to the stages of non-contact and contact between the tool and the workpiece are indicated by t_{nc} and t_c, respectively. Clearly, the time when the tool is not in contact with the workpiece is a small fraction of the whole time.

The chip morphology and the distribution of PENER in CM and UAM process are shown in Fig. 8 for the case of $V_0 = 1.5$ mm/ms. Figure 8a corresponds to the time, at which the cutting velocity in UAM reached a maximum level, whilst Fig. 8b represents the time, at which the cutting tool was at maximum separation

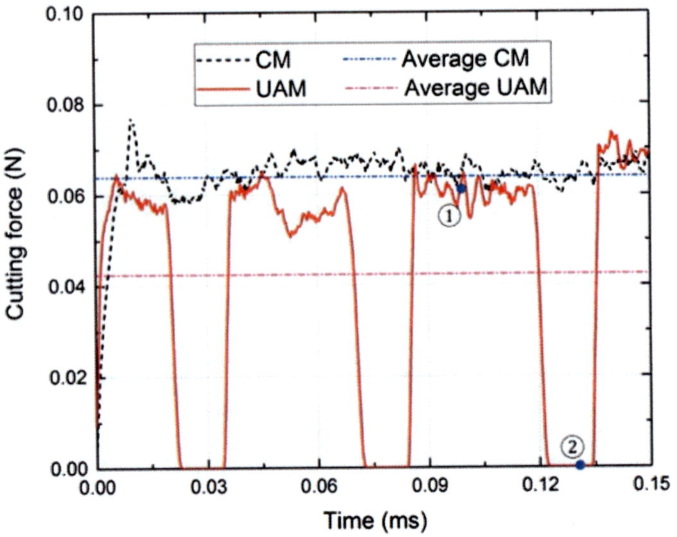

Fig. 7 Comparison of cutting force in CM and UAM for $V_0 = 1.5$ mm/ms

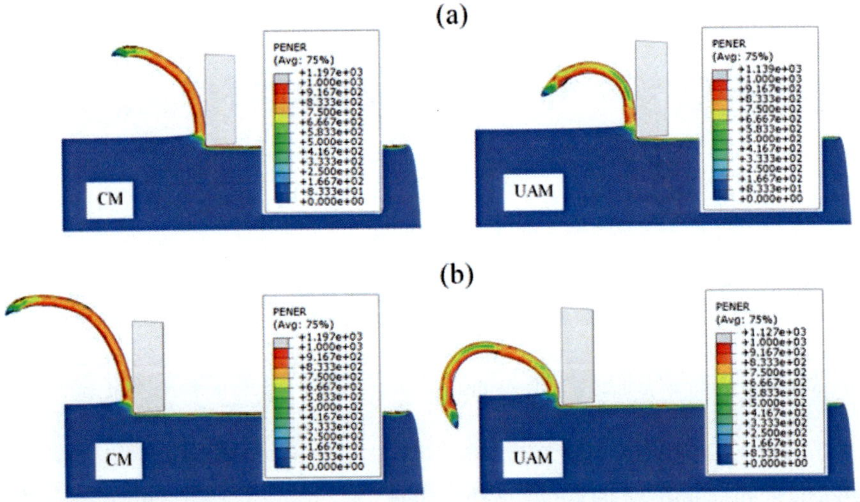

Fig. 8 Comparison of chip morphology for CM and UAM for $V_0 = 1.5$ for moments ① (**a**) and ② (**b**) in Figs. 6 and 7

from the workpiece. The two featured time points are indicated in both Figs. 6 and 7. Apart from chip morphology, a significant difference of PENER fields is observed for CM and UAM in Fig. 8. This field in the chip in UAM exhibits a more inhomogeneous distribution than the one in CM. When the cutting velocity was higher than the nominal velocity in UAM, a chip with higher PENER was observed

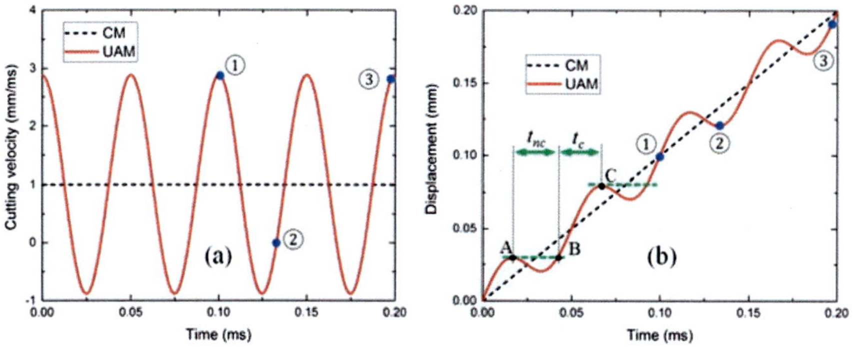

Fig. 9 Comparison of velocity (**a**) and displacement (**b**) of cutting tool for CM and UAM simulations for $V_0 = 1.0$ mm/ms

to form (e.g. Fig. 8a). Consequently, the levels of PENER in the chip were lower when the tool moved with a lower velocity or lost contact with the workpiece (Fig. 8b). In comparison, the highest level in the field of dissipated plastic energy localized in the chip in UAM was comparable to that observed in the CM chip. However, the plastic energy dissipated was significantly lower when the tool completely retracted. Consequently, the total dissipated plastic energy during the UAM process is expected to be lower than that in the CM process. For machining of metallic materials, it is well known that the dissipated plastic energy is one of the primary reasons for heat generation (besides the friction between a tool and a workpiece). Therefore, UAM could potentially reduce the overall temperature during machining of copper, consistent with the conclusions obtained in experiments [1].

The final stage of present research dealt with a significantly reduced cutting velocity: $V_0 = 1.0$ mm/ms; the evolution of respective velocities and displacements of the cutting tool is presented in Fig. 9 for both CM and UAM. Here, V_0 was close to 50% of V_c, implying that the time corresponding to a negative cutting velocity (Fig. 9a) and separation between the cutting tool and the workpiece (Fig. 9b) took a larger proportion in a full vibratory cycle compared to the case of $V_0 = 1.5$ mm/ms. Due to this larger proportion of non-contact as shown in Fig. 9b, the time corresponding to a zero cutting force increased to about 50% of each cycle of tool vibration for the case of $V_0 = 1.0$ mm/ms (Fig. 10). As a result, the average cutting force in UAM was less than half of that in CM.

As before, the formed chips and the PENER fields were compared for CM and UAM (Fig. 11) when $V_0 = 1.0$ mm/ms. At the time corresponding to the maximum cutting velocity (indicated by ① in Fig. 9), locally high levels of the PENER field were observed in the chip in UAM as shown in Fig. 11a. Until the time indicated by ②, the PENER field periodically changed from high levels to low ones in the chip in UAM in contrast to the relatively uniform PENER field in the chip produced by CM. Also, there was a significant difference in the chip morphology for CM and UAM; the chip separation was more difficult in the CM process. In contrast to the case of $V_0 = 1.5$ mm/ms, machining with $V_0 = 1.0$ mm/ms showed a significant variation of the

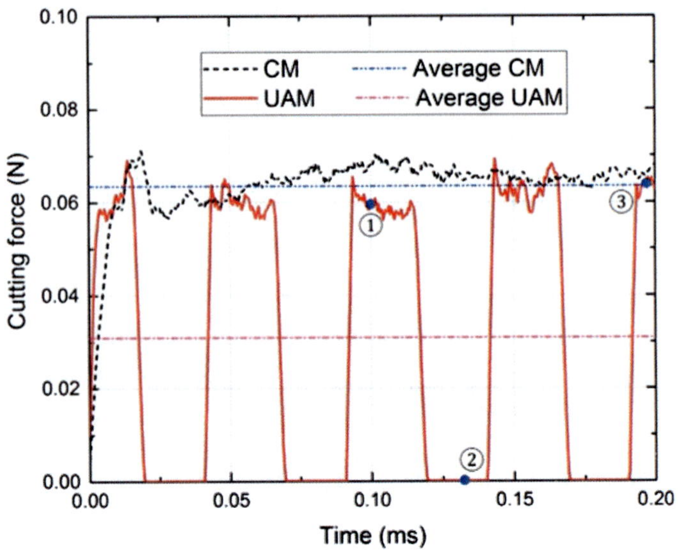

Fig. 10 Comparison of cutting force for CM and UAM for $V_0 = 1.0$ mm/ms

plastic dissipation energy. Therefore, the peak value of PENER in UAM was clearly larger than that in CM. In other words, high plastic deformation, which occurred in UAM corresponding to the chip region with localized high PENER, probably led to fracture of the chip as shown in Fig. 11c. Consequently, the chip formed in UAM was shorter than that in CM, which was also confirmed in the experiment [1].

5 Conclusions

In this paper, the micro-scale CM and UAM processes were investigated for single-crystal copper using FE simulations based on the SCP theory. Compared to the CM process, the effect of ultrasonic vibration in UAM was dependent on the nominal velocity for the fixed amplitude and frequency of ultrasonic vibration. When the nominal cutting velocity was smaller than the critical oscillatory speed induced by ultrasonic vibration, the cutting force and dissipated plastic energy were reduced in UAM. Otherwise, the differences in cutting forces and distributions of plastic-energy density in CM and UAM could be neglected. For the three studied cases, the chip morphology showed significant differences for CM and UAM. The process of separation of the chip from the workpiece was easier in UAM than in CM. In particular, fracture of the chip was observed when the nominal velocity was much smaller than the critical oscillatory speed, leading to shorter chips in UAM. This study provides a fundamental understanding of micromachining of single-crystal copper; it is currently being extended to analysis of HCP systems, e.g. Ti-64 alloy.

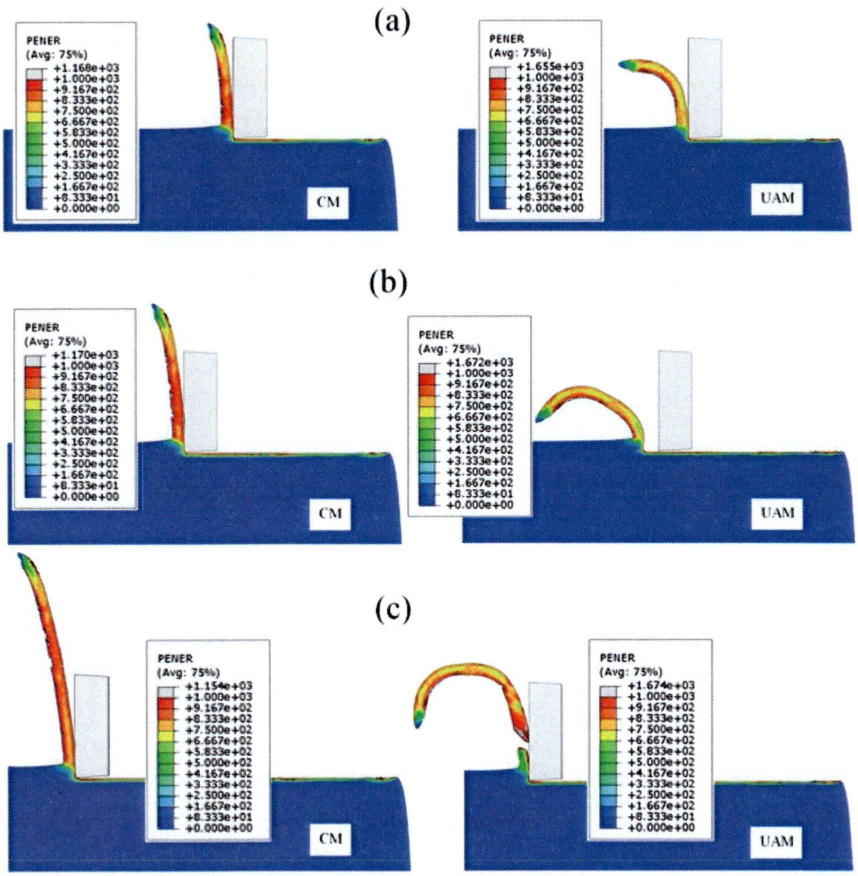

Fig. 11 Comparison of chip morphology for CM and UAM for $V_0 = 1.0$ for moments ① (**a**), ② (**b**) and ③ (**c**) in Figs. 9 and 10

Acknowledgements Funding from the Engineering and Physical Sciences Research Council (UK) through grant EP/K028316/1 and Department of Science and Technology (India), project MAST, is gratefully acknowledged.

References

1. Kumar, M.N., Subbu, S.K., Krishna, P.V., Venugopal, A.: Vibration assisted conventional and advanced machining: a review. Procedia Eng. **97**, 1577–1586 (2014)
2. Muhammad, R., Hussain, M.S., Maurotto, A., Siemers, C., Roy, A., Silberschmidt, V.: Analysis of a free machining $\alpha + \beta$ titanium alloy using conventional and ultrasonically assisted turning. J. Mater. Process. Technol. **214**(4), 906–915 (2014)

3. Kumar, J.: Ultrasonic machining—a comprehensive review. Mach. Sci. Technol. **17**, 325–379 (2013)
4. Singh, R., Khamba, J.: Ultrasonic machining of titanium and its alloys: a review. J. Mater. Process. Technol. **173**, 125–135 (2006)
5. Thoe, T., Aspinwall, D., Wise, M.: Review on ultrasonic machining. Int. J. Mach. Tools Manuf **38**, 239–255 (1998)
6. Babitsky, V., Mitrofanov, A., Silberschmidt, V.: Ultrasonically assisted turning of aviation materials: simulations and experimental study. Ultrasonics **42**, 81–86 (2004)
7. Maurotto, A., Muhammad, R., Roy, A., Silberschmidt, V.V.: Enhanced ultrasonically assisted turning of a β-titanium alloy. Ultrasonics **53**, 1242–1250 (2013)
8. Brehl, D., Dow, T.: Review of vibration-assisted machining. Precis. Eng. **32**, 153–172 (2008)
9. Shi, B., Attia, H.: Current status and future direction in the numerical modeling and simulation of machining processes: a critical literature review. Mach. Sci. Technol. **14**, 149–188 (2010)
10. Paturi, U.M.R., Narala, S.K.R., Pundir, R.S.: Constitutive flow stress formulation, model validation and FE cutting simulation for AA7075-T6 aluminum alloy. Mater. Sci. Eng. A **605**, 176–185 (2014)
11. Zahedi, S.A., Roy, A., Silberschmidt, V.V.: Modelling of vibration assisted machining fcc single crystal. Procedia CIRP **31**, 393–398 (2015)
12. Mian, A., Driver, N., Mativenga, P.: Identification of factors that dominate size effect in micro-machining. Int. J. Mach. Tools Manuf. **51**, 383–394 (2011)
13. Zhang, Y., Mabrouki, T., Nelias, D., Courbon, C., Rech, J., Gong, Y.: Cutting simulation capabilities based on crystal plasticity theory and discrete cohesive elements. J. Mater. Process. Technol. **212**, 936–953 (2012)
14. Kota, N., Ozdoganlar, O.B.: Orthogonal machining of single-crystal and coarse-grained aluminum. J. Manuf. Process. **14**, 126–134 (2012)
15. Zahedi, S.A., Demiral, M., Roy, A., Silberschmidt, V.V.: FE/SPH modelling of orthogonal micro-machining of fcc single crystal. Comput. Mater. Sci. **78**, 104–109 (2013)
16. Tajalli, S., Movahhedy, M., Akbari, J.: Simulation of orthogonal micro-cutting of FCC materials based on rate-dependent crystal plasticity finite element model. Comput. Mater. Sci. **86**, 79–87 (2014)
17. Demiral, M., Roy, A., El Sayed, T., Silberschmidt, V.V.: Numerical modelling of micro-machining of fcc single crystal: Influence of strain gradients. Comput. Mater. Sci. **94**, 273–278 (2014)
18. Pal, D., Stucker, B.: A study of subgrain formation in Al 3003 H-18 foils undergoing ultrasonic additive manufacturing using a dislocation density based crystal plasticity finite element framework. J. Appl. Phys. **113**, 203517 (2013)
19. Huang, Y.: A User-Material Subroutine Incorporating Single Crystal Plasticity in the ABAQUS Finite Element Program. Harvard University (1991)
20. Hutchinson, J.: Bounds and self-consistent estimates for creep of polycrystalline materials. Proc. R. Soc. Lond. Math. Phys. Sci. **348**, 101–127 (1976)
21. Asaro, R.J.: Crystal plasticity. J. Appl. Mech. **50**, 921–934 (1983)
22. Takeuchi, T.: Work hardening of copper single crystals with multiple glide orientations. Trans. Jpn. Inst. Met. **16**, 629–640 (1975)

Laser-Aided Manufacturing: Atom to Application

Jyoti Mazumder

1 Introduction

Laser as a pure form of light offers new form of industrial energy that is being harnessed for a multitude of manufacturing processes. Each form of new energy ushered a new era [1]. Mechanical energy used via tools started the civilizations we know it now. Steam energy started the Industrial Revolution. Table 1 lists different forms of energy and its influence on society.

Lasers are now used for atomic level isotope separation [2] in Atomic Energy and zapping of eyes for corrective eye surgery [3]. Lasers first came as imagination by H. G. Wells in his story "War of the World" serialized in the Pearson magazine in U.K. in 1897. Science of LASER came later. Laser is an acronym for "Light Amplification by Stimulated Emission of Radiation." Einstein [4] first observed the phenomenon of "Stimulated Emissions" in 1916 while studying the interaction of an atom with electromagnetic field.

The concept of generation of laser light includes three main components, i.e., optical cavity, electromagnetic energy for amplification, and atomic media. It was first published by Schawlow and Townes [5]. Theodore Maiman made the first laser operate on May 16, 1960 at the Hughes Research Laboratory in California, by shining a high-power flash lamp on a ruby rod with silver-coated surfaces. In the beginning, CO_2 laser [6] invented by C. K. N. Patel provided high power suitable for manufacturing use. Later on, solid-state lasers were adapted for high power. Now lasers with dimers as medium with ultraviolet wavelength (~192 nm) to far-infrared CO_2 (10.6 μm) are being employed for various manufacturing applications [3]. Lasers are now enjoying revenues exceeding $14 billion (see Fig. 1).

The market segment for the lasers is shown in Fig. 2 [7]. Material processing is the largest segment of laser application. Auto industry has embraced lasers since 1978 in

J. Mazumder (✉)
Department of Mechanical Engineering, University of Michigan, Ann Arbor, USA
e-mail: mazumder@umich.edu

© Springer Nature Singapore Pte Ltd. 2021

U. S. Dixit and S. K. Dwivedy (eds.), *Mechanical Sciences*,
https://doi.org/10.1007/978-981-15-5712-5_11

Table 1 Laser as a new form of industrial energy (*Source* W. M. Steen, Emeritus Professor University of Liverpool)

Form of energy	Usage	Influence on society
Mechanical energy	Tools	Start of civilization
Chemical energy	Furnaces	Iron and bronze ages
Wind energy	Windmills, sailing ships	Start of industry, international trade
Water energy	Water wheels	Industrialization
Steam energy	Steam energy	Industrial revolution
Electrical energy	Dynamo	Distributed power, radio, telephone, motors
Oil/petrol	Compact power	Transport revolution
Nuclear power	Generators, bombs	Unlimited energy changed world politics
Optical energy	Laser	CDs, printers, bar code, communication +++

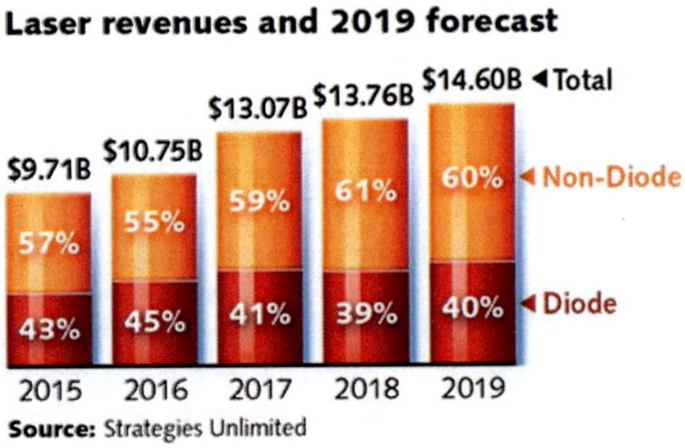

Fig. 1 Global laser market [7]

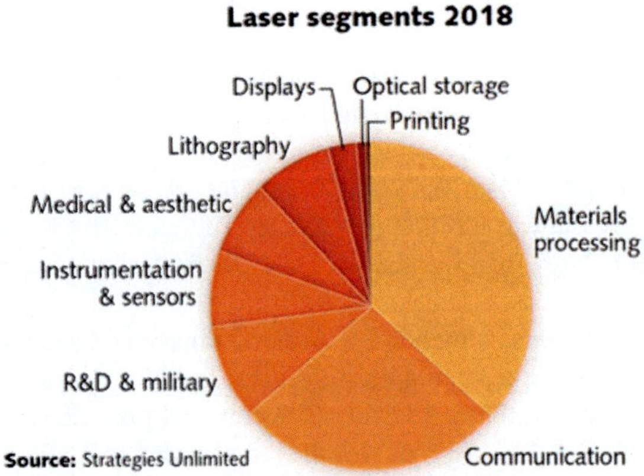

Fig. 2 Laser market segment (*Source* Laser Focus World [7])

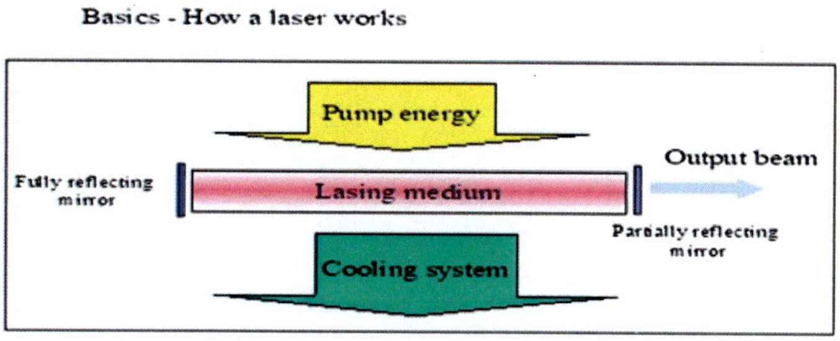

Fig. 3 Basic laser mechanism [1]

GM for laser heat treatment of steering columns. During the past decade, German auto manufacturers have installed hundreds of lasers for body-in-white welding purposes.

2 Why Lasers Are Attractive for Manufacturing?

The basic mechanism of a laser is shown in Fig. 3. Laser light is a stream of photons that interact with surface electrons of the materials it is processing. This mechanism is known as "inverse bremsstrahlung." Laser light, unlike conventional machine tools, does not have any mass or inertia. The beam can be focused to a very small spot concentrating the energy exceeding 10^5 W/cm^2. It is probably one of the few sources of energy with such concentrated energy capable of melting and ablating many of the engineering materials. Energy can be delivered through vacuum, air, or fiber. The beam can be shaped to conform to a particular surface. Shaping can be done both temporally and spatially. This characteristic provides enormous flexibility for manufacturing applications. Energy density can be varied with optics based on the power density needed for the specific manufacturing process. For example, heat treatment requires 10^3 W/cm^2 whereas welding requires 10^6 W/cm^2. Same laser can provide both intensity levels with simple optics. The optical beam can also provide photolytic processes breaking molecules by matching the wavelength of vibration of the molecular bonds. Coherence and spectral purity allow strange non-linear optical effects. Polarization, frequency doubling, diffraction, and multi-photon phenomena all found their way into the manufacturing floor. Moreover, optical energy source is highly amenable to automation.

3 Laser-Aided Manufacturing

Initially, lasers were dubbed as "Solution looking for Problem" but since 1070s, it has been used extensively for various applications from [A to Z] atom splitting to zapping of eyes. However, as shown in Fig. 2, material processing and manufacturing

Table 2 Laser power density versus different manufacturing processes. LCVD stands for Laser Chemical Vapor Deposition

Power intensity (W/cm^2)	Stage of heating	Associated processes
$<10^5$	Thermal conduction	Transformation hard, bending, LCVD
10^5	Melting	Welding, cladding
10^6	Vaporization	Cutting, drilling
10^7	Plasma production	Cleaning, shock hardening
10^9	Solid-state plasma	Shock hardening
$>10^9$	Atomic fusion	Femto second processes

captured the largest market share. Table 2 shows laser energy density needed for various manufacturing processes.

Laser energy density and interaction time have found its way to almost entire manufacturing field to fabricate components from nanometer to several meters. Figure 4 shows a rough map for the energy density, interaction time, and size of the object possible to be fabricated. Holy Grail for manufacturing is to control intensity, interaction time, and cooling rate to modify the microstructure to obtain certain properties for people to use (see Fig. 5). Most of the human endeavors for the laser-aided manufacturing for last five decades have been to develop process-parameter-structure-property relationships both experimentally and by mathematical modeling [3]. Different laser-based manufacturing processes are mentioned in the following sub-sections.

3.1 Laser Heat Treatment

It is used for transformation hardening and annealing of work-hardened materials needing energy density around 10^3 W/cm^2. Laser heat treatment has to follow same time-temperature criteria as conventional heat treatment. One additional criterion is that the substrate should have enough mass compared to heat-treated zone so that conduction rate between treated and untreated zone exceeds the critical rate for hardening. This self-quenching mechanism can also offer a very high cooling rate, sometimes leading to higher hardness than what is available through conventional quenching in water or oil [8].

General Motors has adapted laser heat treatment in 1978 for hardening of their steering column for improved life through wear resistance. Some of the cutting tool companies also apply that for localized hardening. Lasers can focus so precisely that even a tip of teeth of a hacksaw blade can be heat-treated (See Fig. 6). This provides

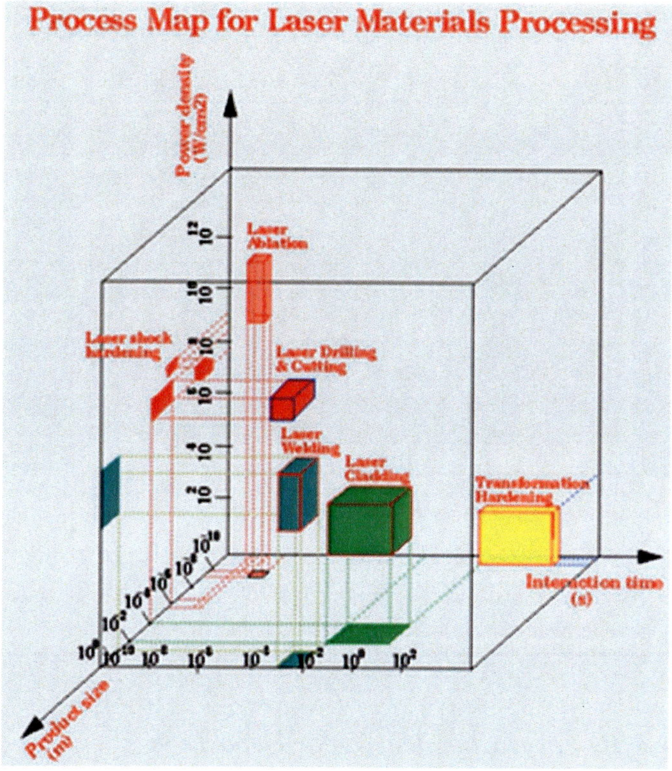

Fig. 4 Intensity and interaction time for different manufacturing processes

Fig. 5 Holy Grail for laser-aided manufacturing

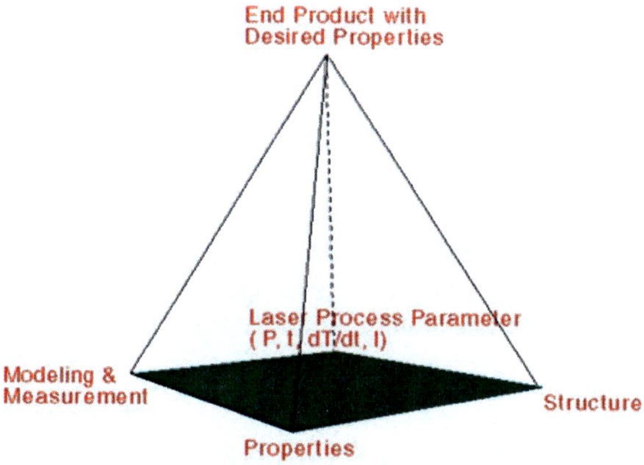

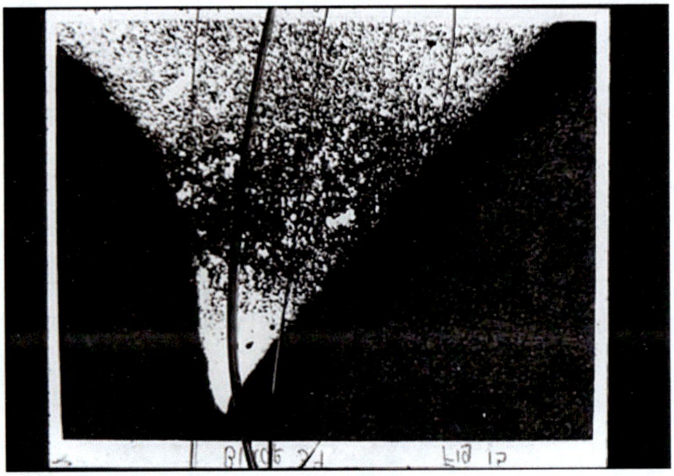

Fig. 6 Laser heat-treated tip of a 12 × 0.5 inch hacksaw blade

a combination of hardness and toughness for the blade. The same concept is also adapted by gear industry.

3.2 *Pyrolytic Laser Chemical Vapor Deposition*

Laser chemical vapor deposition is also another manufacturing application with similar low power density requirements like laser heat treatment. In this process, a substrate is heated in presence of gases which causes endothermic reaction to produce a reaction product. It is deposited on the substrate and another by-product is in the gas form that is pumped out [9]. This process can work with almost all the chemical combinations used by the conventional CVD process, but provides 10,000 times faster depositions [10]. LCVD can also follow photolytic route where laser wavelength is matched with molecular vibration to initiate the reaction [9]. In this case, the laser runs parallel to the substrate. The deposition rate in this route is much slower than that of pyrolytic route. LCVD is used sometimes in integrated circuit industry to repair faulty prints of the circuit.

3.3 *Laser Surface Alloying and Cladding*

It is another laser-aided manufacturing operation that made significant penetration in the market [3]. AT&T adapted laser surface alloying for their telephone connection jacks decades back to achieve longer life through wear resistance. Turbine blades industry employed laser cladding to repair the leading edges of the worn out blades

Fig. 7 Laser cladding with blown powder

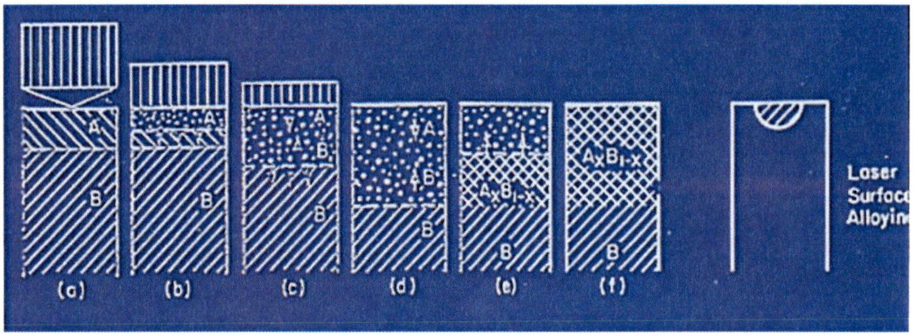

Fig. 8 Stages of laser alloying

since 1980s. These are non-equilibrium processes [11] capable of producing surfaces with tailored properties. In surface alloying, a coating of different material is applied on a surface in many different ways including electroplating, slurry, or powder blown through a nozzle (see Figs. 7 and 8). A high-power laser melts and completely mixes the coating creating a surface with composition that is a mixture of two different alloys. However, in cladding, coating composition is more or less kept with a bit of dilution from the substrate.

3.4 Additive Manufacturing/3D Printing

3D printing has captured the imagination of the manufacturing community worldwide since 2012 [12, 13]. April Issue of Economist magazine dubbed it as 3rd Industrial

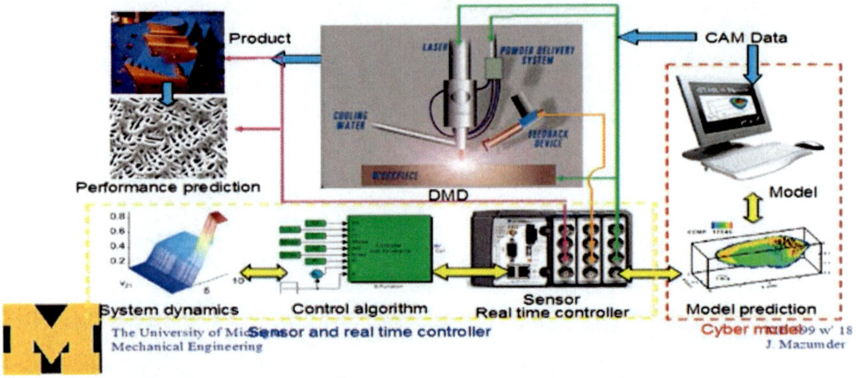

Fig. 9 Metallic additive manufacturing system

Revolution although route of it goes back thousands of years. In 2800 B.C., Egyptians made pyramids layer by layer, which is the basic principle of additive manufacturing. In modern reincarnation, Charles Hull [14] reported 3D printing from polymeric resins with the help of lasers in 1986. Subsequently, Joe Beaman [15] reported solid free form formation by melting plastic granules layer by layer with lasers. Mazumder [16, 17] reported rapid prototyping with metals with closed-loop control of laser cladding (see Fig. 9). Although 3D printing is being developed rapidly, non-laser processes such as extrusion-based resistance heating of polymer are getting adapted more rapidly due to lower cost.

3.5 *Laser Welding*

Laser welding has a wider footprint on the manufacturing floor [3, 18]. Automobile, aerospace, shipping, and chemical industry have adopted laser welding significantly. Keyhole formation without the need for vacuum is the attraction for relative thicker section (see Fig. 10). Keyhole formation means the depth is higher than the width of the weld as shown in Fig. 10. One of the high volume applications is in fabrication of tailored blank for car door as shown in Fig. 11. Laser being an inertia less tool, fixturing is much easier. Narrow heat-affected zone and precision welding are attractive for the users.

3.6 *Laser Cutting*

It is another widely used process. Due to its capability to cut profiles, laser cutting with high-quality edge has become very popular [3]. In laser cutting, highly focused beam with power density exceeding 10^6 W/cm^2 with assist gas is used as shown

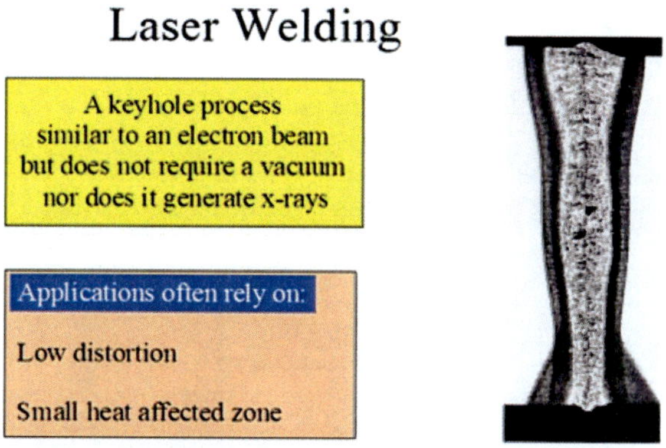

Fig. 10 Keyhole laser welding [1]

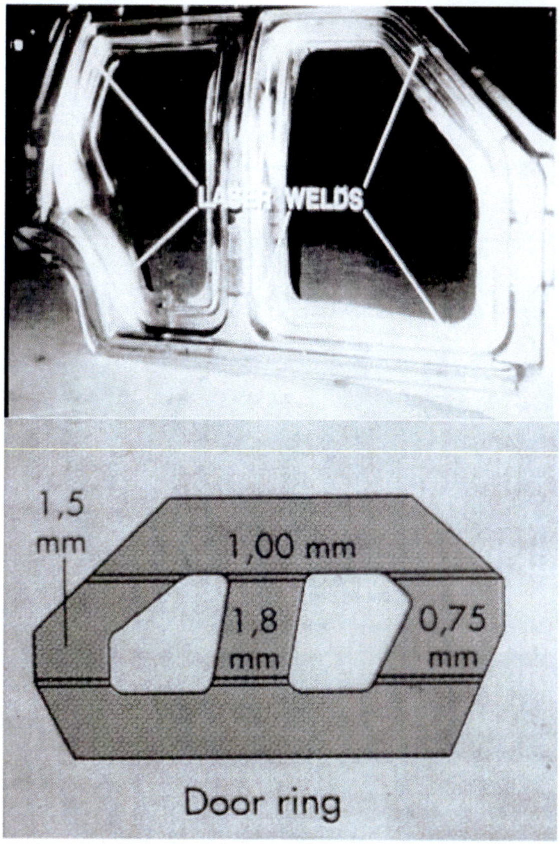

Fig. 11 Tailored blank [1]

in Fig. 12 (adapted from [1]). There are several modes of laser cutting as shown in Fig. 13 (adapted from [1]). Vaporization cutting needs highest power density. Oxygen is often used as an assist gas since it provides additional energy through exothermic reaction. Sometimes gear manufacturers use profile laser cutting to generate the teeth instead of machining. Even greeting card industry uses laser cutting of papers to produce intricate artistic greeting cards.

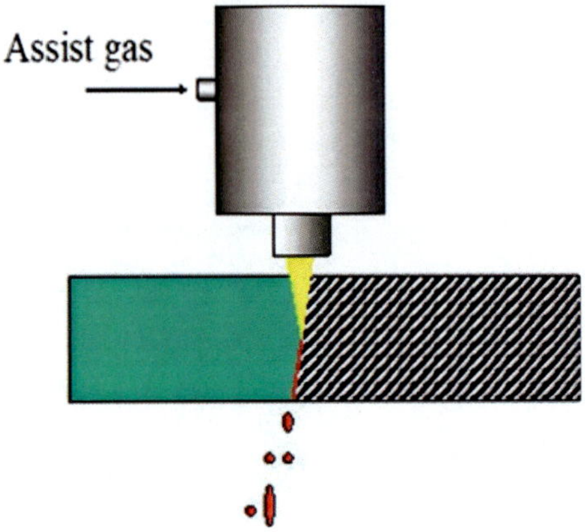

Fig. 12 Laser cutting

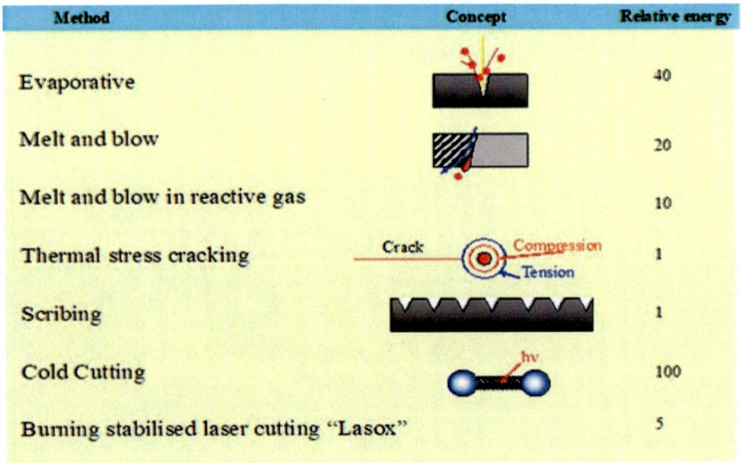

Fig. 13 Modes of laser cutting

3.7 *Laser Drilling*

Laser drilling [3] is used for making nipples for baby bottles. Also, it is used to make electrodes for Li-Ion batteries. Pulsed lasers are more popular for laser drilling. Lasers with 10^{-15} s pulse widths give the cleanest holes but commonly pulse width of 10^{-9} s is used for production. Laser drilling has two different modes. One is percussion drilling where a stationary laser beam evaporates the material to create the whole with a similar size to beam diameter. Another is called "Trepanning" where a laser beam is moved to cut a circular hole. Often oxygen is used as assist gas. For oxidation, sensitive materials like titanium, and inert gas such as argon is used. There are a few more applications such as laser shock hardening, laser cleaning, and laser marking.

3.8 *Laser Shock Hardening or Laser Shock Peening*

This technique is extensively used by aerospace industry [19] to provide compressive residual stress at the surface of aerospace components to increase the hardness and fatigue life. Compressive stress hinders fatigue crack growth. In laser shock hardening, the objects to be treated are often coated with black paint or tape to enhance absorption. Then, a thin layer of water is passed while the substrate is irradiated with a high-power laser pulse as shown in Fig. 14 [1]. Laser pulse width in the order of 10^{-9} s is most suitable for this process. A plasma is formed on the light-absorbing layer and the water confines the expansion of the plasma that creates a shock wave that propagates through the substrate consolidating point and line defects. This results in compressive residual stress.

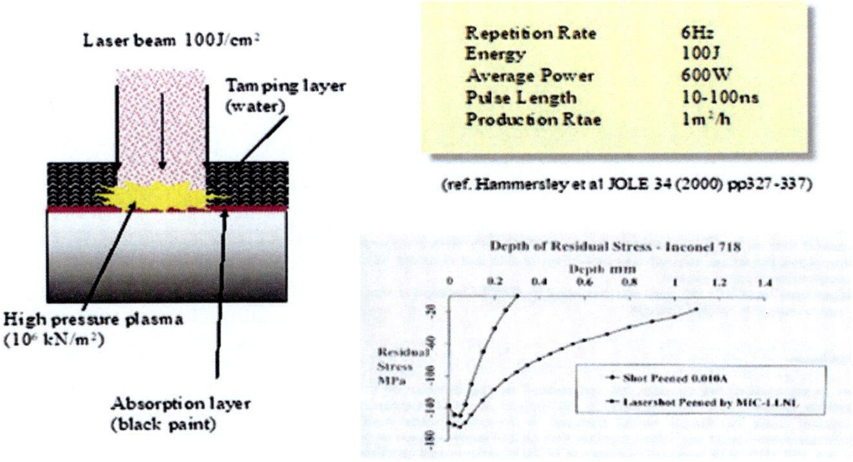

Fig. 14 Laser shock hardening

3.9 Laser Cleaning

It depends on the blast of laser radiation that can remove dirt from the surface. It is a quiet cleaning process without the need for solvent or abrasive. Fume generated by ablation can be easily removed by vacuum. The old church, damaged during WWII, in Dresden Germany was restored and every piece was reassembled after laser cleaning. Old structures with toxic lead paints can also be cleaned by lasers.

4 Summary

Lasers are used for a wide variety of manufacturing processes. The main reason for its popularity is its ease of availability with a wide range of intensity levels from fraction of W/cm^2 to 10^{15} W/cm^2 with precision. It is also amenable to automation, making it a useful tool for modern manufacturing. It is one of the most adaptable forms of energy sources for industry. Its popularity and applications have been continually increasing since its invention.

References

1. Steen, W.M.: Professor Emeritus, University of Liverpool, Fellow, University of Cambridge (Private Communication) (2019)
2. Jensen, R.J., et al.: Laser isotope separation. Sep. Sci. Technol. **15**(3), 1980 (2006)
3. Steen, W., Mazumder, J.: Laser Materials Processing, 4th edn. Springer (2010)
4. Eeinstein, A.: Strahlunggs-emission and absorption nach der Quantenheorie. Verh. Dtsch. Phys. Ges. **18**, 318–323 (1916)
5. Schawlow, A.L., Townes, C.: Infrared and optimal masers. Phys. Rev. **112**, 1940–1949 (1958)
6. Patel, C.K.N.: Continuous-wave laser action on vibrational-rotational transitions of CO_2. Phys. Rev. **136**(5A), A1187–A1193 (1964)
7. Overton, G., Nogee, A., Belforte, D., Wallace, J., Gefvert, B.: Annual Laser Market Review Forecast, 2019. Laser Focus World, Jan 2019
8. Mazumder, J.: Laser heat treatment: the state of the art. J. Met. **35**(5), 18–26 (1983)
9. Mazumder, J., Kar, A.: Theory and Application of Laser Chemical Vapor Deposition. Prentice Hall (1995)
10. Chen, X., Mazumder, J.: Laser chemical vapor deposition of titanium nitride. Phys. Rev. B **52**(8), 5947–5952 (1995)
11. Mazumder, J.: Non-equilibrium synthesis by laser for tailored surfaces. In: Mazumder, J., Conde, O., Vilar, R., et al. (eds.) NATO Advanced Study Institute entitled "Laser Processing: Surface Treatment and Film Deposition". Series E: Applied Sciences, vol. **307**, pp. 47–75, Sesimbra, Portugal, 3–16 July 1994
12. Economist Magazine, April 2012 Cover page
13. Gibson, I., Rosen, D., Stucker, B.: Additive Manufacturing Technologies: 3D Printing, Rapid Prototyping, and Direct Digital Manufacturing, 2nd edn. (2015)
14. Hull, C.W.: Apparatus for production of three-dimensional objects by stereolithography. US Patent 4575330, 11 Mar 1986

15. Beaman, J.J., Deckard, C.R.: Selective laser sintering with assisted powder handling. US Patent 4,938,816, 3 July 1990
16. Koch, J., Mazumder, J.: Rapid prototyping by laser cladding. In: McKay, T.D., Matsanawa, A., Hugel, A. (eds.) Proceedings of the ICALEO '93. Laser Synthesized Films and Materials, vol. 77, pp. 556–565, Orlando, FL, 17–20 Oct 1993
17. Koch, J., Mazumder, J.: Apparatus and methods for monitoring and controlling multi-layer laser cladding. US Patent 6,122,564 19 Sept 2000
18. Mazumder, J.: Laser welding: state of the art review. JOM **34**, 16–24 (1982)
19. Mazumder, J.: Lasers in aerospace industry manufacturing. In: Encyclopedia of Aerospace Engineering, Online © 2010 Wiley